Current Topics in Microbiology and Immunology

Volume 443

The reviews series *Current Topics in Microbiology and Immunology* publishes cutting-edge syntheses of the latest advances in molecular immunology, medical microbiology, virology and biotechnology. Each volume of the series highlights a selected timely topic, is curated by a dedicated expert in the respective field, and contains a wealth of information on the featured subject by combining fundamental knowledge with latest research results in a unique manner.

Nigel Paneth · Michael J. Joyner ·
Arturo Casadevall
Editors

Convalescent Plasma and Other Antibody Therapies for Infectious Diseases

Lessons learned from COVID-19 and Future Prospects

Responsible Series Editor: Arturo Casadevall

Springer

Editors
Nigel Paneth
Department of Epidemiology
and Biostatistics
Michigan State University
East Lansing, MI, USA

Michael J. Joyner
Department of Anesthesiology
and Perioperative Medicine
Mayo Clinic
Rochester, MN, USA

Arturo Casadevall
Department of Molecular Microbiology
and Immunology
Johns Hopkins University
Baltimore, MD, USA

ISSN 0070-217X ISSN 2196-9965 (electronic)
Current Topics in Microbiology and Immunology
ISBN 978-3-031-87166-5 ISBN 978-3-031-87167-2 (eBook)
https://doi.org/10.1007/978-3-031-87167-2

This Springer imprint is published by the registered company Springer Nature Switzerland AG
The registered company address is: Gewerbestrasse 11, 6330 Cham, Switzerland

Preface

This book is the product of the collaborative efforts of a group of several hundred physician-scientists who gathered in the early days of the COVID-19 pandemic to consider whether and how convalescent plasma might be most effectively deployed in management of this new disease. The effort—which we refer to as the National Convalescent Plasma Project—emerged in March 2020, and its leadership team are the editors of this volume. Our goal in writing this book was to provide an overview of the several forms of passive antibody therapy used in COVID management, and also to provide a background to these therapies, reviewing their use since the first tentative, but effective, steps that began with the development of antitoxins in the 1890s.

We here provide documentation of the effectiveness of convalescent plasma and other forms of passive antibody therapy in COVID-19 and further demonstrate that this success was built on a firm historical foundation of effectiveness against a range of serious infectious diseases as varied as diphtheria, polio and plague. The three basic forms of passive antibody treatment for infectious diseases—antitoxins, antisera and convalescent plasma; hyperimmune globulins; and monoclonal antibodies—developed steadily as scientific knowledge grew, with a new technology evolving about every half-century. Antitoxins and antisera were first used in the last decade of the 19th century and convalescent plasma soon followed; the technology behind hyperimmune globulins was developed in the 1940s; and monoclonal antibodies have been used to treat and prevent infections since the 1990s. But as this volume also shows, no form of passive antibody therapy has become outdated. Each form was used successfully in COVID-19, but their efficacy was apparent only when they were used appropriately.

The historical evidence is set out in the first three chapters, with special attention to antitoxins and antisera (chapter "A Brief History of Polyclonal Antibody Therapies Against Bacterial and Viral Diseases Before COVID-19") and to monoclonal antibodies (chapter "Monoclonal Antibody Therapies for Infectious Diseases"). After considering the importance of antibody measurement (chapter "The Importance of Antibody Titer Determination to the Effective Use of Convalescent Plasma"), we review the signals of effectiveness of convalescent plasma in COVID-19. The US Expanded Access Program (EAP) of 2020 provided early results on mortality reduction with early treatment (chapter "Convalescent Plasma and the US Expanded Access Program: A Personal Narrative") on safety (chapter "The Safety Profile of COVID-19 Convalescent Plasma") and on the importance of donor-recipient proximity (chapter "The Importance of Geographic Proximity of Convalescent Plasma Donors"). An overview of virtually every study of convalescent plasma that assessed mortality is provided in chapter "Evidence for the Efficacy of COVID-19 Convalescent Plasma", and the critical importance of convalescent plasma in the immunocompromised is discussed in chapter "Convalescent Plasma for Immunocompromised Patients". The value of monoclonal antibodies and of hyperimmune globulins in COVID-19 are reviewed in chapters "Monoclonal Antibody Therapies Against SARS-CoV-2: Promises and Realities" and "Hyperimmune Globulins in COVID-19", respectively.

We then turn to the mechanics and logistics of administering antibody therapies, emphasizing patient advocacy (chapter "The Role of the Patient Advocate During a Pandemic: The Case of Convalescent Plasma"), blood banking in low-and-middle income countries (chapter "Blood Banking Capacity in Low-and Middle-Income Countries: Covid-19 Convalescent Plasma in Context") and cost-effective means of obtaining plasma (chapter "HemoClear: A Practical and Cost-Effective Alternative to Conventional Convalescent Plasma Retrieval Methods").

The final section looks to the future. We will need to prepare systems for evaluating any new passive antibody product (chapter "Generating the Evidence Base for Convalescent Plasma Use for a New Infectious Disease"), develop technologies for efficient outpatient use of intravenous treatments (chapter "Convalescent Plasma and Other Antibody Therapies for Infectious Diseases—Lessons Learned from COVID-19 and Future Prospects") and find the best ways to use monoclonal antibodies and hyperimmune globulins (chapter "Monoclonal Antibodies and Hyperimmune Immunoglobulins in the Next Pandemic"). The final (chapter "Using Passive Antibody Therapies in the Next Pandemic") considers the use of passive antibody therapies in a future pandemic, and how we can—indeed must—prepare ourselves to be able to use them properly.

It is not unusual for important and effective means of controlling disease to be supplanted by simpler, more efficient and more effective treatments. But in pandemic situations, many of our modern tools prove ineffective and turning to earlier forms of treatment may be essential. But all too often it is not only the treatment that has been forgotten, but the manner of its use. This lesson was learned in COVID-19, where the absolute requirement for early treatment with antibody rich plasma was ignored in many settings. But when used appropriately, passive antibody therapies can be extremely valuable additions to the preventive and therapeutic armamentaria.

East Lansing, USA Nigel Paneth
Rochester, USA Michael J. Joyner
Baltimore, USA Arturo Casadevall

Acknowledgements

We acknowledge the assistance in the editing of this book of the four other members of the leadership team of the National Convalescent Plasma Project—Brenda Grossman MD and Jeffrey Henderson MD, both of Washington University, St. Louis; Liise-Anne Pirofski MD, Albert Einstein College of Medicine; and Shmuel Shoham MD, Johns Hopkins University.

Contents

Passive Antibody Therapies Prior to Covid

Monoclonal Antibody Therapies for Infectious Diseases

Arturo Casadevall and Nigel Paneth

Contents

Abstract In contrast to therapy in oncology and immune-related diseases, where dozens of monoclonal antibodies (mAbs) have been introduced, often in transformative fashion, the use of mAbs for infectious diseases is generally underdeveloped, with fewer than a dozen mAbs currently licensed for the treatment of microbial diseases. This situation is paradoxical given that antibodies are major products of the immune system for protecting against infectious diseases. The underdevelopment of mAbs for infectious diseases has several causes including the availability of effective therapy against many microbial diseases, the fact that many pathogenic microbes are antigenically diverse and thus all strains are not covered by a single mAb, and the high expense of mAb therapies. Despite these hurdles the

A. Casadevall (✉)
Department of Molecular Microbiology and Immunology, Johns Hopkins Bloomberg School of Public Health, Baltimore, MD, USA
e-mail: acasade1@jhu.edu

N. Paneth
Departments of Epidemiology & Biostatistics and Pediatrics & Human Development, College of Human Medicine, Michigan State University, East Lansing, MI, USA
e-mail: paneth@msu.edu

Current Topics in Microbiology and Immunology (2025) 443: 3–13
https://doi.org/10.1007/82_2024_265

Published Online: 14 June 2024

number of mAbs licensed for infectious disease indications is slowly increasing and there are numerous opportunities for the development of mAbs in the prevention and treatment of microbial diseases.

The story of monoclonal antibody therapy for infectious diseases is one of underdevelopment. Almost a half-century after the discovery of hybridoma technology by Kohler and Milstein in 1975 (Kohler and Milstein 1975), fewer than a dozen mAbs have been fully licensed for use against infectious diseases (mAbs employed in the COVID pandemic were licensed only for emergency use). Including all forms of licensing, in 2022, 122 mAbs were approved for clinical use in the United States of which only 11.7% targeted an infectious disease. The COVID pandemic has stimulated the development of several mAbs for COVID (see Chap. 3, P3), and this indication alone constitutes about half of all antimicrobial mAbs (Lyu et al. 2022). In contrast, 78% of mAbs target cancer or immune-related diseases (Lyu et al. 2022).

One of your authors has bemoaned this state of affairs for decades and argued for greater use of mAb therapies in infectious diseases (Casadevall and Scharff 1995; Casadevall 1996). The topic of mAbs for infectious disease is diverse, including an enormous literature, and we note other recently published reviews that may be of interest to the reader (Motley et al. 2019; Pecetta et al. 2020). In this chapter, we review the current use of mAbs in infectious diseases, their advantages and disadvantages, and the economic and biological issues that have prevented their greater use as antimicrobial agents. We also identify areas of potential development for anti-infective mAbs.

1 Monoclonal Antibodies in Brief

The first monoclonal antibodies were made by cell biological techniques that involved the fusion of an immortal cell, often of myeloma origin, to an antibody-producing B cell resulting in a hybridoma. Initially, it was much easier to make murine hybridomas producing mouse mAbs than human mAbs. Consequently, the first licensed mAb to find widespread clinical use was a murine mAb—anti-CD3 OKT3 (muromonab-CD3)—which was licensed by the FDA in 1986 (Lyu et al. 2022). In recent decades, the technology for making mAbs has matured greatly, and the application of recombinant DNA techniques provides highly versatile technologies, making it possible today to make fully human antibodies of many types including bi-specific, isotope-labeled, drug-labeled, antibodies and their fragments.

mAb-based therapies differ from other antibody treatments such as hyperimmune globulins and convalescent plasma in that they consist of homogenous preparations of antibodies, in which the protein structure of all molecules is identical. All antibody molecules in a mAb preparation recognize a single epitope or binding determinant in their target antigen and all have the same constant region. Since the constant region confers the isotype, determines serum half-life, and the interaction with Fc receptors and complement activation, mAb preparations are homogenous in their biological properties. This homogenous composition leads to high levels of activity per protein content but with functionality reduced to a single isotype.

2 Monoclonal Versus Polyclonal—Advantages and Disadvantages

In contrast to mAbs, in both hyperimmune globulins and convalescent plasma, the amount of specific antibody is only a small fraction of the total amount of antibody present, and these preparations usually include antibodies of all isotypes. This means that these antibody preparations are polyclonal, with low activity per protein content, but having multiple forms of antibody, such as IgM, IgG, and IgA, each with different effector functions.

In trying to understand the relative paucity of mAb therapies for infectious diseases it is important to contrast their advantages and disadvantages in comparison to polyclonal preparations. mAbs are produced by cell culture, and the process requires several purification steps. Overall, the process is expensive with high production costs relative to polyclonal antibody preparations such as convalescent plasma that usually come from donors. In contrast to their widespread use in oncology and rheumatology where they are used in conditions often difficult to treat effectively, the development of mAbs for infectious diseases must contend with the fact that, for many bacterial, viral, parasitic, and fungal diseases, there is already effective antimicrobial therapy in the form of small molecule drugs. This means that for mAbs to be successfully developed into clinically useful drugs they must show value added over existing therapy, which can be difficult when less costly effective antimicrobial therapy is available. As noted before, mAbs are biological agents produced in tissue culture making them much more expensive than antimicrobial drugs.

Moreover, the exquisitely fine specificity of mAbs is a disadvantage for targeting antigenically diverse microbes since these cannot all be covered by a single antibody. One strategy to overcome this limitation is to design mAb cocktails—in effect moving away from the single clone idea of "monoclonal" to something more like bi and triclonal. Successful mAb cocktails include agents targeting such antigenically diverse viruses and toxins as SARS-CoV-2 (Aleem and Vaqar 2023), rabies (Kim et al. 2021; de Melo et al. 2022), and botulinum toxin (Guptill et al. 2021). One area where mAbs have a distinct advantage over polyclonal preparations is in situations where the immune response is inconsistent because of simultaneous production of protective and non-protective mAbs. For example, polyclonal responses to *M. tuberculosis* and fungal pathogens are often inconsistent with regards to protective efficacy, but it is possible to make mAbs that are consistently protective (reviewed in (Casadevall and Pirofski 2012)). For example, in the era of serum therapy, when sera were produced against several pneumococcal serotypes, no serum protective against type 3 pneumococcus was ever developed. But it is possible to make protective mAbs to this pneumococcal serotype (Babb et al. 2021).

3 Monoclonal Antibodies in Current Clinical Use for Infectious Diseases

The relative paucity of mAb use against infectious diseases is evident from the fact that, of the more than one hundred fully licensed mAbs for clinical use, only ten have antimicrobial indications. This low proportion is paradoxical when one considers that the field of infectious disease introduced antibody therapies in the 1890s (see Chap. 5, P5) and that antibodies are immune molecules whose natural function is to protect against infectious diseases. The ten mAbs are palivizumab (first mAb licensed for an infectious disease in 1996) and nirsevimab for respiratory syncytial virus; raxibacumab and obiltoxaximab for inhalational anthrax; bezlotoxumab for recurrent C. difficile colitis; ibalizumab for multiple drug-resistant HIV infection; rabishield and babimabs for prevention of rabies; and inmazeb and ebanga for Ebola virus infection (Whaley and Zeitlin 2022). Of these ten mAbs, seven are antiviral and three target bacterial toxins.

This selection of commercially viable pathogen targets is informative, illustrating the value of the development of mAbs against viruses for which there is no effective antiviral therapy or bacterial toxins that cause disease. For HIV, there are more than two dozen effective small molecule antiviral agents, but ibalizumab targets the niche created by patients who are infected with multiple drug-resistant viral strains. For bacterial infections largely mediated by toxins such as *C. difficile* colitis and anthrax, and where there are no antitoxin therapies available, mAbs have found another niche. It is likely that the mAb arsenal against infectious diseases will expand in future years as there are ongoing efforts to develop mAb therapy against a variety of infectious diseases including influenza, COVID-19, and malaria.

4 COVID-19 Illustrates the Strengths and Weaknesses of mAb Therapies

When the pandemic caused SARS-CoV-2 struck, there were no specific therapies against this coronavirus. Many companies raced to develop mAbs and within months these entered clinical trials, as reviewed in Chap. 3 (P6). President Trump was treated with a cocktail of two mAbs made by Regeneron in October of 2020, which was about 10 months after SARS-CoV-2 was described, demonstrating a major advantage of mAbs—the rapidity with which they can be developed. Three sole-use monocolonals (bebtelovimab, bamlanivimab, and sotrovimab) and three combination therapies of two monocolonals (tixagevimab and cilgavimab; casirivimab and imdevimab; bamlanivimab and etesevimab) were found effective and granted Emergency Use Authorization by the Food and Drug Administration for COVID-19 (Aleem and Vaqar 2023). These mAbs proved highly effective when given to individuals with early COVID-19, preventing hospitalization in outpatients, and having an excellent safety record (Aleem and Vaqar 2023). However, unlike convalescent plasma, these mAbs were not associated with reduced mortality when given to hospitalized patients.

Unfortunately, SARS-CoV-2 evolved rapidly in human populations and generated many antigenically different variants such that by January 2023, when the omicron variant was finally declining, all had lost efficacy and were withdrawn from clinical use (Focosi et al. 2023). It is noteworthy that preparations that combined two mAbs recognizing different epitopes in the SARS-CoV-2 spike protein lasted longer in clinical use, since it was more difficult for the emerging variants to defeat both antibodies. This experience highlights a major weakness of mAbs for antigenically variable infectious disease agents—they can be defeated by changes in the target antigen.

The remarkable specificity of mAbs becomes a weakness because this specificity relies on precise interactions between the antibody paratope and the epitope and any changes to the protein structure can reduce the binding affinity and the mAb activity. For palivizumab, a mAb licensed for the prevention of respiratory syncytial virus in neonates a single amino acid change on the viral protein renders it useless (Adams et al. 2010), but fortunately, this mutation has not become prevalent among clinical isolates and the mAb remains clinically useful. Not surprisingly then, when mAb-resistant viruses emerge in immunocompromised individuals, physicians have resorted to again using high titer COVID-19 convalescent plasma—a polyclonal product—to help eradicate the infection.

In fact, it is possible that some of these COVID virus variants may have actually been generated by mAb antibody selection in immunosuppressed patients who were being treated to prevent or suppress infection (Casadevall and Focosi 2023). Such patients harbor large viral populations that include naturally emerging variants and the combination of mAb selection and a poorly functioning immune system creates propitious conditions for variant selection (Gliga et al. 2022). The temporal emergence of variants resistant to mAbs occurring shortly after the introduction of such mAbs into clinical use raises the concern that mAb therapies in immunocompromised patients may have contributed to the persistence of the pandemic through facilitating the emergence of variants that are more resistant to vaccine immunity (Focosi et al. 2023).

5 Enhanced mAbs for Infectious Diseases

mAbs can be enhanced by modifications to carry radioactive isotopes and drugs. The attachment of a radioactive isotope to an mAb can deliver microbicidal radiation to the site of infection and the feasibility and efficacy of this approach have been established against fungi, bacteria, and viral-infected cells (Dadachova and Casadevall 2014). Radiolabeled antibodies have been used for the treatment and diagnosis of certain types of cancer, but their clinical use has been limited by the additional logistical complexity of dealing with a radioactive compound that can have a short half-life.

Similarly, the linking of an antimicrobial drug to a monoclonal antibody takes advantage of the great specificity of immunoglobulins to allow its delivery directly to the microbial target (Cavaco et al. 2022). Such compounds are called antibody–antibiotic conjugates and have been demonstrated to promote the killing of such troublesome bacteria as *Staphylococcus aureus* (Lehar et al. 2015) and *Pseudomonas aeruginosa* (Kajihara et al. 2021). Although conceptually simple, antibody–antibiotic conjugates require sophisticated chemistry and knowledge of pathogenesis for efficacy. For example, an antibody–antibiotic conjugate targeting *S. aureus* links the drug vancomycin to an antibacterial antibody that promotes phagocytosis and the drug is released from the conjugate in the acidic environment of the phagosome (Lehar et al. 2015). The antibody–vancomycin conjugate was more effective than vancomycin alone against staphylococcal experimental infection in mice (Lehar et al. 2015).

Although *S. aureus* is not considered a traditional intracellular pathogen, there is evidence that intracellular residence contributes to persistence and spread, and the antibody–antibiotic conjugate approach could provide physicians with yet another tool for treating difficult-to-treat staphylococcal disease, although it will be necessary to identify the subset of individuals with staphylococcal disease who could benefit from antibody–vancomycin conjugates and test the combination in them in randomized trials.

The process of thinking through how the clinical testing of such a novel compound would be done illustrates some of the hurdles in developing mAbs for infectious diseases where effective therapy is already available in the form of small-molecule antimicrobial agents. Unlike other conditions for which there is no therapy, the fact that vancomycin alone is effective against *S. aureus* would imply that it would be ethically difficult if not impossible to test antibody–vancomycin conjugates as a stand-alone therapy versus vancomycin alone. Hence, clinical trials would likely involve its initial testing in combination with the unconjugated drug. Establishing additivity or synergy will be challenging.

6 The Future of mAb Therapies in Infectious Diseases

The experience with mAb therapies in infectious diseases shows that these compounds can be highly effective when used properly and are generally well tolerated. In the almost three decades since palivizumab was first licensed for the prevention of Respiratory Syncytial Virus (RSV), a handful of other mAbs have joined the therapeutic armamentarium with indications for infectious diseases. The paucity of mAb therapies in infectious diseases relative to other fields such as oncology and immune-mediated diseases is a consequence of the challenges posed by the variable nature of microbial antigens, the existence of effective therapies for many infectious diseases, and the economics of drug development and usage. In contrast to host components such as interferons and cytokines, microbial antigens are often antigenically variable, and this antigenic variation can greatly complicate the development of antibody therapies.

In the era of serum therapy (Chap. 5, P10), it was necessary to develop serotype-specific sera for each of the dozens of pneumococcal serotypes and the ongoing COVID-19 pandemic has shown how excellent mAbs can be rapidly defeated by the emergence of new viral variants. Another hurdle for mAbs is the uncertainty and variability of demand for many infectious diseases, as epidemics are unpredictable. For example, mAbs developed for lethal viruses such as the Ebola virus would be invaluable during outbreaks but find no use during periods when the human disease is quiescent. For other infectious diseases, demand can wane and remove the rationale for development. In this regard, one of your authors worked for many years to develop mAb therapy for cryptococcosis, which reached epidemic proportions in the early 1990s as a consequence of the AIDS epidemic and that work led to the Phase I testing of a mAb to the capsular polysaccharide (Larsen et al. 2005). However, it took more than a decade from bench to bedside, and by the time the mAb was ready for clinical use, the prevalence of the disease had waned because of improvements in anti-retroviral therapy and there was little or no economic incentive for further development of this therapy. While this result was perhaps disappointing, the effort spawned new insights into basic immunology, showing how a strictly oriented clinical quest can lead to new basic science understanding.

A new development which may enhance opportunities for the development of mAbs against infectious diseases is the ability to genetically modify mAbs to greatly increase their serum half-life in humans. The half-life of a mAb is determined by its constant region and given that IgG has the longest serum half-life of about 20 days, it is the preferred isotype for clinical use. mAbs can be engineered for longer half-life by introducing amino acid changes into the constant region and several such antibodies are in clinical use. This functionality may allow mAbs to elicit immune responses that provide durable protection against infectious diseases much like vaccination, but via the passive immunization mechanism. This approach was proven to be successful with mAbs for the prevention of COVID-19 (Levin et al. 2022; Abraham 2022) and a new mAb against RSV (Talha and Ali 2023) has recently been licensed for long-lasting prophylaxis. A long-acting mAb to *S. aureus* is in clinical development for the prevention of staphylococcal pneumonia (Yu et al. 2017). Given the increasing prevalence of immunosuppressed individuals in our society who are susceptible to many infectious diseases and who often respond poorly to vaccines because of their underlying immune impairment, this approach could have application for the prevention of many conditions.

Another potential opportunity is the development of catalytic mAbs against infectious diseases. Catalytic mAbs have the added property that their binding site is capable of catalysis such as proteolysis. Catalytic antibodies can destroy antigens and could find use in diseases where microbial antigens accumulate in tissues, such as in cryptococcosis, where polysaccharide antigens shed from fungal cells accumulate and contribute to pathogenesis by impairing the immune response (Bowen et al. 2017a). Catalytic mAbs that digest the *C. neoformans* capsule (Bowen et al. 2017b) and hydrolyze the spike protein of SARS-CoV-2 (McConnell et al. 2023) have been described, which brings the new dimension of catalysis-mediated destruction of microbial antigens and pathogens to the anti-infective armamentarium.

The existence of effective therapies for many infectious diseases is a disincentive to develop mAb therapies since the latter are expensive and would have to either show superiority or some advantage over existing therapies to find clinical use. McConnell has argued that the continued existence of small molecule antimicrobial agents against bacteria is likely to be the major disincentive holding up the development of mAbs against multiply drug-resistant bacteria (McConnell 2019). Nevertheless, promising mAbs are under clinical and pre-clinical development against Gram-negative bacteria that are often highly drug-resistant, such as *Pseudomonas aeruginosa* (Que et al. 2014), *Klebsiella pneumoniae* (Diago-Navarro et al. 2018), and *Acinetobacter baumannii* (Slarve et al. 2023).

Another area of opportunity for mAb development is in the generation of human antibodies to replace older equine immune sera preparations currently in use, such as the development of mAbs for the treatment of diphtheria (Smith et al. 2016). One can anticipate that, as the drug resistance problem worsens for certain bacterial pathogens, and the available drug options are reduced, the development of mAb therapies for bacterial infections will be incentivized.

On the other hand, it must be conceded that the nature of infectious diseases creates disincentives for mAb development. Most infectious diseases resolve quickly, and successful therapy is generally brief, making anti-infective mAbs less attractive to companies and investors than the development of mAbs for use in chronic diseases, which can require prolonged, and often, lifelong usage. Yet infectious diseases will continue to provide opportunities for the development of new anti-infective mAbs (Table 1). As we have learned repeatedly in recent years, new infectious diseases for which there is no therapy appear on a nearly regular schedule. In addition, the efficacy of established antimicrobial therapy is declining

Table 1 Opportunities for the development of anti-infective mAbs

Target	Rationale	Example(s)
New viruses and major viral outbreaks with known viruses	mAbs with antiviral activity can be developed quickly	mAbs targeting Ebola virus (Tshiani Mbaya et al. 2021)
Multiple-drug resistant bacteria	Experience from age of serum therapy shows that antibodies can be effective antibacterial agents	mAbs in development for highly drug resistant bacteria such as *S. aureus* (Lehar et al. 2015), *P. aeruginosa* (Que et al. 2014), *K. pneumoniae* (Diago-Navarro et al. 2018), and *A. baumannii* (Slarve et al. 2023)
Toxins that cause disease	Antibodies are premier agents for the neutralization of toxins	mAb cocktails being developed for the treatment of botulism (Guptill et al. 2021)
Fungal diseases	Mycotic diseases are notoriously difficult to treat and adjunctive mAb immunotherapy could be clinically useful	mAbs have been developed that protect against *C. neoformans*, *A. fumigatus*, *H. capsulatum* and *C. albicans*
Long lasting passive immunization	mAbs can be genetically modified for long serum half-life and thus provide long-term protection	Long-half-life mAbs have been developed for prevention of COVID-19 (Aleem and Vaqar 2023) and RSV

due to widespread drug resistance, creating a constant need for new and better anti-infective therapies. In this environment, mAb therapies for infectious diseases are likely to see continued development and we can anticipate that the number of licensed immunoglobulins will increase with time.

Declarations
Neither author has any conflicts to declare nor received any funding for this work.

References

Abraham J (2022) Monoclonal antibodies with extended half-life to prevent Covid-19. N Engl J Med 386(23):2236–2238

Adams O, Bonzel L, Kovacevic A, Mayatepek E, Hoehn T, Vogel M (2010) Palivizumab-resistant human respiratory syncytial virus infection in infancy. Clin Infect Dis off Publ Infect Dis Soc Am 51(2):185–188

Aleem A, Vaqar S (2023) Monoclonal antibody therapy for high-risk coronavirus (COVID 19) patients with mild to moderate disease presentations. StatPearls. Treasure Island (FL) ineligible companies. Disclosure: S. Vaqar declares no relevant financial relationships with ineligible companies. StatPearls Publishing. Copyright © 2023, StatPearls Publishing LLC

Babb R, Doyle CR, Pirofski LA (2021) Isolation and characterization of human monoclonal antibodies to pneumococcal capsular polysaccharide 3. Microbiol Spectrum 9(3):e0144621

Bowen A, Wear MP, Cordero RJ, Oscarson S, Casadevall A (2017b) A monoclonal antibody to *Cryptococcus neoformans* glucuronoxylomannan manifests hydrolytic activity for both peptides and polysaccharides. J Biol Chem 292(2):417–434

Bowen A, Wear M, Casadevall A (2017a) Antibody-mediated catalysis in infection and immunity. Infect Immun 85(9)

Casadevall A (1996) Antibody-based therapies for emerging infectious diseases. Emerg Infect Dis 2:200–208

Casadevall A, Pirofski LA (2012) A new synthesis for antibody-mediated immunity. Nat Immunol 13(1):21–28

Casadevall A, Scharff MD (1995) Return to the past: the case for antibody-based therapies in infectious diseases. Clin Infect Dis 21:150–161

Casadevall A, Focosi D (2023) SARS-CoV-2 variants resistant to monoclonal antibodies in immunocompromised patients constitute a public health concern. J Clin Invest 133(6)

Cavaco M, Castanho MA, Neves V (2022) The use of antibody-antibiotic conjugates to fight bacterial infections. Front Microbiol 13:835677

Dadachova E, Casadevall A (2014) Radiolabeled antibodies for therapy of infectious diseases. Microbiol Spectrum 2(6)

de Melo GD, Hellert J, Gupta R, Corti D, Bourhy H (2022) Monoclonal antibodies against rabies: current uses in prophylaxis and in therapy. Curr Opin Virol 53:101204

Diago-Navarro E, Motley MP, Ruiz-Peréz G, Yu W, Austin J, Seco BMS et al (2018) Novel, broadly reactive anticapsular antibodies against carbapenem-resistant *Klebsiella pneumoniae* protect from infection. mBio 9(2)

Focosi D, McConnell S, Sullivan DJ, Casadevall A (2023) Analysis of SARS-CoV-2 mutations associated with resistance to therapeutic monoclonal antibodies that emerge after treatment. Drug Resist Updates: Rev Commentaries Antimicrobial Anticancer Chemother 71:100991

Gliga S, Luebke N, Killer A, Gruell H, Walker A, Dilthey AT, et al (2022) Rapid selection of sotrovimab escape variants in SARS-CoV-2 Omicron infected immunocompromised patients. Clin Infect Dis: Off Publ Infect Dis Soc Am

Guptill JT, Raja SM, Juel VC, Walter EB, Cohen-Wolkowiez M, Hill H et al (2021) Safety, tolerability, and pharmacokinetics of NTM-1632, a novel mixture of three monoclonal antibodies against botulinum toxin B. Antimicrob Agents Chemother 65(7):e0232920

Kajihara KK, Pantua H, Hernandez-Barry H, Hazen M, Deshmukh K, Chiang N et al (2021) Potent killing of Pseudomonas aeruginosa by an antibody-antibiotic conjugate. Mbio 12(3). https://doi.org/10.1128/mbio.00202-21

Kim PK, Ahn JS, Kim CM, Seo JM, Keum SJ, Lee HJ et al (2021) A broad-spectrum and highly potent human monoclonal antibody cocktail for rabies prophylaxis. PLoS ONE 16(9): e0256779

Kohler G, Milstein C (1975) Continuous cultures of fused cells secreting antibody of predefined specificity. Nature 256:495–497

Larsen RA, Pappas PG, Perfect JR, Aberg JA, Casadevall A, Cloud GA et al (2005) A phase I evaluation of the safety and pharmacodynamic activity of a murine-derived monoclonal antibody 18B7 in subjects with treated cryptococcal meningitis. Antimicrob Agents Chemother 49:952–958

Lehar SM, Pillow T, Xu M, Staben L, Kajihara KK, Vandlen R et al (2015) Novel antibody–antibiotic conjugate eliminates intracellular *S. aureus*. Nature 527(7578):323–328

Levin MJ, Ustianowski A, De Wit S, et al (2022) Intramuscular AZD7442 (tixagevimab–cilgavimab) for prevention of Covid-19. N Engl J Med 386:2188–2200

Lyu X, Zhao Q, Hui J, Wang T, Lin M, Wang K et al (2022) The global landscape of approved antibody therapies. Antibody Therapeutics 5(4):233–257

McConnell MJ (2019) Where are we with monoclonal antibodies for multidrug-resistant infections? Drug Discovery Today 24(5):1132–1138

McConnell SA, Sachithanandham J, Mudrak NJ, Zhu X, Farhang PA, Cordero RJB et al (2023) Spike-protein proteolytic antibodies in COVID-19 convalescent plasma contribute to SARS-CoV-2 neutralization. Cell Chem Biol 30(7):726–38.e4

Motley MP, Banerjee K, Fries BC (2019) Monoclonal antibody-based therapies for bacterial infections. Curr Opin Infect Dis 32(3):210–216

Pecetta S, Finco O, Seubert A (2020) Quantum leap of monoclonal antibody (mAb) discovery and development in the COVID-19 era. Semin Immunol 50:101427

Que YA, Lazar H, Wolff M, François B, Laterre PF, Mercier E et al (2014) Assessment of panobacumab as adjunctive immunotherapy for the treatment of nosocomial *Pseudomonas aeruginosa* pneumonia. Eur J Clin Microbiol Infect Dis: off Publ Eur Soc Clin Microbiol 33 (10):1861–1867

Slarve M, Reyna Z, Burk E, Ruiz-Delgado J, Li R, Yan J et al (2023) Therapeutic, humanized monoclonal antibody exhibits broad binding and protective efficacy against *Acinetobacter baumannii*. Antimicrob Agents Chemother 67(3):e0008623

Smith HL, Cheslock P, Leney M, Barton B, Molrine DC (2016) Potency of a human monoclonal antibody to diphtheria toxin relative to equine diphtheria anti-toxin in a guinea pig intoxication model. Virulence 7(6):660–668

Talha M, Ali MH (2023) Latest FDA-approved drug nirsevimab-alip (Beyfortus): a gamechanger for treatment of respiratory syncytial virus. J Med Virol 95(10):e29169

Tshiani Mbaya O, Mukumbayi P, Mulangu S (2021) Review: insights on current FDA-approved monoclonal antibodies against ebola virus infection. Front Immunol 12:721328

Whaley KJ, Zeitlin L (2022) Emerging antibody-based products for infectious diseases: Planning for metric ton manufacturing. Hum Vaccin Immunother 18(2):1930847

Yu XQ, Robbie GJ, Wu Y, Esser MT, Jensen K, Schwartz HI et al (2017) Safety, tolerability, and pharmacokinetics of MEDI4893, an investigational, extended-half-life, anti-*Staphylococcus aureus* alpha-toxin human monoclonal antibody, in healthy adults. Antimicrobial Agents Chemother 61(1)

Nigel Paneth is a pediatrician and perinatal and developmental epidemiologist who is an emeritus professor at Michigan State University and founding chair of the Department of Epidemiology and Biostatistics. He is the author or editor of 4 books and 390 peer-reviewed papers and has an H-index of 95.

A Brief History of Polyclonal Antibody Therapies Against Bacterial and Viral Diseases Before COVID-19

Nigel Paneth, Madison Walsh, Breanna Kornatowski, and Arturo Casadevall

Contents

Abstract The use of the serum or plasma of patients or animals who have recovered from an infectious disease, or had been immunized with a relevant antigen, to treat or prevent the same infection in others began in the late 1880s when French and German scientists uncovered, one step at a time, several of the elements of the immune system's response to infection. A key finding was that the damage caused by some bacteria depends upon their secreted toxins which can be

The original version of the chapter has been revised: The incorrect text 'a survival rate of 24.5' has been corrected as 'a mortality rate of 24.5%'. A correction to this chapter can be found at https://doi.org/10.1007/82_2025_295

N. Paneth (✉)
Departments of Epidemiology & Biostatistics and Pediatrics & Human Development, College of Human Medicine, Michigan State University, East Lansing, MI, USA
e-mail: paneth@msu.edu

N. Paneth · M. Walsh · B. Kornatowski
College of Human Medicine, Michigan State University, East Lansing, MI, USA

A. Casadevall
Department of Molecular Microbiology and Immunology, Johns Hopkins University Bloomberg School of Public Health, Baltimore, MD, USA

Current Topics in Microbiology and Immunology (2025) 443: 15–39
https://doi.org/10.1007/82_2024_279

Published Online: 30 August 2024

neutralized by biologic agents. Antitoxins to diphtheria and tetanus began to be manufactured in large animals in France, Germany, and the US in the 1890s and were soon being used worldwide. The impact of diphtheria antitoxin on childhood mortality was profound. Shortly after the development of antitoxins, convalescent serum began to be used for its anti-bactericidal properties thus addressing serious infections caused by non-toxin-producing organisms. The effectiveness of antitoxins and antisera was demonstrated by examining mortality rates in hospitals before and after the introduction of antitoxins, by comparisons of treated and untreated patients, by comparing early and late treatment and dosage, by examining vital data mortality trends, and by several randomized and alternate assignment trials. Antitoxins continue to have a role in the rare cases of diphtheria and other conditions largely eradicated by immunization, but serum therapy nearly disappeared from the medical armamentarium with the development of antibiotics in the 1940s. Inasmuch as new human pathogens are now emerging with unprecedented regularity as seen in the recent COVID-19 pandemic, and because specific therapies are unlikely to be available for them, plasma-based antibody therapies are likely to again carve out a niche in infectious disease control.

1 Introduction

For a period roughly corresponding to the first half of the twentieth century, the major mode of control of many infectious diseases was through the use of passive antibody treatment in the form of antitoxins and antisera. The first part of this period (1890–1930) was dominated by the use of antitoxins, a form of treatment restricted to infections with microorganisms that produced toxins. Soon after, the passive antibody principle was extended to organisms that did not produce toxins, and the word that came to be used to describe these agents was "antisera". Paradoxically, as the Swiss immunologist Jean Lindenmann has pointed out, terms like antibody or antiserum reverse the logic of the term antitoxin. Lindemann remarked that "the antitoxin is *something* directed against a toxin; the antibody is a *body* directed against something" (Lindenmann 1984).

In approaching the historical development of both forms of passive antibody treatment, we recognize that these forms of therapy—operating at the dawn of modern medicine—consumed an enormous amount of medical energy and are represented by a vast medical literature. Any summary of those momentous decades in medical history must necessarily be incomplete. Our approach here is to highlight key lessons of the past by summarizing the most salient studies, understanding how medical authorities of the time assessed the effectiveness of therapies, and weighing the value of serum or plasma-based antibody therapies for the present day. The uses of antibody therapies during the COVID-19 emergency are covered in other chapters of this volume.

2 The Antitoxin Era (1890–1930)

The discovery, by Emil Behring and Kitasato Shibasaburō, that serum from animals that had recovered from experimental infection with diphtheria and tetanus microbes contained antitoxins that could prevent or cure these infections in animals newly exposed to these microbes was revolutionary (Behring and Kitasato 1890; Behring 1890). Until then, no agent had been found that reliably cured any systemic infectious disease in animals or humans. In 1901 Emil von Behring received the first Nobel Prize in Medicine or Physiology for his discovery that immunity could be transferred by immune serum (Kaufmann 2017). Antitoxins—which came to be produced in large animals in bulk—were later found effective in curing bacterial diseases, but only those whose major pathogenicity was owed to exotoxins, such as diphtheria and tetanus, and only if given early in the disease course. Antitoxins did not attack organisms, but the toxic products of organisms.

Behring and Kitasato's classic 1890 paper demonstrating the effect of the tetanus antitoxin in rabbits (Behring and Kitasato 1890) and Behring's paper published a week later showing the same for diphtheria in guinea pigs (Behring 1890) were preceded by much important work. Kitasato had isolated and cultured the tetanus bacillus in 1889 (Kitasato 1889) and Frederick Loeffler identified the diphtheria bacillus in 1884 (Loeffler 1884). Loeffler fulfilled all of his teacher Koch's stringent postulates for ascribing causality to a putative pathogen (Koch 1884, 1890) in the process. His classic paper explained that the bacillus is not present at sites of serious pathology distant from the inoculation, suggesting that some product of the bacterium must be responsible. Then two students of Pasteur—Emile Roux and Alexandre Yersin—produced the full pathology of diphtheria in guinea pigs via the injection of sterile filtrates of *Cornybacterium diphtheriae*, showing that a sterile, blood-borne, bacterially produced substance could produce damage at a distant site (Roux and Yersin 1888).

The translation of this momentous discovery to humans was soon to take place. Behring in Germany and Roux in France worked independently of each other to manufacture large amounts of antitoxin in large animals (usually horses) and to purify, as best they could, the preparations (see Fig. 1). A year after Behring's December 1890 paper, diphtheria antitoxin was used to effect a miraculous Christmas Day 1891 cure of a comatose child in Berlin, although the veracity of that unpublished anecdote has been questioned (Nakayama 2022). But in April 1894, Behring and colleagues published the results of antitoxin use in 30 children with diphtheria of whom 24 survived (Behring et al. 1893). Yet more influential was the report that same year by Emile Roux and colleagues of a mortality rate of 24.5% in a much larger sample of 428 diphtheritic children in Paris, which compared favorably to mortality rates of 50% or more without the use of antitoxin (Roux et al. 1894). Diphtheria antitoxin was rapidly adopted on both sides of the Atlantic, and by 1911, was being produced by public health agencies such as the New York City Health Department, and by major pharmaceutical companies such as Parke Davis, Lederle, and Burroughs Wellcome (Rosenau 1913). As recently as

Fig. 1 Inoculation of diphtheria toxin to horses producing antidiphtheria serum, at the stables of Marne la Coquette. Engraving after a drawing by Alexis Lemaistre (1896). From Cavaillon J-M: From Bacterial Poisons to Toxins: The Early Works of Pasteurians. Toxins (Basel). 2022 Nov; 14 (11):759

February 2024, the World Health Organization issued a recommendation for using diphtheria antitoxin as the standard of care for the treatment of diphtheria cases, estimating that its use is associated with a 76% reduction in mortality (Clinical Management of Diphtheria 2024).

Tetanus antitoxin was first conclusively shown effective in animals by Nocard in 1897 (Nocard 1897), but, given the rarity of tetanus, was used only occasionally in humans. Much of the early use was by veterinarians to treat tetanus in domestic animals. But the high frequency of tetanus after wounds in World War 1 led to the antitoxin being widely used by both sides in the conflict with strong evidence of effectiveness (Bruce 1917).

The very first use of tetanus antitoxins in humans is hard to pinpoint. Kneass assembled 68 antitoxin-treated human tetanus cases from the literature in 1896, but was not convinced of its effectiveness (Kneass 1896). But a review by MacConkey in 1914 found evidence of effectiveness in several published case series in humans dating back to 1902 (MacConkey 1914). The antitoxin was administered by various routes, including directly into the spinal fluid (Andrewes 1917). A 1941 review of tetanus treatments at Cook County Hospital in Chicago found that at least 250 tetanus patients had been treated with antitoxin in the previous 18 years (Spaeth 1941). A major development was the switch from horse to human-derived antitoxin (Nation et al. 1963), thereby reducing the at times severe impact of serum sickness.

The next antitoxin for which we can find evidence of use in the nineteenth century was to Brucellosis (Aldridge 1898), the name assigned to what was originally called Maltese or Mediterranean Fever after David Bruce discovered the organism in 1887. Both bovine and human preparations were used sporadically in the early twentieth century. Carpenter and Boak, reviewing virtually the entire literature until 1936, concluded that while clinical reports of serum use were generally favorable, "Spectacular cures have not occurred, while disappointments were not infrequent" (Carpenter and Boak 1936).

Tetanus was not the only member of the Clostridial family of organisms to become the subject of antitoxin treatment. The organism causing gas gangrene, like tetanus, often the result of war injury, was identified in 1892 by Welch and Nuttall (1892). Bull and Pritchett identified the toxin and produced an antitoxin successful in animals in 1917 (Bull and Pritchett 1917), which, along with antitoxins developed by French (Weinberg and Seguin 1918) and German scientists (Linton 2000), began to be used, with some apparent success, in the very last year of WW I (Van Beuren 1919).

The antitoxin was used more broadly during WW II, after an antitoxin to the three related agents capable of causing gas gangrene—*Clostridium perfringens Type A*; *C. septicum;* and *C. oedematiens* had been developed, and was found to reduce the extremely high mortality in soldiers with gas gangrene, especially if used early (Macfarlane 1943). The antitoxin is now also used to treat rare cases of liver or lung abscess caused by *C. perfringens* (Hifumi et al. 2013, 2018) and the production of an antitoxin has been proposed for the treatment of sepsis from this organism (Hifumi 2020).

A third *Clostridial species, C. botulinum,* produces the single most potent toxin known and causes some 100 cases of disease a year in the US, mostly infant botulism, with wound botulism and botulism acquired from contaminated food also contributing cases (https://www.cdc.gov/botulism/hcp/clinical-overview/index.html#:~:text=Botulism%20Antitoxin%20Heptavalent%20and%20BabyBIG,of%20paralysis%20and%20consequent%20complications). An equine antitoxin to the seven known botulinum toxin types is available from the CDC (Centers for Disease Control and Prevention 2010), as well as a human-derived antitoxin, whose use is restricted to infant botulism (Arnon et al. 2006). The organism was discovered in 1896 (Ermengem 1897), and many attempts were made to produce an antitoxin, but evidence of effectiveness in humans was hard to come by in light of the scattered and occasional incidence of the disease. Thus, although Lederle Laboratories began producing an equine antitoxin in 1940 and making it freely available to the public (Cooper 1964), no firm evidence of successful use in humans was obtained until 1963 (Iida 1963; Dolman and Iida 1963; O'Horo et al. 2017), findings confirmed by a CDC review of treated cases in the US (Tacket et al. 1984).

The application of antitoxins to another common scourge—infections by the Group A hemolytic streptococcus—became useful rather late in the antitoxin era.

Scarlet fever and erysipelas are toxin-mediated complications of infection with this organism. Scarlet fever in the nineteenth and early twentieth centuries was an important cause of death, and even after a considerable decline in the case-fatality rate, 2,500 US deaths were annually attributed to scarlet fever as late as 1929–1930 (Collins 1938). While attempts to create an antiserum for scarlet fever occurred as early as 1895 in Paris (Mamorek 1895) and in 1902 in Vienna (Moser 1902) these products were hampered by the fact that the specific causative streptococcus—and its toxin—were not identified until the work of Dick and Dick in 1925. Among the several accomplishments of this remarkable medical couple was the production of a specific antitoxin to treat the disease, which they showed was valuable both for treatment and for prophylaxis after exposure (Dick and Dick 1924, 1925). From then until the end of the antitoxin era, streptococcal antitoxins were used with considerable success in hospitalized patients (Eley 1928; Veldee et al. 1931; Hunt 1933) and for the prophylaxis of local outbreaks (Banks 1929).

More complex was the antitoxin to the bacterial cause of Shigellosis. Attempts to create either an antiserum or an antitoxin (Todd 1903) were made within just a few years of the discovery of the bacillus by Shiga in 1898 (Shiga 1898). Kruse found some promising results (Kruse 1903) and Graham reported success in treating soldiers in WW I if antiserum was given early (Graham 1918), but results in the US were described as "disappointing" (Holt et al. 1904). In addition to the usual problem of late treatment of patients already in extremis, it came to be understood that more than one species of the Shigella organism causes bacterial dysentery (e.g. *S. dysenteriae*, *S. flexneri*, *S. boydii*, *S. sonnei)* and that they collectively produce a multitude of toxins (both endo- and exo-) or no detectable toxin. Simon Flexner, the founding and long-term (1901–1935) director of the Rockefeller Institute in New York, whose work with colleagues was central to the development of several forms of antiserum therapy, eventually found that it was possible to produce a polyvalent antiserum by injecting horses with more than one type of Shigella bacillus and toxin (Flexner and Amoss 1915). With better standardization of the several preparations and careful attention to dosage, Shiga reported favorable results with antitoxin in Japan in the 1920s, but only if treatment began before the patient was very ill (Shiga 1936). Efforts to develop serum therapy for tuberculosis found inconsistent results (Glatman-Freedman and Casadevall 1998).

Antitoxins have been used to treat non-infectious toxins, such as snake bites, ricin, and other chemical agents, but these topics are outside the scope of this review. Table 1 shows the earliest uses of the major antitoxins as far as we have been able to ascertain.

Table 1 Major antitoxins and their earliest human use

Disease	Organism	Year of first use in humans
Diphtheria	*Cornybacterium Diphtheriae*	1894 (Behring and Boer 1893; Roux et al. 1894)
Scarlet Fever	*Group A Streptococcus*	1895 (Mamorek 1895)
Tetanus	*Clostridium Tetani*	1896 (Kneass 1896)
Brucellosis	*Brucella Abortus*	1898 (Aldridge 1898)
Shigella	*Shigella species*	1903 (Kruse 1903)
Gas gangrene	*Clostridium Perfringens*	1919 (Beuren 1919)
Botulism	*Clostridium Botulinum*	1963 (Dolman and Iida 1963)

3 The Antiserum Era (1910–1945)

3.1 Antiserum for Bacterial Diseases

As early as 1891, Klemperer demonstrated that serum could transfer immunity to *Streptococcus pneumonia* in experimental animals thus extending the principle of passive immunity to bacterial infections whose pathophysiology, unlike diphtheria and tetanus, was not dependent on a bacterial toxin (Klemperer and Klemperer 1891). However, unlike the diphtheria antitoxin, which was greeted throughout the world as a life-saving development, antiserum treatment of pneumococcal pneumonia did not become widely used until the 1920s, in part because the type-specificity of the several types of pneumococcal infections required a match of the antibodies in convalescent plasma (CP) to the infecting serotype. Occasional uses of anti-pneumococci sera are recorded as early as the late nineteenth century (Washbourn 1897), and Anders, in 1904, was able to assemble many small series of cases treated with antiserum but concluded that "*the very slight reduction in the mortality percentage does not warrant its general introduction*" (Anders 1904). The first large series of cases in which typing of the pneumococcus strain was employed was described by Rufus Cole of the Rockefeller Institute Hospital in 1913, whose antiserum targeted the type 1 strain (Cole 1913).

Pioneering work in the use of pneumococcal antiserum was also carried out by Maxwell Finland and colleagues at the Boston City Hospital, where pneumonia accounted for about 50% of deaths, and where a cases series of more than 1,000 bacterially confirmed cases of pneumococcal pneumonia was assembled in less than 3 years (Sutliff and Finland 1934).

Using his knowledge of the capsular polysaccharides of the pneumococcus, Finland created an archive of type-specific sera from his own patients, which enabled his use of antiserum to produce a substantial reduction in mortality in the most serious complication, pneumococcal sepsis, which was nearly always fatal at the time (Finland 1930). Finland further showed, by comparing treated and untreated cases, reductions in mortality from pneumonia for all four serotypes of the pneumococcus studied (1–4) that averaged 31%. If treatment was provided in the first 3 days of illness, mortality was reduced by 48%.

In the late 1930s, Horsfall et al. (1937) and Wood (1939), concerned by the high frequency of serum sickness with horse-based preparations, revisited the use of rabbit serum. Horsfall et al. showed apparent efficacy in four pneumococcal serotypes (1,2,7,8) with many fewer reactions than with horse serum. Wood found that rabbit serum lowered mortality in both mice and people, required smaller volumes, and cost less, but unlike Horsfall et al., found adverse reactions commoner than in patients treated with horse serum. By 1940, serum and sulfonamides were being used as a pneumonia therapeutic in combination (Finland et al. 1941), successes were soon overshadowed by the discovery of penicillin as a treatment for pneumococcal infection.

Just a year after discovering the plague bacillus, Yersin et al. (1895) developed an antiserum to the agent, but results in humans using this serum and several other early sera were not encouraging (Rowland 1912). Technical difficulties in preparing the serum seem to have plagued many investigators (Platzer 1945). However, decades later, the great Indian scientist Sir Sahib Singh Sokhey developed antisera in lambs and horses that lowered plague mortality by about 50% (Sokhey 1936). Later, the same group again showed, in a 4-arm alternate assignment trial, that serum nearly halved mortality, but that sulfathiozole reduced mortality by 70% (Wagle et al. 1941). Thus, as with many antiserum preparations, effective antibiotics ended their use.

Among the earliest uses of antiserum in bacterial disease was in the treatment of meningococcal meningitis. A key figure in this effort was once again Simon Flexner, who began developing anti-meningococcal serum in horses in the first decade of the twentieth century. In 1908 Flexner and Jobling published evidence of the effectiveness of its use during several meningococcal epidemics in Ohio, Pennsylvania, New York, Scotland, and Ireland (Flexner and Jobling 1908). As with tetanus, another disease with central nervous system effects, the serum was injected directly into the spinal canal, with the removal of an amount of spinal fluid equal in volume to the injected serum. By 1913, Flexner was able to describe results in some 1,300 patients, reporting that in this previously nearly always fatal disease, mortality in serum recipients was 31%, and 18% if serum was provided in the first three days of illness (Flexner 1913).

Meningitis caused by another organism, *Haemophilus Influenza* (so named because the agent was mistakenly thought by its discoverer, Pfeiffer, to cause influenza (Pfeiffer 1893)), was treated with antiserum as early as 1916 (Torrey 1916), and its effectiveness was assumed in a major pediatric text of 1939 in its chapter on *Haemophilus influenza* (H flu) (Alexander 1940). The author of that chapter was perhaps the leading US authority on the management of this form of meningitis, Hattie Alexander. Her many papers on the subject included notable animal work on antisera to the organism with the immunologist Michael Heidelberger (Alexander and Heidelberger 1940), showing, among other things, that *H. influenza* antibodies had many similarities to pneumococcal antibodies. She strongly advocated for rabbit serum over horse serum in treatment (Alexander 1943a), and favored combining intravenous antiserum therapy with sulfa drug treatment, while reserving intrathecal therapy for non-responders to these agents (Alexander 1943b). Mortality from *H. influenza* meningitis was very nearly 100%, and early attempts to use serum were of dubious value (Torrey 1916), but by the 1930s serum treatment was found to be moderately useful, as it seems to have lowered mortality to 86% (Fothergill 1937) or to 72% (Silverthorne et al. 1937) according to investigators in Boston and Toronto, respectively.

Tularemia was treated with serum prepared in goats in 1932 (Foshay 1932), and by 1940, Foshay was able to assemble a series of 600 patients treated with serum prepared in various animal species, with some evidence of effectiveness (Foshay 1949).

As for other bacterial organisms, we find a somewhat mixed picture of the effectiveness of antiserum in Pertussis. While Scheinblum (Scheinblum and Bullowa 1944) and colleagues reported clinical improvement with antiserum treatment, and Lapin (1945) and McGuiness et al. (1940) described effectiveness in prevention in exposed family members, Frank et al. (1942) and colleagues did not see benefits. Much of the discordance may have been due to the source of the serum (human or animal), mode of preparation, and dosage and timing of treatment.

The first antiserum against anthrax was produced in 1895 by Marchoux (1895), but was used largely in animals. More widely used in humans was an anti-anthrax serum produced by Sclavo (1903) in 1903, which apparently reduced mortality from 24 to 6% in Italy. Regan, in New York City, pioneered the use of injecting anti-anthrax serum locally into the lesion with good results in a single case report (Regan 1919), later recommending adding more extensive systemic (intravenous or intramuscular) serum treatment for severe cases (Regan 1921). A 1941 review of anthrax treatment advised against local use but continued recommending the use of systemic serotherapy in combination with neoarsphenamine (Lucchesi and Gildersleeve 1941).

3.2 *Antisera for Viral Illnesses*

Viruses do not make toxins, and they thus became key targets for antiserum therapy for prophylaxis. However, the use of animals to develop antisera or antitoxins for bacterial diseases relies on the capacity to grow the pathogen of interest in the laboratory followed by immunization, a practice not available for viruses until the development of cell culture in the 1940s. Thus at first only human convalescent serum or plasma was used for the treatment of viral diseases.

Convalescent serum was used quite widely during the 1918–1920 influenza epidemic in the US, and several accounts at the time pointed to favorable results with the use of CP (summarized in Marano et al. (2016) and Luke et al. (2006)). In 2004–5, an H5N1 influenza A virus of avian origin created outbreaks, largely in China and Southeast Asia. Although the number of individuals affected was modest, mortality was very high compared to most influenza epidemics. CP was used with some apparent success, but the only published evidence was from individual case reports (Kong 2006; Zhou et al. 2007). An H1N1 influenza A virus thought to be similar to the agent of the 1918–1920 pandemic spread through much of the world in 2009, and CP from survivors of that pandemic, fractionated down to its immunoglobulins, was compared to ordinary immunoglobulin in a randomized trial in Hong-Kong when that influenza strain returned in the winter of 2010–11. If given in the first 5 days of illness, but not if given later, this product apparently reduced mortality (Hung et al. 2013).

Another virus that stimulated the use of convalescent serum was the poliovirus, a consistent public concern because of its epidemic outbreak potential that caused deaths and left many children with severe motor deficits. While the threats of diphtheria, scarlet fever, and childhood tuberculosis had to some extent faded by the 1950s, wards full of children with polio in iron lungs were a constant reminder of the limitations of medicine and public health. The most famous and dramatic response to the public concern was the development of vaccines against polio (Francis et al. 1957), while the iron lungs played an important role in initiating the era of intensive respiratory care (Wunsch 2023).

The first documented use of serum therapy in poliomyelitis was by Netter and colleagues in France in 1911 (Netter and Gendron 1911). But the results of this effort and several subsequent uses of serum were that treatment was inevitably delayed by the time it took to diagnose the disease. In experimental monkey infections, serum worked, but only if provided very soon after the virus had been introduced into the animal (Flexner and Stewart 1928). An attempt to use convalescent serum applied both intravenously and intrathecally in the pre-paralytic phase of polio during the New York City 1931 epidemic was deemed unsuccessful (Park 1932), but the following year, during a Philadelphia outbreak, Stokes et al. injected uninfected children with parental whole blood to prevent polio (Stokes et al. 1932). More than 2,000 healthy children received 60cc of whole blood (30 cc in each buttock!), 11 of whom contracted polio, and while no firm comparative data were available, the authors concluded that the cases were milder than average, with only one instance of paralysis.

The most significant effort to prevent polio by passive immunization was a pioneering randomized trial led by pediatric and public health physicians whose work in this and other areas of infectious diseases was notable.[1] The leader of the

[1] Hammon was a recipient of the Medal of Freedom (later termed Presidential Medal of Freedom) in 1947 and served as the first chair of Epidemiology and Microbiology at the University of Pittsburgh. Coriell's work on growing polio virus in tissue culture proved fundamental to the Nobel prize winning work of Enders, Weller and Robbins on that topic and he founded the Coriell Institute for Medical Research (https://www.coriell.org/). Stokes and Wehrle were both long-time chairs of pediatrics (Stokes at Children's Hospital of Philadelphia [1939–1963] and Wehrle at the University of Southern California [1961–1986]) and were two of the most significant figures in US pediatric infectious disease in the 20th century. The Children's Hospital Research Institute carries Stokes name and Wehrle played a key role in the WHO smallpox eradication campaign.

effort, William Hammon, worried that a vaccine would be difficult to support, even if were one to be developed, in light of the relative rarity and epidemic nature of polio, but felt that passive immunization had potential (Hammon 1950). In a series of five papers (Hammon et al. 1952a, b, c, 1953, 1954) in JAMA, Hammon and colleagues described their randomized, placebo-controlled trial in several states, beginning with a pilot study in Utah (Hammon et al. 1952). The intervention consisted of pooled gamma globulin obtained from thousands of WW II servicemen and women that was shown to have antibodies to all three major poliovirus strains. Pooled gamma globulin derived from human plasma was at that time separated by fractionation processes without chromatographic methods, using water-soluble organic solvents or neutral salts. The product was thus in effect, partially purified CP, unlike the more heavily treated and processed contemporary IgG (Maor and Zimhony 2024).

Over the course of a year, over 54,000 children were inoculated (Hammon et al. 1952b) with no major adverse effects (Hammon et al. 1952c). In the final analyses, protection against disease occurring between 3 days and 8 weeks after receipt of gamma globulin was 75.5%. For cases occurring from one week to five weeks post-injection, protection was 88% (Hammon et al. 1953, 1954). Despite this impressive achievement, the success of the Salk polio vaccine trial, announced just seven months after the publication of the final paper from the gamma globulin trial, led to the abandonment of passive antibody therapy for polio which thus experienced only a short period of use (Rinaldo 2005; Paul 1971).

Of the commoner childhood diseases, mumps, chickenpox, and measles have been subject to prophylaxis and/or treatment with convalescent serum. In mumps, the evidence for the prevention of the disease using human convalescent serum during epidemics and for lowering the incidence of the common complication of orchitis seems fairly strong (Rambar 1946). Hess, in New York City, appeared to have stemmed an outbreak of mumps in an orphanage by injecting whole blood from recovered children into exposed children (Hess 1915). Most of the later usage occurred in outbreaks in the military, since orchitis is thought more likely in pubertal than pre-pubertal males. Wesselhoeft reviewed several studies in 1942 and, while uncertain of the value of serum in the treatment of disease, claimed that human mumps convalescent serum "is of established value" for post-exposure prophylaxis (Wesselhoeft 1942). Despite the overall mildness of varicella, prophylactic use of convalescent serum was undertaken to prevent secondary cases in hospitals which could lead to the quarantining of wards. Successes in small case series were reported by Blackfan et al. (1923) and by Weech (1924).

Measles prevention using human convalescent serum was shown in Italy as early as 1907 (Cenci 1907) and by French scientists in 1918 (Nicolle and Conseil 1918) and was employed to stem a school outbreak in the US (Gallagher 1935) and to prevent disease in rubber plantations in Malaya (Kingsbury 1927). Animal convalescent serum was also occasionally used (Barenberg et al. 1930). Evidence for effectiveness in treating the disease was, however, mixed.

More recently, antiserum has been used in the management of newly occurring or less familiar viral infections. Antiserum effectiveness in lowering mortality in Argentine Hemorrhagic Fever was documented by a randomized trial (Maiztegui et al. 1979), and a comparison of more than 4,000 patients with the condition treated with antiserum over several decades found a mortality drop from 43% prior to antiserum use to 3% with treatment (Ruggiero et al. 1986). Small observational studies have appeared to show a favorable effect of CP on mortality in Lassa Fever (Frame et al. 1984) and Ebola Hemorrhagic Fever (Mupapa et al. 1999). In Ebola, the WHO supported the use of CP as a treatment modality (World Health Organization 2014).

Two coronavirus outbreaks occurred in the decades prior to the COVID-19 pandemic. SARS-CoV-1, then called just SARS-CoV, caused an epidemic that spread from China to the Western Hemisphere in 2002–2004. Treatment with CP was at times undertaken, with a 2015 review identifying 8 reports of use (Mair-Jenkins et al. 2015). Most studies were small, but some reasonably persuasive evidence of effectiveness came from an observational study of 80 treated patients in Hong-Kong, showing better survival and clinical improvement with early CP treatment (Cheng et al. 2005).

The outbreak of the Middle East Respiratory Syndrome in 2012, an infection caused by another beta-coronavirus very closely related to SARS-CoV-1 (Wong and Saier 2021), prompted some use of CP, and a prospectus for a trial in Saudi Arabia was proposed (Arabi et al. 2015). However, likely because of the difficulty in procuring the product, and the sporadic nature of cases, the trial seems not to have been conducted (Arabi and Hajeer 2016). A 2020 meta-analysis (Aviani et al. 2021) could identify only two case reports of CP use in MERS (Chun et al. 2016; Ko et al. 2018), both from the 2015 Korean outbreak of the disease, the largest outside of the Middle East.

Table 2 shows the earliest uses of the major antisera as far as we can determine.

4 Methods Used in Early Studies of Antitoxin and Antiserum Effectiveness

When antitoxins and antisera were first introduced into medical practice, methods for the evaluation of effectiveness were yet in their infancy. Most assessments made of the clinical effectiveness of these agents would not pass muster in today's medical world, which sees the randomized clinical trial as the gold standard in evaluating treatments. In some instances, rapid recovery as shown by reduction of fever and improvement of symptoms seemed to the authors sufficient evidence of effectiveness, as in a report of nearly 600 treated cases of scarlet fever analyzed by the NY State Health Department Laboratories (Wadsworth 1929).

Below we provide examples of the most frequently used forms of evaluation on the effectiveness of antitoxins and antisera. Despite the imperfections of the

Table 2 Major antisera and their earliest use in humans

Disease	Organism	Year of first use in humans
Bacteria		
Meningococcal Meningitis	*Neisseria meningitidis*	1906 (Jochmann 1906)
Bubonic Plague	*Yersinia pestis*	1912 (Rowland 1912)
Pneumococcal Pneumonia	*Streptococcus pneumoniae*	1890s (Klemperer and Klemperer 1891; Washbourn 1897)
Tularemia	*Bacillus tularense*	1932 (Foshay 1932)
Haemophilus Meningitis	*Haemophilus influenzae*	1916 (Torrey 1916)
Viruses		
Measles	*Measles virus (MeV)*	1907 (Cenci 1907)
Poliomyelitis	*Poliovirus*	1911 (Netter and Gendron 1911)
Mumps	*Mumps orthorubulavirus*	1915 (Hess 1915)
Influenza	*Influenza A virus*	1919 (Redden 1919)
Chickenpox	*Varicella virus*	1923 (Blackfan et al. 1923)
Argentine Hemorrhagic Fever	*Junin virus*	1959 (Martinez Pintos et al. 1961)
Lassa Fever	*Lassa virus (LASV)*	1984 (Frame et al. 1984)
Ebola Hemorrhagic Fever	*Ebola virus*	1999 (Mupapa et al. 1999)
SARS	SARS-associated coronavirus (*SARS-Cov1*)	2005 (Cheng et al. 2005)
MERS	Middle East respiratory syndrome coronavirus (*MERS-CoV*)	2016 (Chun et al. 2016)

designs, which most authors acknowledged, the very large differences favoring antitoxins and antisera found in many studies proved convincing to the medical world, as indicated by the widespread use of antitoxins and antisera before the advent of antibiotics.

1. **Comparisons of the outcomes of individuals treated with these products to individuals treated in the past when they were not available.**

Since CP or serum was often employed in epidemic situations, many evaluations of effectiveness were derived from comparisons to similar cases of disease treated in the past before the product was available. An early example is from Graham-Smith (1902) who described an epidemic of diphtheria in Colchester, England, where early in the epidemic, antitoxin was not provided in hospitals or given only to very severe cases, but on a specific date, public health authorities began to insist on antitoxin treatment of all hospitalized cases. The case-fatality rate before the general use of antitoxin was 25.9% and afterward, 5.8%. Bruce (1917), describing the use of tetanus antitoxin in British soldiers during WW I, noted that in September and October of 1914, when antitoxin was not used routinely in the wounded, the ratio of tetanus cases to wounded soldiers ranged from 15.9 to 31.8 per thousand. In the following 25 months, when antitoxin was routinely given to the wounded, the ratio ranged from 0.2 to 2.8 per thousand. These very large differences seemed to the authors unlikely to be accounted for by changes in virulence of the organism or selection of milder cases for treatment.

2. **Comparisons of individuals treated and not treated in the same period of time.**

A popular approach to the evaluation of antitoxin and antisera was to compare treated to non-treated cases, generally in the same setting. Nearly all authors gave consideration to the fact that treated and untreated cases were not exactly comparable, but in general treated cases tended to be more severely ill. Thus Eley compared 235 recipients of scarlet fever antitoxin to 250 non-recipients, showing similar rates of complications notwithstanding that antitoxin was given only to the more severe cases (Eley 1928). Hunt showed that among 2,303 scarlet fever patients, 882 of whom received antitoxin, complications were substantially reduced even though treated cases were more likely to have been described as moderately or severely ill prior to treatment (Hunt 1933). In a more recent study, a CDC assessment of the value of tetanus antitoxin, 11 potentially confounding variables were examined in a comparison of cases treated and untreated with tetanus immune globulin, but none could explain the 22% reduction in mortality with treatment (Blake et al. 1976).

3. **Comparisons of results of treatment by time in the course of illness the product was administered.**

A very frequent form of analysis was to examine the effect of time of treatment. Writing in the then-leading textbook of Pediatrics (Hodes 1939), Hodes showed a geometric increase in mortality from diphtheria in relation to the day of treatment with antitoxin. Among more than 4,000 patients reported by the Chicago Board of Health, mortality was 0.27% if treatment was on day 1 of illness, 1.67% for day 2 treatment, 3.77% for day 3, 11.39% for day 4, and 25.37% if antitoxin was given later. Similarly, Hunt (1933) showed that among antitoxin-treated cases of scarlet fever, complications ranged from 2.5% if treated on the first day of illness to 8.3% if treated on day 4 of illness or later, to 10.5% if untreated. Restriction of attention to severe cases showed even larger differences (Hunt 1933). In pneumococcal pneumonia, Cecil and Plummer (1930) and Finland (1930) both showed more than a halving of mortality from treatment if serum was provided in the first three days of illness. Convincingly, alternate assignment controls in these studies (cited below)

Table 3 Declines in diphtheria mortality in US: 1890–1924

City	Diphtheria death rate per 100,000 1890–94	Diphtheria death rate per 100,000 1920–24	Percent reduction (%)
New York City	134.4	14.0	89.6
Boston	112.3	20.2	82.0
Philadelphia	119.4	16.7	86.0
Chicago	117.3	17.5	85.1

whose non-serum treatment was begun in the first 3 days of illness showed no mortality advantage compared to later-treated controls.

4. **Dose–Response Findings**

Park and Freeman treated 1,500 contacts of measles cases in New York City with plasma obtained from convalescing cases and found measles was fully prevented in 91% of recipients of 4–10 cc of plasma,[2] but in only 72% if 3 cc was used (Park and Freeman 1926). Cole, writing in 1929, attributed the relatively low mortality rate from type 1 pneumococcus (10.2%) at the Rockefeller Institute Hospital, where convalescent serum had been used nearly universally since 1913, to their practice of using large doses of serum (Cole 1929).

5. **Examinations of vital data to note mortality trends after the introduction of these products**

Diphtheria was the commonest highly lethal disorder treated with passive antibody and thus the condition most examined in vital data. Inasmuch as diphtheria vaccine was licensed in the US in 1923, the period from 1890 to 1920 is where vital data provide insight into the effects of diphtheria antitoxin. Figures reported in JAMA in 1926 (Anon 1925) reviewed diphtheria death rates in 5-year intervals from 1890 to 1894 (before antitoxin use) to 1920–1924 in 63 US cities that had data for the full period of time (vital data then was restricted to cities, because death registration was as yet incomplete in the rural US). The four major cities tabulated above (Table 3) experienced more than an 80% reduction in diphtheria mortality. Similar trends were seen throughout the US.

It is important to recognize, however, that the incidence of diphtheria was also in decline in this period, perhaps due to improvements in hygiene and reductions in crowding. Lucia and Welke, examining data from San Francisco, a city with substantial access to antitoxin, were not convinced that the decline in diphtheria mortality prior to 1930 could be attributed solely to antitoxin (Lucia and Welke 1939). Yet there is evidence that case-fatality rates declined earlier and more rapidly

[2] Notwithstanding the paper's title, the authors state clearly that plasma was used, explaining their methodology for obtaining it.

than incidence rates in the very early years of this century, suggesting an effect of treatment (Dauer 1950).

It is sobering to consider that a diphtheria death rate of 100/100,000, a figure exceeded by all the cities tabulated above in the early 1890s, would translate to more than 300,000 deaths a year from diphtheria in the US now.

6. Randomized Trials

At least three randomized controlled trials (RCTs) have been used to assess the effectiveness of antitoxins and antisera in addition to seven quasi-randomized trials that used alternate assignment of patients to receive or not receive the treatment. The alternate assignment trials were done about a century ago (one in diphtheria (Fibiger 1898), one in plague (Wagle et al. 1941), and five in pneumococcal pneumonia (Finland 1930; Sutliff and Finland 1931; Cecil and Plummer 1930; Bullowa 1928; Medical Research Council 1934) and the three conventionally randomized trials were done in more modern times (in polio (Hammon et al. 1952a, b, c, 1953, 1954), in Argentine Hemorrhagic Fever (Maiztegui et al. 1979), and in influenza (Hung et al. 2013)).

The trial of diphtheria antitoxin was conducted in Denmark by Johannes Fibiger, who assigned treatment or non-treatment on alternating days from May 1896 to May 1897 (Fibiger 1898). The results were striking—30 of 250 untreated patients died (12%) while just 8 of 239 treated died (3.3%). In 1930, Finland described an alternate assignment trial in the Boston City Hospital that began in 1924. Patients with lobar pneumonia were alternately assigned to serum treatment or not (Finland 1930). From 1927 to 1929, serum was provided in two of the four hospital medical services and withheld in the other two, a design we would now refer to as a quasi-experiment (Shadish et al. 2002). In both experiments, mortality was lower and disease duration shorter in the serum treated. Notwithstanding the positive results of these trials, Sutcliffe and Finland undertook a further alternate assignment trial of convalescent serum in type 1 pneumococcal pneumonia, this time using concentrated serum (Sutliff and Finland 1931). Conducted in 1929–30, the study showed similar findings to their 1930 report, but with more details on the clinical course, including more rapid resolution of fever, clinical symptoms and blood culture positivity, and a lowered rate of extension of pulmonary disease. Four of the five alternate assignment trials of CP in pneumonia (two in Boston by Finland and his team (Finland 1930; Sutliff and Finland 1931) and two in New York by Cecil and Plummer (Cecil and Plummer 1930) and by Bullowa (1928) of convalescent serum in pneumococcal pneumonia reported reductions in mortality of about a third, from about 30% to about 20%. The fifth, conducted in the United Kingdom by the Medical Research Council (1934), came under considerable criticism by statistical authorities as unreliable due to poor protocol compliance (Bryder 2011). As noted above, the trial in bubonic plague, conducted in India, confirmed the value of serum, but showed that sulfathiazole was yet more effective (Wagle et al. 1941).

We have reviewed above the very large Hammon et al. RCT (Hammon et al. 1952a, b, c, 1953, 1954) of convalescent serum in polio in some detail; treatment reduced the risk of polio disease by 75.5%. The trial of antiserum prepared from patients recovered from Argentine Hemorrhagic Fever (Maiztegui et al. 1979) showed, in 188 participants, a mortality of 1.1% in the treated arm and 16.5% in the placebo arm. The small Influenza trial in Hong-Kong (17 treated, 17 controls) was conducted in very sick ICU patients 94% of whom were on ventilators and whose overall mortality was 26%. Compared to immune plasma prepared prior to the pandemic, plasma from convalescents produced the same mortality (5 deaths in the treatment arm, 4 among controls), but lower viral loads and cytokine levels. When treatment was initiated early, no deaths took place in the treatment arm (Hung et al. 2013).

5 Concluding Thoughts

Much of the history of CP is the history of forgetting. Periodic reminders of the value of CP have appeared in the medical literature (Casadevall and Scharff 1994, 1995; Casadevall 2002; Casadevall et al. 2004), as well as occasional suggestions for its use in disorders as varied as cryptococcal meningitis (Gordon and Casadevall 1995), chronic ECHO virus infection (Bodensteiner et al. 1979), and hepatitis B in the immunosuppressed (McGory et al. 1996). Yet medical leadership has by and large felt that antitoxins and antisera are therapeutics of the past and has not prioritized their use, even in pandemic situations (Paneth et al. 2022). No less an authority than Russell L. Cecil, whose 1927 *Textbook of Medicine* is now in its 27th edition, stated "*It is a fundamental principle in all serum therapy that to obtain the best results the serum must be given early in the disease*" (Cecil 1937). Sadly, that knowledge was forgotten nearly a century later when many RCTs for COVID-19 convalescent plasma were designed that tested antibody therapy in conditions that precluded its effectiveness (Casadevall et al. 2023).

We have reviewed here only a fraction of the enormous literature on the use of antitoxins and antisera in the great age of their use from 1890 to 1945, and the sparser literature of the past 80 years or so. But even this fraction of the extensive medical experience with passive immunization using convalescent serum or plasma or its derivatives of either human or animal origin points to a highly effective medical intervention that saved hundreds of thousands, perhaps millions of lives. This effectiveness was recognized by several governmental agencies. The State of Michigan (DeKruif 1949), and the cities of New York (Park 1931) and San Francisco (Lucia and Welke 1939)—to cite but a few examples—manufactured and/or funded the purchase of antitoxins, providing them without charge to people with few resources.

This effective form of intervention was relegated to the sidelines by the development of antibiotics but remains alive in the therapeutic armamentarium. Medical practitioners should be careful not to forget the lessons of the past—not just the value of using passive antibody therapy, but the importance of using it properly—treating early in the disease course with a product of known high antibody content. As antibiotic resistance increases and as new viral and bacterial agents emerge, we must use all the means at our disposal to counter these threats. History teaches us that passive antibody therapies, used appropriately, will likely be powerful tools when we confront, as we inevitably must, the next pandemic.

Declarations None of the authors has any conflicts to declare and no funding was obtained for this work.

References

Aldridge AR (1898) Note on the serum reaction of Mediterranean fever and its treatment by antitoxic plasma. Lancet 1:1394–1395

Alexander HE (1940) Influenza bacillus infections. In: Jr Holt LE, McIntosh R (eds) Diseases of infancy and childhood, 11th edn. Appleton-Century, New York

Alexander HE (1943a) Treatment of *Haemophilus influenzae* infections and of meningococcic and pneumococcic meningitis. Am J Dis Child 66:172–187

Alexander HE (1943b) Experimental basis for treatment of *Haemophilus infiuenzae* infections. Am J Dis Child 66:160–171

Alexander HE, Heidelberger M (1940) Chemical studies on bacterial agglutination. V. Agglutinin and precipitin content of antisera to *Haemophilus influenza*. J Exp Med 71(1): 1–11

Anders JM (1904) Serum treatment of pneumonia. JAMA 1777–1781

Andrewes FW (1917) On the intrathecal route for the administration of tetanus antitoxin. Lancet 189:682–686

Anon (1926) Diphtheria mortality in large cities of the United States in 1925. JAMA 86:1005–1007

Arabi YM, Hajeer AH (2016) Luke T er al: Feasibility of using convalescent plasma immunotherapy for MERS-CoV infection, Saudi Arabia. Emerg Infect Dis 22(9):1554–1561

Arabi Y, Balkhy H, Hajeer AH et al (2015) Feasibility, safety, clinical, and laboratory effects of convalescent plasma therapy for patients with Middle East respiratory syndrome coronavirus infection: a study protocol. Springerplus 19(4):709

Arnon SS, Schechter R, Maslanka SE, Jewell NP, Hatheway CL (2006) Human botulism immune globulin for the treatment of infant botulism. N Engl J Med 354:462–471

Aviani JK, Halim D, Soeroto AY et al (2021) Current views on the potentials of convalescent plasma therapy (CPT) as Coronavirus disease 2019 (COVID-19) treatment: a systematic review and meta-analysis based on recent studies and previous respiratory pandemics. Rev Med Virol 31:e2225

Banks H (1929) Intravenous antitoxin in an outbreak of scarlet fever in a public school. Brit Med J 2:50

Barenberg LH, Lewis JM, Messer WH (1930) Measles prophylaxis: comparative results with the use of adult blood, convalescent serum and immune goat serum. JAMA 95(1):4–8

Behring E (1890) Untersuchungen über das Zustandekommen der Diphtherie-Immunitat bei Thieren (Investigations into the development of diphtheria immunity in animals). Deutsche Medicinishe Wochenschrift. 50:1145–1148

Behring E, Kitasato S (1890) über das Zustandekommen der Diphtherie-Immunitat und der Tetanus-Immunitat bei Thieren. (On the development of diphtheria immunity and tetanus immunity in animals). Deutsche Medicinische Wochenschrift 49:1113–1114

Behring EA, Boer O, Kossel H (1893) Zur Behandlung diphtheriekranker Menschen mit Diphtherieheilserum (Treatment of diphtheria sufferers with diphtheria healing serum). Deutsche Medicinische Wochenschrift 17:389–93
Blackfan KD, Peterson MF, Conroy FC (1923) Use of convalescent serum as a prophylaxis in measles and chickenpox. Ohio State Med Jnl 19:97
Blake PA, Feldman RA, Buchanan TB et al (1976) Serologic therapy of tetanus in the United States, 1965–1971. JAMA 235:42–44
Bodensteiner JB, Morris HH, Howell JT et al (1979) Chronic ECHO type 5 virus meningoencephalitis in X-linked hypogammaglobulinemia: treatment with immune plasma. Neurology 29:815–819
Bruce D (1917) Note on the incidence of tetanus among wounded soldiers. Br Med J 2926:118–119
Bryder L (2011) The Medical Research Council and clinical trial methodologies before the 1940s: the failure to develop a 'scientific' approach. J Royal Soc Med 104:335–343
Bull CB, Pritchett IW (1917) Toxin and antitoxin and protective inoculation against B. Welchii. J Exp Med 26:119–138
Bullowa JGM: The use of antipneumococcic refined serum in lobar pneumonia: data necessary for a comparison between cases treated with serum and cases not so treated, and the importance of a significant control series of cases. JAMA 90:1354–1358
Carpenter CM, Boak RA (1936) The treatment of human brucellosis. A review of current therapeutic methods. Medicine (Baltimore) 15:103–128
Casadevall A (2002) Passive antibody administration (immediate immunity) as a specific defense against biological weapons. Emerg Infect Dis 8:833–841
Casadevall A, Scharff MD (1994) Serum therapy revisited: animal models of infection and development of passive antibody therapy. Antimicrob Agents Chemother 38:1695–1702
Casadevall A, Scharff MD (1995) Return to the past: the case for antibody-based therapies in infectious diseases. Clin Infect Dis 21:150–161
Casadevall A, Dadachova E, Pirofski LA (2004) Passive antibody therapy for infectious diseases. Nat Rev Microbiol 2:695–703
Casadevall A, Joyner MJ, Pirofski LA, Senefeld JW, Shoham S, Sullivan D, Paneth N, Focosi D (2023) Convalescent plasma therapy in COVID-19: unravelling the data using the principles of antibody therapy. Exp Rev Respir Med 17(5):381–395
Cecil RL (1937) Use of very early serum treatment in pneumococcus type I pneumonia. JAMA 108:689–692
Cecil RL, Plummer N (1930) Pneumococcus type I pneumonia—a study of eleven hundred and sixty-one cases, with especial reference to specific therapy. JAMA 95:1547–1553
Cenci F (1907) Alcune esperienze di sieroimmunizzazione e sieroterapia nel morbillo (Some experiences of seroimmunization and serotherapy for Measles). Rivista di Clinica e Pediatrica 5:1017–1025
Centers for Disease Control and Prevention (2010) Investigational heptavalent botulinum antitoxin (HBAT) to replace licensed botulinum antitoxin AB and investigational botulinum antitoxin E. MMWR 59:299
Cheng Y, Wong R, Soo YOY et al (2005) Use of convalescent plasma therapy in SARS patients in Hong Kong. Eur J Clin Microbiol Infect Dis 24(1):44–46
Chun S, Chung CR, Ha YE (2016) Possible transfusion-related acute lung injury following convalescent plasma transfusion in a patient with Middle East Respiratory Syndrome. Ann Lab Med 36(4):393–395
Clinical management of diphtheria: guideline, 2 February 2024, World Health Organization, Geneva, 2024 (WHO/DIPH/Clinical/2024.1). Licence: CC BY-NC-SA 3.0 IGO
Cole R (1913) Treatment of pneumonia by use of specific serums. JAMA 61:663
Cole R (1929) Serum treatment in type 1 lobar pneumonia. LAMA 93:741–747
Collins SD (1938) History and frequency of clinical scarlet fever cases and of injections for artificial immunization among 9,000 families, based on nation-wide periodic canvasses, 1928–31. Public Health Rep 53:409–427

Cooper MS (1964) Antitoxins to *C. botulinum.* In: Lewis KH, Cassel K (eds) Botulism: proceedings of a symposium. USDHEW, Cincinnati
Dauer CC (1950) Trends in age distribution of diphtheria in the United States. Public Health Rep 38:1209–1218
DeKruif P (1949) Life among the doctors. Harcourt, Brace, New York
Dick GF, Dick GH (1924) A scarlet fever antitoxin. JAMA 82:1246–1247
Dick GF, Dick GH (1925) Therapeutic results with concentrated scarlet fever antitoxin. JAMA 85:1693–1695
Dolman CD, Iida H (1963) Type E botulism: its prevention and specific treatment. Can J Pub Health 54:293–308
Eley RC (1928) Scarlet-fever antitoxin—report of four hundred and sixty-five cases. Am J Dis Child 35:14–17
Fibiger J (1898) Om Serumbehandling af Difteri (On the serum treatment of diphtheria). Hospitalstidende 6:309–325, 337–350
Finland M (1930) The serum treatment of lobar pneumonia. N Engl J Med 202(26):1244–1247
Finland M, Lowell FC, Strauss E (1941) Treatment of pneumococcic pneumonias with sulfapyridine, sulfathiazole and serum: analysis of the results of specific therapy at the Boston City Hospital from July 1939 through June 1940. Ann Intern Med 14(7):1184–1185
Flexner S (1913) The results of the serum treatment in 1300 cases of epidemic meningitis. J Exp Med 17(5):553–576
Flexner S, Jobling JW (1908) Serum treatment of epidemic cerebro-spinal meningitis. J Exp Med 10(1):141–203
Flexner S, Stewart FW (1928) Protective action of convalescent poliomyelitis serum. JAMA 91:383–384
Flexner S, Amoss HL (1915) The rapid production of antidysenteric serum. J Exp Med 19:515–524
Foshay L (1932) Serum treatment of tularemia. JAMA 98:552
Foshay L (1949) Tularemia: a summary of certain aspects of the disease including methods for early diagnosis and the results of serum treatment in 600 patients. Medicine 19:1–84
Fothergill LD (1937) Hemophilus influenzae (*Pfeiffer bacillus*) meningitis and its specific treatment. New Engl J Med 216:587–590
Frame JD, Verbrugge GP, Gill RG, Pinneo L (1984) The use of Lassa fever convalescent plasma in Nigeria. Trans R Soc Trop Med Hyg 78(3):319–324
Francis T Jr, Napier JA, Voight RB et al (1957) Evaluation of the 1954 field trial of poliomyelitis vaccine. Edwards Brothers, Ann Arbor
Frank WP, Patton EF, Hamilton PM (1942) Treatment of whooping cough bronchopneumonia. J Pediatr 20:720–722
Gallagher JR (1935) Use of convalescent measles serum to control measles in a preparatory school. Am J Public Health 25:595–598
Glatman-Freedman A, Casadevall A (1998) Serum therapy for tuberculosis revisited: reappraisal of the role of antibody-mediated immunity against *Mycobacterium tuberculosis.* Clin Microbiol Rev 11:514–532
Gordon MA, Casadevall A (1995) Serum therapy for cryptococcal meningitis. Clin Infect Dis 21:477–479
Graham D (1918) Some points in the diagnosis and treatment of dysentery occurring in the British Salonika force. Lancet 1:51–55
Graham-Smith GA (1902) The measures taken to check the diphtheria outbreak of 1901 at Colchester. Epidemiol Infect 2:170–193
Hammon WM (1950) Possibilities of specific prevention and treatment of poliomyelitis. Pediatrics 6:696–705
Hammon WM, Coriell LL, Stokes J Jr (1952a) Evaluation of Red Cross gamma globulin as a prophylactic agent for poliomyelitis. I. Plan of controlled field tests and results of 1951 pilot study in Utah. J Am Med Assoc 150(8):739–749

Hammon WM, Coriell LL, Wehrle PF, Klimt CR, Stokes J Jr (1952b) Evaluation of Red Cross gamma globulin as a prophylactic agent for poliomyelitis. 3. Preliminary report of results based on clinical diagnosis. J Am Med Assoc 150(8):757–760
Hammon WM, Coriell LL, Stokes J Jr (1952c) Evaluation of Red Cross gamma globulin as a prophylactic agent for poliomyelitis. 2. Conduct and early follow-up of 1952 Texas and Iowa-Nebraska studies. J Am Med Assoc 150(8):750–756
Hammon WM, Coriell LL, Wehrle PF, Stokes J Jr (1953) Evaluation of red cross gamma globulin as a prophylactic agent for poliomyelitis. IV. Final report of results based on clinical diagnoses. J Am Med Assoc 151(15):1272–1285
Hammon WM, Coriell LL, Ludwig ED, McAllister RM, Greene AE, Sather GE, Wehrle PF (1954) Evaluation of red cross gamma globulin as a prophylactic agent for poliomyelitis. 5. Reanalysis of results based on laboratory-confirmed cases. J Am Med Assoc 156(1):21–27
Hess AE (1915) A protective therapy for mumps. Am J Dis Child 10:99–103
Hifumi T (2020) Spontaneous non-traumatic *Clostridium perfringens* Sepsis. Jpn J Infect Dis 73 (3):177–180
Hifumi T, Nakano D, Chiba J et al (2018) Combined therapy with gas gangrene antitoxin and recombinant human soluble thrombomodulin for *Clostridium perfringens* sepsis in a rat model. Toxicon 141:112–117
Hifumi T, Koido Y, Takahashi M (2013) Antitoxin treatment for liver abscess caused by *Clostridium perfringens*. Clin Mol Hepatol 19:97–98
Hodes H (1939) Diphtheria. In: Holt LE Jr, McIntosh R (eds) Holt's diseases of infancy and childhood, 11th edn. Appleton-Century, New York
Holt LE (1904) Clinical conclusions. In: Flexner S, Holt LE (eds) Bacteriological and clinical studies of the diarrheal diseases of infancy with reference to the *Bacillus dysinteriae* (Shiga). Rockefeller Institute, New York, p 189
Horsfall FL, Goodner K, MacLeod CM et al (1937) Serum as a therapeutic agent in lobar pneumonia. JAMA 108(18):1483–1490
https://www.cdc.gov/botulism/hcp/clinical-overview/index.html#:~:text=Botulism%20Antitoxin%20Heptavalent%20and%20BabyBIG,of%20paralysis%20and%20consequent%20complications
Hung IFN, To KKW, Lee C-K et al (2013) Hyperimmune iv immunoglobulin treatment: a multicenter double-blind randomized controlled trial for patients with severe 2009 influenza a (H1N1) infection. Chest 144:464–473
Hunt LW (1933) The treatment of scarlet fever with antitoxin. JAMA 101:1444–1447
Iida H (1963) Specific antitoxin therapy in type E botulism. Japan J Med Sci Biol. 16:311–313
Jochmann G (1906) Versuche zur Serodiagnostik und Serotherapie der epidemischen Genickstarre (Experiments on the serodiagnosis and serotherapy of epidemic nuchal rigidity). Deutsche Medicinische Wochenschrift 1:788–793
Kaufmann SHE (2017) Emil von Behring: translational medicine at the dawn of immunology. Nat Rev Immunol 17:341–343
Kingsbury AN (1927) Watson: the serum prophylaxis of measles. Epidemiol Infect 27:1–23
Kitasato S (1889) Uber dem Tetanusbacillus (On the tetanus bacillus). Zeitschr Hyg Infekt 7:225–234
Klemperer G, Klemperer F (1891) Versuche uber immunisirung und heilung bei der pneumokokkeninfection (Experiments on immunization and therapy for pneumococcal infection). Berl Klin Wchnschr. 28:833–835
Kneass SS (1896) The antitoxin of tetanus. JAMA 27:134–138
Ko J-H, Seok H, Cho SY (2018) Challenges of convalescent plasma infusion therapy in Middle East respiratory coronavirus infection: a single centre experience. Antivir Ther 23(7):617–622
Koch R (1884) Die aetiologie der tuberculose (The etiology of tuberculosis). Mittheilungen Aus Dem Kaiserlichen Gesundheitsamt 2:1–88
Koch R (1890) Ueber bakteriologische Forschung (On bacteriological research). Verhandlungen des X Internationalen Medichinischen Congresses, Berlin, 1, 35, August Hirschwald, Berlin
Kong LK (2006) Successful treatment of avian influenza with Convalescent plasma. Hong Kong Med J 12:489

Kruse W (1903) Die Blutserum therapie bei der Dysenterie (Serum therapy for dysentery). Deutsche Medicinische Wochenschrift, 1:6–8

Lapin JH (1945) Serum in the prophylaxis of contacts and the treatment of whooping cough. J Pediatr 26:555–559

Lindenmann J (1984) Origins of the terms “antibody” and “antigen.” Scand J Immunol 19:281–285

Linton DS (2000) The obscure object of knowledge: German military medicine confronts gas gangrene during world war I. Bull Hist Med 74:291–316

Loeffler F (1884) Untersuchungen über die Bedeutung der Mikroorganismen für die Entstehung der Diphtheric beim Menschen, bei der Taube und der Kalbe (Studies on the significance of bacteria in causing diphtheria in man, pigeons and calves). Mitteilung Klinische Gesundheit, Berlin 2:451–499

Lucchesi PF, Gildersleeve N (1941) The treatment of anthrax. JAMA 116:1506–1508

Lucia EL, Welke HF (1939) Trends in the prevalence of diphtheria. J Infect Dis 58:306–317

Luke TC, Kilbane EM, Jackson JL, Hoffman SL (2006) Meta-analysis: convalescent blood products for Spanish influenza pneumonia: a future H5N1 treatment? Ann Intern Med 145:599–609

MacConkey A (1914) Tetanus: its prevention and treatment by means of antitetanic serum. BMJ 2 (2806):609–614

Macfarlane MG (1943) Therapeutic value of gas-gangrene antitoxin. Br Med J 2(4324):636–640

Mair-Jenkins J, Saavedra-Campos M, Baillie K et al (2015) The effectiveness of convalescent plasma and hyperimmune immunoglobulin for the treatment of severe acute respiratory infections of viral etiology: a systematic review and exploratory meta-analysis. J Infect Dis 211:80–90

Maiztegui JI, Fernandez NJ, de Damilano AJ (1979) Efficacy of immune plasma in treatment of Argentine haemorrhagic fever and association between treatment and a late neurological syndrome. Lancet 2(8154):1216–1217

Mamorek A (1895) Le streptocoque et le serum antistreptococique (The streptococcus and antistreptococcal serum). Ann de l'Inst Pasteur 9:592–620

Maor Y, Zimhony O (2024) Hyperimmune globulins in COVID-19. Curr Top Microbiol Immunol. https://doi.org/10.1007/82_2024_277

Marano G, Vaglio S, Pupella S et al (2016) Convalescent Plasma: new evidence for an old therapeutic tool. Blood Transfus 14(2):152–157

Marchoux E (1895) Serum anticharbonneux (Antianthrax serum). Annales de l’institut Pasteur 9:785

Martinez Pintos IF et al (1961) Nuestra experiencia en el tratamiento del “mal de los rastrojos” (Our experience with the treatment of “stubble disease”). Sem Med (B Aires) 118:856–864

McGory RW, Ishitani MB, Oliveira WM et al (1996) Improved outcome of orthotopic liver transplantation for chronic hepatitis B cirrhosis with aggressive passive immunization. Transplantation 61:1358–1364

McGuinness AC, Bradford WL, Armstrong JC (1940) The production and use of hyperimmune human whooping cough serum. J Pediatr 16:21–29

Medical Research Council (1934) The serum treatment of lobar pneumonia. A report of the therapeutic trials committee of the Medical Research Council. Lancet 1:290–295

Moser P (1902) Ober die Behandlung des Scharlachs mit einer Schar lachstreptokkenserum (On the treatment of scarlet fever with an anti-streptoccal serum) Wien. Klin Wchnschr 15:1053–1055

Mupapa K, Massamba M, Kibadi K, Kuvula K, Bwaka A, Kipasa M et al (1999) Treatment of Ebola hemorrhagic fever with blood transfusions from convalescent patients. International scientific and technical committee. J Infect Dis 179(Suppl 1):S18–S23

Nakayama DK (2022) The Christmas miracle of Emil von Behring. Am Surg 23:31348221129515

Nation NS, Pierce NF, Adler SJ, Chinnock RF, Wehrle PF (1963) Tetanus: the use of human hyperimmune globulin in treatment. Calif Med 98:305–307

Netter A, Gendron A, Touraine (1911) Serotherapie de la poliomyelite anterieure aigue (serotherapy of acute anterior poliomyelitis). Comptes Rendu Des Séance De Societe De Biologie 70:625

Nicolle C, Conseil E (1918) Pouvoir preventif du serum d'un malade convalescent de rougeole (Preventive power of serum from a patient convalescing from measles). Bulletins et Mémoires de la Société Médicale des Hôpitaux de Paris 42:336

Nocard, E (1897) Sur la sérothérapie du tétanos (On the serotherapy of tetanus). Receuil Méd. Vét. 8e série, tome 4:481–490

O'Horo JC Harper EP, El Rafei A (2017) Efficacy of antitoxin therapy in treating patients with foodborne botulism: a systematic review and meta-analysis of cases, 1923–2016. Clin Infect Dis 66(suppl_1):S43–S66

Paneth N, Casadevall A, Pirofski L-A et al (2022) WHO covid-19 drugs guideline: reconsider using convalescent plasma. BMJ 8(376):o295

Park WH (1931) The history of diphtheria in New York City. Am J Dis Child 42:1439–1445

Park WH (1932) Therapeutic use of antipoliomyelitis serum in preparalytic cases of poliomyelitis. JAMA 99:1050–1053

Park WH, Freeman RG (1926) The prophylactic use of measles convalescent serum. JAMA 87:556–558

Paul JR (1971) A history of poliomyelitis. Yale University Press, New Haven

Pfeiffer R (1893) Die Atiologie der Influenza (The etiology of influenza). Zeitschrift für Hygiene und Infektionskrankheiten 18:357

Platzer RF (1945) Evaluation of therapeutic agents in plague. A review of the literature. US Naval Med Bull 46:1674–1688

Rambar AC (1946) Mumps: use of convalescent serum in the treatment and prophylaxis of orchitis. Am J Dis Child 71(1):1–13

Redden WR (1919) Treatment of influenza-pneumonia by use of convalescent human serum. Boston Med Surg J 161:688–691

Regan J (1919) The local use of antianthrax serum in the treatment of anthrax. JAMA 72:1724–1725

Regan JC (1921) The local and general treatment of cutaneous anthrax. JAMA 77:1044–1048

Rinaldo CR (2005) Passive immunization against poliomyelitis. Am J Pub Health 95:790–799

Rosenau MJ (1913) Preventive Medicine and Hygiene. Appleton, New York, p 383

Roux E, Martin L, Chaillou A (1894) Trois cent cas de diphtherie traite par le serum antidiphtherique (Three hundred cases of diphtheria treated using anti-diphtheria serum). Annales De L'institut Pasteur 8:640–661

Roux E, Yersin A (1888) Contribution à l'étude de la diphtérie (Contributions to the study of diphtheria). Annales de l' Institut Pasteur 2:629–661

Rowland S (1912) Serum treatment of human plague in: Seventh report on plague investigations in India. J Hyg 12 (Plague Supplement lvi):326–339

Ruggiero HA, Pérez Isquierdo F, Milani HA, Barri A, Val A, Maglio F et al (1986) Treatment of Argentine hemorrhagic fever with convalescent's plasma. 4433 cases. Presse Med 15 (45):2239–2242

Scheinblum IE, Bullowa JGM (1944) The treatment of pertussis with lyophile hyperimmune human pertussis serum. J Pediatr 25:49–55

Sclavo S (1903) Sullo stato presente della serioterapia anticarbonchiosa (On the present state of antianthrax serotherapy). Rivista d'Igiene e di Santa Publica 14:519

Shadish WR, Cook TD, Campbell DT (2002) Experimental and quasi-experimental designs for generalized causal inference. Houghton-Mifflin, Boston

Shiga K (1898) Ueber den Erregerder Dysenterie in Japan (On the causative agents of dysentery in Japan). Vorlaufige Mitteilung. Zentralbl Bakteriol Microbiol Hyg 23:599–600

Shiga K (1936) The trend of prevention, therapy and epidemiology of dysentery since the discovery of its causative organism. NEJM 15:120–211

Silverthorne N, Fraser DT, Snelling CE (1937) Influenzal meningitis. J Pediatr 10:228–323

Sokhey SS (1936) Un nouveau serum antipesteux (A new antiplague serum). Bull Office Internat. d'hygiene Publique e 28:1097–1100

Spaeth R (1941) Therapy of tetanus. A study of two hundred and seventy-six cases. Arch Intern Med 68:1133–1160

Stokes J Jr, Wolman IJ, Carpenter HC et al (1935) Prophylactic use of parents' whole blood in anterior poliomyelitis: Philadelphia epidemic of 1932. Am J Dis Child 50(3):581–595

Sutliff WD, Finland M (1931) Type 1 lobar pneumonia treated with concentrated pneumococcic antibody (Felton)—the clinical course. JAMA 96:1465–1469

Sutliff WD, Finland M (1934) Type 1 pneumococcic infection with especial reference to specific serum treatment. NEJM 210:237–245

Tacket CO, Shandera WX, Mann JM, Hargrett NT, Blake PA (1984) Equine antitoxin use and other factors that predict outcome in type A foodborne botulism. Am J Med 76:794–798

Todd C (1903) On a dysentery antitoxin. Br Med J 2(2240):1456–1458

Torrey RG (1916) Influenzal meningitis with report of a case. Am J Med Sci 152:403

Van Beuren FT (1919) The treatment of gas bacillus infection. JAMA 73:239–242

Van Ermengem E (1897) Ueber einen neuen anaeroben Bacillus und seine Beziehungen zum Botulismus (On a new anaerobic bacillus and its relationship to botulism). Zeitschrift fur Hygiene und Infektionskrankheiten 26:1–56

Veldee MV, Stevenson FE, Mitchell EG (1931) Scarlet fever streptococcus antitoxin in the treatment of scarlet fever. Public Health Rep 51:3023–3050

Wadsworth AB (1929) The hemolytic streptococci and anti streptococcus serum in scarlet fever. Am J Public Health 19:1287–1302

Wagle PM, Sokhey SS, Dikshit BB et al (1941) Chemotherapy in Plague. Ind Med Gaz 76:29–32

Washbourn JM (1897) Anti-pneumoccic serum. BMJ 1:510–514

Weech AA (1924) The prophylaxis of varicella with convalescent's serum. JAMA 82:1245–1246

Weinberg M, Seguin P (1918) La Gangrene Gazeuse (Gas Gangrene). Masson et Cie, Paris

Welch WH, Nuttall GHE (1892) A gas producing bacillus (*Bacillus aerogenes capsulatus*, nov. spec.) capable of rapid development in the blood vessels after death. Bull Johns Hopkins Hosp 3:81

Wesselhoeft C (1942) Mumps. New Engl J Med 226:530–534

Wong NA, Saier MH (2021) The SARS-coronavirus infection cycle: a survey of viral membrane proteins, their functional interactions and pathogenesis. Int J Mol Sci 22(3):1308

Wood WB (1939) Treatment of pneumococcic pneumonia with concentrated antipneumococcus rabbit serum. JAMA 113(9):745–749

World Health Organization (2014) Use of convalescent whole blood or plasma collected from patients recovered from Ebola virus disease for transfusion, as an empirical treatment during outbreaks: interim guidance for national health authorities and blood transfusion services. World Health Organization

Wunsch H (2023) The autumn ghost: how the battle against a polio epidemic revolutionized modern medical care. Greystone Books, Vancouver

Yersin A, Calmette A, Borrel A (1895) La peste bubonique. Deuxieme note. (Bubonic Plague, 2nd note). Ann Inst Pasteur 9:589–592

Zhou B, Zhong N, Guan Y (2007) Treatment with convalescent plasma for influenza A (H5N1) infection. N Engl J Med 357:1450–1451

Nigel Paneth is University Distinguished Professor Emeritus and founding chair of the Department of Epidemiology & Biostatistics at MSU. His principal research is in perinatal and child developmental epidemiology and he is a member of the leadership team of the National COVID-19 Convalescent Plasma Project (CCPP19). He is the author or editor of five books and 378 peer-reviewed publications.

Madison Walsh is a Research Assistant at Michigan State University in the Department of Obstetrics, Gynecology and Reproductive Biology. Her past work includes serving as a graduate assistant for CCPP19. Her current work involves substance use disorder, treatment, and subsequent health outcomes.

Breanna Kornatowski is a research study manager, responsible for follow up assessments in the Michigan State University center of the NIH funded "Environmental Influences on Child Health Outcomes (ECHO)" project. She has been involved in CCPP.19 from its inception, helping to manage its website, CCPP19.org. Her interests include maternal and child public health, health administration, and efficiency in observational research.

Arturo Casadevall Ph.D. is a Bloomberg Distinguished Professor and Chair of the Department of Molecular Microbiology and Immunology at the Johns Hopkins School of Public Health. He has been involved in the development of antibody therapies for infectious diseases for more than three decades and chaired the CCPP19 group during the COVID-19 crisis.

The Logic and History of Passive Immunity and Antibody Therapies

Arturo Casadevall and Nigel Paneth

Contents

Abstract This volume takes a broad overview of antibody-based therapies prior to and during the COVID pandemic and examines their potential use in future pandemics. Passive antibody therapy was the first effective antimicrobial treatment and its development in the early twentieth century helped catalyze immunological and microbiological research. During the era of serum therapy (1890–1940) antibody-based therapies were developed against both viral and bacterial diseases. Effective treatment required an understanding of how to quantify antibodies, how to develop serotype-specific sera and recognition of the need to treat early in disease. Thus, although the era of serum therapy essentially ended with the development of small molecule antimicrobial therapy in the 1940s, antibody-based therapies led to

A. Casadevall (✉)
Department of Molecular Microbiology and Immunology, Johns Hopkins School of Medicine, Baltimore, MD, USA
e-mail: acasade1@jhu.edu

N. Paneth
Departments of Epidemiology and Biostatistics and Pediatrics and Human Development, College of Human Medicine, Michigan State University, East Lansing, MI, USA
e-mail: paneth@msu.edu

Current Topics in Microbiology and Immunology (2025) 443: 41–52
https://doi.org/10.1007/82_2024_267

Published Online: 22 May 2024

important new scientific understanding, while remaining in use for some toxin and venom-caused diseases and in the prevention of outbreaks of viral hepatitis. A renewed interest in antibody-based therapies was seen in the widespread deployment of convalescent plasma and monoclonal antibodies during the COVID-19 pandemic. Convalescent plasma will likely be the first specific therapy during outbreaks with new pathogens for which there is no other therapy. For all forms of antibody-based therapies, effectiveness relies on the key principles of antibody therapy, namely, treatment early in disease with preparations containing sufficient antibody specific to the microbe in question.

For many infectious diseases, the presence of antibody to the specific causative microbe is synonymous with immunity. Having sufficient antibody to Hepatitis B virus, diphtheria toxin, and Hemophilus influenza type B means that the individual is protected against disease caused by those pathogens. However, this rule is not universal, as serum antibody to infections by the Human Immunodeficiency Virus (HIV) and *Mycobacterium tuberculosis* does not confer protection against disease. However, in these diseases, as in others, antibodies are always a marker of prior infection, prior immunization, or both.

The logic of passive immunity and antibody therapies is based on the concept that transfer of specific antibodies from an immune host to a non-immune host confers immunity in the latter. Such passive transfer of immunity was first demonstrated for toxins in 1890 by von Behring and Kitasato (Borrell 2008), and extended to non-toxin-producing organisms by Klemperer et al. (Ivens 1918), as reviewed in greater detail in Chap. 5 (P3). Von Behring's discoveries were recognized by the first Nobel Prize in Medicine or Physiology, awarded in 1901 (Watson et al. 1993).

1 The Age of Serum Therapy

The Age of Serum therapy spans roughly 50 years from 1890 to 1940. It was a time when the major antimicrobial agents were specific antibodies found in the serum of immune hosts, both animal and human. For microbes that could be grown in vitro, such as bacteria, serum was raised by immunizing large animals such as horses or cows and using the immune serum thus produced for human therapy. For microbes that could not be grown in the laboratory, such as measles, passive antibody therapy used serum obtained from humans who had recently recovered from such viral diseases.

The age of serum therapy coincided with the beginning of modern medicine and its therapeutic power left its mark on the general language as in the phrase 'truth serum'. Although 'truth serum' refers to drugs and not to serum, the acceptance of the efficacy of serum at the time gave name to a family of drugs that were thought to be useful in interrogation (Borrell 2008). The age of serum therapy also left a mark on the national heritage. Today, the Iditarod Trail Sled Dog Race in Alaska commemorates a great effort in 1925 to bring diphtheria antitoxin to Nome to help stay a diphtheria epidemic. Serum therapy contributed to the war effort since it was used widely in World War I for the prophylaxis and therapy of both tetanus (caused by *Clostridium tetani*) and gas gangrene (caused by *Clostridium perfringens*) since both organisms were frequent contaminants of shrapnel wounds (Ivens 1918).

The finding that serum was therapeutic catalyzed a tremendous interest in, and research into, the field of immunology. In the first half of the twentieth century, lobar pneumonia was a major killer of adults, and much of serum therapy research was devoted to better and more detailed understanding of the causative agent, *Streptococcus pneumoniae* (Watson et al. 1993). This in turn led to such major biological discoveries as the 'transforming principle' by Griffith (1928), whereby pneumococcus regained virulence by what was ultimately shown to be DNA. Furthermore, it is worthwhile to note that the finding that DNA was responsible for passage of genetic information came from experiments related to the pneumococcal capsule and virulence of the organism (Avery et al. 1944), where Avery and collaborators at the Rockefeller Institute had been working for decades on various aspects of the capsular polysaccharide. The discovery of pneumococcal serotypes allowed the generation of serotype-specific serum to match the infecting organism. By the early 1930s the therapy of pneumococcal pneumonia had been largely perfected with tools for rapid diagnosis such as the capsular swelling reaction (Quellung) and the availability of various type-specific serums. The age of serum therapy thus served as an important catalyst for the fuller understanding of both microbial pathophysiology and humoral immunity in humans.

Despite successes against toxin-mediated diseases such as diphtheria and tetanus, and also against a variety of bacterial and viral pathogens (Table 1), a consistently effective serum against *Mycobacterium tuberculosis* could not be found (Glatman-Freedman and Casadevall 1998). Likewise, efforts to generate effective serum therapy against cancers were not successful. The realization that pathogenic microbes differed in their susceptibility to therapeutic serums was an important catalyst in the search for effective immunity against these organisms, which eventually identified the concept that organisms like *M. tuberculosis* were contained by tissue immunity, the intellectual precursor of cell-mediated immunity. Incidentally, this finding catalyzed a split in immunology between considering that antibody- and cell-mediated immunity were responsible for extracellular and intracellular pathogens, respectively, a view that remained paradigmatic until the 1990s, although eventually shown to be erroneous (Casadevall and Fang 2020). The problem was not that humoral immunity could not be effective, but that some microbes tended to elicit non-effective antibody responses that were sometimes comprised of a mix of protective and non-protective antibodies such that their sum effect was not protective or curative. In the 1990s, investigators used hybridoma technology to generate monoclonal antibodies (see below) and when these were tested in animal models of infection, protective antibodies were identified for

Table 1 Diseases that have been treated with immune or convalescent serum[a]

Disease	Causative organism	References[b]
Anthrax	*Bacillus anthracis*	Lucchesi and Gildersleeve (1941)
Botulism	*Clostridium botulinum*	Tacket et al. (1984)
Brucellosis	*Brucella abortus*	Carpenter and Boak (1936)
Diphtheria	*Corynebacterium diphtheriae*	Wesselhoeft (1994)
Dysentery	*Shigella* spp.	Manson-Bahr (1939)
Gas gangrene	*Clostridium perfringens*	Ivens (1918)
Hemophilus meningitis	*Hemophilus influenzae*	Fothergill (1937)
Meningococcal meningitis	*Neisseria meningitidis*	Casadevall and Scharff (1994)
Mumps	Mumps virus	Rambar (1946)
Pneumococcal pneumonia	*Streptococcus pneumoniae*	Casadevall and Scharff (1994)
Poliomyelitis	Polio virus	Park (1932)
Scarlet fever and Erysipelas	Group A streptococcus	Wadsworth (1929)
Tetanus	*Clostridium Tetani*	Wever and Bergen (2012)
Tuberculosis	*Mycobacterium tuberculosis*	Glatman-Freedman and Casadevall (1998)
Tularemia	*Francisella tularensis*	Foshay (1940)
Whooping cough	*Bordetella pertussis*	Frank et al. (1942)

[a]Not a complete list
[b]Representative reference. For many of these diseases extensive literature is available on the use of serum therapy

pathogens such as *M. tuberculosis* (Teitelbaum et al. 1998), *Histoplasma capsulatum* (Nosanchuk et al. 2003), and *Listeria monocytogenes* (Edelson et al. 1999) among others, establishing the efficacy of antibodies provided that these expressed the right specificity and effector function.

For viral diseases, it was not possible to generate therapeutic sera in animals since the technology of viral growth in cell culture followed by virion purification did not exist, and serum therapy thus relied on convalescent human serum. This kind of serum therapy was used during the 1918 influenza pandemic, and a retrospective analysis estimated that its use reduced mortality by 20% (Luke et al. 2006). For measles and mumps, convalescent plasma was reportedly effective when used early in disease. However, for polio the results with serum therapy were inconsistent, possibly because antibody was used too late and had no efficacy during the paralytic phase of disease. The use of human convalescent plasma for viral diseases was sometimes associated with serum hepatitis and this phenomenon led to the discovery of viral hepatitis.

2 Abandonment of Serum Therapy

Serum therapy was largely abandoned with the development of small molecule antimicrobial therapy. With the introduction of sulfonamide in 1935 and penicillin in 1942, physicians had effective antimicrobial therapy that was considered superior to serum therapy. In contrast to serum therapy, where physicians had to match the type of serum to the infective organism and had to make a specific diagnosis, antimicrobial therapy was much less specific and a single agent could often treat many types of infectious diseases without the need for pathogen identification. Casadevall has argued that this shift in the ability to treat bacterial infections without making a specific microbial diagnosis was to have long-term consequences on the development of the infectious disease specialty and could have contributed to a culture of empiricism and the emergence of antimicrobial resistance (Casadevall 2017). Antimicrobial therapy was easier to administer and did not require careful observation for the allergic reactions that were associated with the use of horse serum, such as serum sickness, a self-limited illness characterized by rash and arthralgias occurring days or weeks after serum administration. This disease was produced by antigen–antibody complexes as patients mounted antibody responses to other horse proteins. There were some attempts to determine whether the combination of serum therapy and antimicrobial therapy was more effective than either, but these showed little or no additional benefit over antimicrobial therapy, which hastened abandonment of serum therapy (Plummer et al. 1941).

It is a historical paradox that serum therapy was abandoned at precisely the same time when great progress was being made in the purification of immunoglobulins. The Cohn fractionation method, developed in the early 1940s, allowed the purification of immunoglobulins by differential precipitation with alcohol (Cohn and Strong 1946), resulting in highly purified antibody preparations that had many fewer side effects than animal sera. One is left to wonder whether, had antimicrobials been developed a few years later, or Cohn fractionation a few years earlier, antibody therapies would have been maintained as anti-infective therapies.

3 Passive Antibody Therapy Retains a Niche

Despite being abandoned for the treatment of bacterial diseases, serum therapy has retained a niche for the treatment of some viral diseases, toxin-mediated diseases, and envenomation from the bites of poisonous animals. Serum therapy remains first-line therapy for the treatment of tetanus (Blake et al. 1976) and bites from rabid animals (using antibody preparations derived from human donors (Hattwick et al. 1976) or immunized horses (Lang et al. 1998). Over subsequent decades a variety of hyperimmune globulins derived from human donors have been developed for the treatment or prevention of Hepatitis A and B, varicella zoster virus, respiratory syncytial virus, and the toxigenic bacteria *C. tetani* and *Clostridium difficile*

(Pati et al. 2023). Hyperimmune globulins are preparations of purified human antibodies where the amount of specific antibody is increased by obtaining material from many convalescent or vaccinated individuals, providing more specific antibody than is possible with single-donor convalescent or vaccinated serums or plasmas. The term gamma globulin is often used to describe such preparations since the material is predominantly IgG. It is noteworthy that even though childhood diphtheria is largely under control in vaccinated populations, horse antiserum preparations are still being used for unvaccinated cases and in regions that lack vaccines (Pati et al. 2023).

4 The Monoclonal Antibody Revolution Largely Bypasses Infectious Diseases

In 1975, the development of hybridoma technology by Kohler and Milstein (1975) allowed for the production of homogenous preparations of monoclonal antibodies (mAbs). When first introduced, there was high hope for their use against infectious diseases, but this hope was unrealized. mAbs by virtue of being a homogenous preparation of a single immunoglobulin recognize only a single epitope and were all composed of a single isotype. This precise specificity has proven a weakness for the use of mAbs against pathogenic microbes, which are usually antigenically diverse. Expectations were raised for using monoclonals in the treatment of endotoxin-mediated sepsis, with some promising early reports (Ziegler et al. 1991). However, only patients who were endotoxin positive benefitted from monoclonal antibody therapy (Wortel et al. 1992), and the approach was not pursued clinically because of the difficulties involved in identifying those subgroups and the expense (Barriere 1992). The specificity of mAbs encountered the difficulty that endotoxin is heterogeneous, and, moreover, that not all sepsis episodes are endotoxin-mediated.

However, some limited successes were found for mAbs. In 1998, Palivizumab became the first licensed mAb for use against an infectious disease—to prevent respiratory syncytial virus (RSV) infection in some high-risk newborns. Later, the mAb bezlotoxumab was introduced for prevention of *C. difficile* colitis and Raxibacumab and Obiltoxaximab were authorized to prevent anthrax (Pati et al. 2023). The COVID-19 pandemic saw the successful use of several monoclonal antibody preparations to reduce progression of disease, but these were eventually withdrawn when SARS-CoV-2 variants emerged that evaded neutralization (Focosi et al. 2023). In contrast to their limited use against infectious diseases, mAbs have begun to play an important role in the therapy of cancer and inflammatory diseases where their targets are host molecules that, unlike infectious agents, are rarely antigenically variable.

5 Use of Antibody Therapies in Infectious Disease Emergencies and Pandemics

Since the 1918 influenza pandemic, antibody-based therapies in the form of convalescent serum or plasma have been regularly deployed for infectious disease emergencies because these are available as soon as there are survivors who can provide convalescent serum. As we have noted, there is evidence of effectiveness of convalescent serum in the 1918 influenza pandemic, and human convalescent serum was used successfully to prevent the spread of measles and mumps (Park and Freeman 1926; Gallagher 1935; Rambar 1946). For polio, however, the results were mixed (Brodie 1932; Park 1932; Wesselhoeft 1933).

It is noteworthy that for none of the therapeutic studies of convalescent serum against viral disease in the early twentieth century was the presence of specific antibody known, since virology was in its infancy and serologic assays for antiviral antibodies had yet to be developed. By contrast, a measure of antibody to bacterial agents in serum was established through capsular swelling (Quellung effect) (Sabin 1933), agglutination, and the mouse protection test (Felton 1928). While convalescent serum was used with some success during the 1918 influenza pandemic, it was virtually unused during the 1957 influenza pandemic, a reflection of the extent to which this option had fallen out of favor, possibly due to concerns about the risk of serum hepatitis.

A renaissance of serum therapy occurred in the late 1970s when it was shown to be effective in a randomized clinical trial in Argentina for the treatment of Argentine hemorrhagic fever (Maiztegui et al. 1979). By the late twentieth century, advances in blood banking together with the increased ability to test for hepatitis viruses and HIV led to renewed interest in this therapeutic option. In recent decades, convalescent plasma has been deployed during several infectious disease emergencies including the outbreak of the syndrome of acute respiratory syndrome (SARS) in 2003, the middle east respiratory syndrome (Mair-Jenkins et al. 2015), and the Ebola epidemic in West Africa (Sahr et al. 2017; Griensven et al. 2016). However, it was the COVID-19 pandemic that most clearly stimulated the use of convalescent plasma in the twenty-first century. As described in Chap. 7 (P9), COVID-19 convalescent plasma was first deployed at scale under the Expanded Access Program initiated in March 2020 (Senefeld et al. 2021) and then, after evidence of both safety and efficacy emerged, CPP began to be very widely used under the Emergency Access Authorization of August 2020. Within 6 months of this authorization, at least a half million hospitalized COVID patients in the US had received this therapy (Casadevall et al. 2021a). The practice diminished after some large clinical trials reported lack of efficacy, but unfortunately these trials focused on patients with advanced disease and sometimes did not have plasma with sufficient antibody (Casadevall et al. 2023). By the end of the acute phase of the pandemic, strong evidence emerged indicating that CPP reduced hospitalization in outpatients and reduced mortality in hospitalized patients, especially when provided early (Senefeld et al. 2023). These studies are reviewed in detail in Chap. 10 (P10).

6 The Principles of Antibody Therapy

In 1937, Russell Cecil, one of the giants of early 20th medicine wrote '*It is a fundamental principle in all serum therapy that to obtain the best results the serum must be given early in the disease. This statement holds true regardless of whether one is using antitoxic or antibacterial*' (Cecil 1937). In 2021, Casadevall and his collaborators, frustrated that the lessons from the age of serum therapy were being ignored or forgotten—so that COVID-19 convalescent plasma was being evaluated too late to make a difference and often without sufficient antibody—enumerated the three key principles of antibody therapy known as the specificity, temporality, and quantitative principles (Casadevall et al. 2021b). These are

1. SPECIFICITY: The antibody preparation must contain antibody *specific* for the pathogen being treated.
2. TEMPORALITY: The antibody preparation should be administered as *early* as possible in the course of disease.
3. QUANTITY: The antibody preparation should contain *sufficient* specific immunoglobulin to mediate a biological effect.

However, it is noteworthy that in immunocompromised patients the temporal principle is suspended since these individuals behave very differently than immunocompetent patients where the lack of efficacy with time reflects the inability of antibody to affect a progressive inflammatory condition that includes an endogenous antibody response (Casadevall et al. 2023). This topic is reviewed in detail in Chap. 11 (P11).

7 Summary

This book attempts to provide a detailed overview of the use of the several forms of passive antibody therapy highlighted briefly in this introductory chapter. The chapters that follow will delve into many of the topics introduced here in greater detail addressing past and present use of these modalities, especially in COVID-19, but with a particular focus on anticipating their uses in any future pandemic.

Paneth and Casadevall review the origins and early history of convalescent serum and antitoxins. The effectiveness of convalescent plasma in COVID-19 (CPP) is addressed in chapters by Senefeld and Klassen, while the safety of CPP is reviewed by Joyner. A chapter by Shoham reviews the use of CPP in immuno-suppressed patients, and Kunze and Johnson's chapter emphasizes the importance of donor proximity. Sullivan identifies lessons learned about antibody therapies from the COVID experience for future pandemics and Pirofski provides a chapter specifically devoted to the topic of convalescent plasma use in future pandemics. Some very practical matters in the logistics of obtaining, storing, and delivering convalescent plasma are reviewed in chapters by Bloch, by Nunez and colleagues, by Nierich and by Lebowits. The extraordinary Expanded Access Program at the Mayo Clinic that initiated widespread CPP use in the COVID pandemic is described by Joyner. Marks addresses the important issue of antibody determination in convalescent plasma treatment.

Casadevall and Paneth discuss the development of monoclonal antibodies for the prevention and treatment of infectious diseases, and their use in COVID is reviewed by Focosi, while their likely use in future pandemics is addressed by Franchini and Focosi. Maor and Zimhoni assess the use of hyperimmune globulin in COVID.

One hundred and thirty-three years after Behring's discovery of antitoxins, antibody therapy remains a critical component of the armamentarium of medicine and is the first available therapy during epidemics and infectious disease outbreaks. The COVID-19 pandemic saw perhaps the largest scale deployment of antibody-based therapies in human history, using both convalescent plasma and molecular antibodies. Furthermore, these antibody therapies were rigorously tested in COVID-19 and the principles of antibody therapy derived from the lessons with serum therapy in the early twentieth century were validated using randomized controlled trials and other study designs. A key scientific and public health legacy of the COVID-19 pandemic is the knowledge that antibody therapies are effective if used properly. This knowledge is critically important to humanity as it deals with the inevitability of future infectious disease emergencies.

Declarations
Neither author has any conflicts to declare nor received any funding for this work.

References

Avery OT, Macleod CM, McCarty M (1944) Studies on the chemical nature of the substance inducing transformation of pneumococcal types: Induction of transformation by a desoxyribonucleic acid fraction isolated from pneumococcus type IIII. J Exp Med 79(2):137–158

Barriere SL (1992) The economic impact of HA-1A (Centoxin) against endotoxin. Pharmacoeconomics 2(5):408–413

Blake PA, Feldman RA, Buchanan TM, Brooks GF, Bennett JV (1976) Serologic therapy of tetanus in the United States. JAMA 235:42–44

Borrell B (2008) What is truth serum? Scientific American

Brodie M (1932) A comparison between convalescent serum and non-convalescent serum in poliomyelitis. J Exp Med 55:507–519

Carpenter CM, Boak RA (1936) The treatment of human brucellosis. A review of current therapeutic methods. Medicine 15:103–125

Casadevall A (2017) Crisis in infectious diseases: 2 decades later. Clin Infect Dis off Publ Infect Dis Soc Am 64(7):823–828

Casadevall A, Fang FC (2020) The intracellular pathogen concept. Mol Microbiol 113(3):541–545

Casadevall A, Scharff MD (1994) Serum Therapy" revisited: animal models of infection and the development of passive antibody therapy. Antimicrob Agents Chemotherap 38:1695–1702

Casadevall A, Dragotakes Q, Johnson PW, Senefeld JW, Klassen SA, Wright RS et al (2021a) Convalescent plasma use in the USA was inversely correlated with COVID-19 mortality. Elife 10

Casadevall A, Pirofski LA, Joyner MJ (2021b) The Principles of antibody therapy for infectious diseases with relevance for COVID-19. mBio 12(2)

Casadevall A, Joyner MJ, Pirofski LA, Senefeld JW, Shoham S, Sullivan D et al (2023) Convalescent plasma therapy in COVID-19: Unravelling the data using the principles of antibody therapy. Exp Rev Respir Med 1–15

Cecil RL (1937) Effects of very early serum treatment in pneumococcus Type I pneumonia. JAMA 108:589–692

Cohn EJ, Strong LE et al (1946) Preparation and properties of serum and plasma proteins; a system for the separation into fractions of the protein and lipoprotein components of biological tissues and fluids. J Am Chem Soc 68:459–475

Edelson BT, Cossart P, Unanue ER (1999) Cutting edge: paradigm revisited: antibody provides resistance to *Listeria* infection. J Immunol 163(8):4087–4090

Felton LD (1928) The units of protective antibody in anti-pneumococcus serum and antibody solution. J Infect Dis 43:531–542

Focosi D, McConnell S, Sullivan DJ, Casadevall A (2023) Analysis of SARS-CoV-2 mutations associated with resistance to therapeutic monoclonal antibodies that emerge after treatment. Drug Resist Updates Rev Commentaries Antimicrob Anticancer Chemother 71:100991

Foshay L (1940) Tularemia: a summary of certain aspects of the disease including methods for early diagnosis and the results of serum treatment in 600 patients. Medicine 19:1–81

Fothergill LRD (1937) *Hemophilus influenzae* (*Pfeiffer bacillus*) meningitis and its specific treatment. N Engl J Med 216:587–590

Frank WP, Patton EF, Hamilton PM (1942) Treatment of whooping cough bronchopneumonia. J Pediatr 20:720–722

Gallagher JR (1935) Use of convalescent measles serum to control measles in a preparatory school. Am J Public Health 25:595–598

Glatman-Freedman A, Casadevall A (1998) Serum therapy for tuberculosis revisited: a reappraisal of the role of antibody-mediated immunity against *Mycobacterium tuberculosis*. Clin Microbiol Rev 11:514–532

Griffith F (1928) The significance of pneumococcal types. J Hyg 27(2):113–159

Hattwick MA, Corey L, Creech WB (1976) Clinical use of human globulin immune to rabies virus. J Infect Dis 133(Suppl):A266–A272

Ivens F (1918) The preventive and curative treatment of gas gangrene by mixed serums. BMJ 2 (3016):425

Kohler G, Milstein C (1975) Continuous cultures of fused cells secreting antibody of predefined specificity. Nature 256:495–497

Lang J, Attanath P, Quiambao B, Singhasivanon V, Chanthavanich P, Montalban C et al (1998) Evaluation of the safety, immunogenicity, and pharmacokinetic profile of a new, highly purified, heat-treated equine rabies immunoglobulin, administered alone or in combination with a purified, Vero-cell rabies vaccine. Acta Trop 70:317–333

Lucchesi PF, Gildersleeve N (1941) The treatment of anthrax. JAMA 14:1506–1508

Luke TC, Kilbane EM, Jackson JL, Hoffman SL (2006) Meta-analysis: convalescent blood products for Spanish influenza pneumonia: a future H5N1 treatment? Ann Intern Med 145 (8):599–609

Mair-Jenkins J, Saavedra-Campos M, Baillie JK, Cleary P, Khaw FM, Lim WS et al (2015) The effectiveness of convalescent plasma and hyperimmune immunoglobulin for the treatment of

severe acute respiratory infections of viral etiology: a systematic review and exploratory meta-analysis. J Infect Dis 211(1):80–90
Maiztegui JI, Fernandez NJ, de Damilano AJ (1979) Efficacy of immune plasma in treatment of Argentine haemorrhagic fever and association between treatment and a late neurological syndrome. Lancet 2(98154):1216–1217
Manson-Bahr P (1939) The dysenteric disorders. The diagnosis and treatment of dysentery, sprue, colitis, and other diarrheas in general practice. Baltimore: The Williams & Wilkins Co.
Nosanchuk JD, Steenbergen JN, Shi L, Deepe GS Jr, Casadevall A (2003) Antibodies to a cell surface histone-like protein protect against *Histoplasma capsulatum*. J Clin Invest 112 (8):1164–1175
Park WH (1932) Therapeutic use of antipoliomyelitits serum in preparalytic cases of poliomyelitis. JAMA 99:1050–1053
Park WH, Freeman RG (1926) The prophylactic use of measles convalescent serum. JAMA 87:556–561
Pati I, Cruciani M, Candura F, Massari MS, Piccinini V, Masiello F et al (2023) Hyperimmune globulins for the management of infectious diseases. Viruses 15(7)
Plummer N, Liebmann J, Solomon S, Kammerer W, Kalkstein M, Ensworth HK (1941) Chemotherapy versus combined chemotherapy and serum in the treatment of pneumonia: a study of 607 alternated cases. J Am Med Assoc 116(21):2366–2371
Rambar AC (1946) Mumps: use of convalescent serum in the treatment and prophylaxis of orchitis. Am J Dis Child 71(1):1–13
Sabin AB (1933) Immediate pneumococcus typing directly from sputum by the Neufeld reaction. JAMA 100:1584–1586
Sahr F, Ansumana R, Massaquoi TA, Idriss BR, Sesay FR, Lamin JM et al (2017) Evaluation of convalescent whole blood for treating Ebola virus disease in Freetown, Sierra Leone. J Infect 74(3):302–309
Senefeld JW, Johnson PW, Kunze KL, Bloch EM, van Helmond N, Golafshar MA et al (2021) Access to and safety of COVID-19 convalescent plasma in the United States expanded access program: a national registry study. PLoS Med 18(12):e1003872
Senefeld JW, Gorman EK, Johnson PW, Moir ME, Klassen SA, Carter RE et al (2023) Rates among hospitalized patients with COVID-19 treated with convalescent plasma: a systematic review and meta-analysis. Mayo Clinic Proc: Innov Qual Outcomes 7(5):499–513
Tacket CO, Shandera WX, Mann JM, Hargrett NT, Blake PA (1984) Equine antitoxin use and other factors that predict outcome in Type A foodborne botulism. Am J Med 76:794–798
Teitelbaum R, Glatman-Freedman A, Chen B, Robbins JB, Unanue ER, Casadevall A et al (1998) A monoclonal antibody recognizing a surface antigen of *Mycobacterium tuberculosis* enhances host survival. Proc Natl Acad Sci 95:15688–15693
van Griensven J, Edwards T, de Lamballerie X, Semple MG, Gallian P, Baize S et al (2016) Evaluation of convalescent plasma for Ebola virus disease in guinea. N Engl J Med 374 (1):33–42
Wadsworth AB (1929) The hemolytic streptococci and antistreptococcus serum in scarlet fever. Am J Public Health 19:1287–1302
Watson DA, Musher DM, Jacobson JW, Verhoef J (1993) A brief history of the pneumococcus in biomedical research: a panoply of scientific discovery. Clin Infect Dis 17:913–924
Wesselhoeft C (1933) The present status of serum treatment for acute poliomyelitis. J Pediatr 3:330–351
Wesselhoeft C (1994) Treatment of scarlet fever and diphtheria. Med Clin N Am 19:1389–1429

Wever PC, van Bergen L (2012) Prevention of tetanus during the First World War. Med Humanit 38:78–82

Wortel CH, von der Möhlen MA, van Deventer SJ, Sprung CL, Jastremski M, Lubbers MJ et al (1992) Effectiveness of a human monoclonal anti-endotoxin antibody (HA-1A) in gram-negative sepsis: relationship to endotoxin and cytokine levels. J Infect Dis 166 (6):1367–1374

Ziegler EJ, Fisher CJ, Sprung CL, Straube RC, Sadoff JC, Foulke GE et al (1991) Treatment of gram-negative bacteremia and septic shock with HA-1A human monoclonal antibody against endotoxin. N Engl J Med 324:429–436

Nigel Paneth is a pediatrician and perinatal and developmental epidemiologist who is Emeritus Professor at Michigan State University and Founding Chair of the Department of Epidemiology and Biostatistics. He is author or editor of four books and 390 peer-reviewed papers and has an H-index of 95.

Passive Antibody Use in Covid

The Importance of Antibody Titer Determination to the Effective Use of Convalescent Plasma

Peter W. Marks

Contents

Abstract Convalescent Plasma (CP) has been used prophylactically and therapeutically over the past century to address a variety of infectious threats. Two tenets of the use of CP were clear from prior experience in the setting of other infectious outbreaks: (1) best results are obtained when CP is given early in the course of the disease, and (2) plasma containing high-titer neutralizing capacity is necessary to achieve optimal results. The magnitude of the COVID-19 pandemic along with the initial lack of effective therapeutic alternatives, combined with the relative safety of the approach of administration of CP, led to the initiation of an expanded access program (EAP) that ultimately provided CP to tens of thousands of individuals. When the program was initiated, no high-throughput assay was available for the determination of antibody titers, so antibody positive units were administered without regard to titer. With foresight regarding the need to ultimately determine such titers, samples from the CP units administered were retained and titers were determined retrospectively. An automated live-virus neutralization assay was ultimately selected for this purpose based on an evaluation of its accuracy and precision. Ultimately, an analysis performed in 13,794 individuals from the EAP for which clinical outcomes were known following the administration of single units of

P. W. Marks (✉)
Center for Biologics Evaluation and Research, U.S. Food and Drug Administration, Silver Spring, MD, USA
e-mail: Peter.Marks@fda.hh.gov

Current Topics in Microbiology and Immunology (2025) 443: 55–62
https://doi.org/10.1007/82_2024_281

Published Online: 7 December 2024

COVID-19 CP between the period of April and August 2020 indicated that higher titer COVID-19 CP was associated with a modest reduction in absolute mortality. The benefit observed was confined to individuals who were not intubated, and there was a trend toward a greater reduction in mortality using the highest SARS-CoV-2 neutralizing antibody-containing CP units. This experience during the COVID-19 pandemic is instructive for the future. To facilitate the production of CP that is likely to be most effective, high-throughput assays to determine neutralizing antibody titers need to be developed and implemented early during an outbreak to facilitate the identification and early administration of high-titer units.

1 Introduction

In January 2020, as the potential magnitude of the spread of SARS-CoV-2 was becoming apparent, based upon decades of prior practical experience, thoughts were given to the use of convalescent plasma (CP) as a therapeutic to try to reduce the observed morbidity and mortality of the illness that became known as COVID-19. At that time, two tenets of its use were completely clear from prior experience with CP in the setting of seasonal influenza, SARS-CoV-1, MERS, and Argentine Hemorrhagic Fever (Xu et al. 2020; Mo and Fisher 2016; Ruggiero et al. 1986): (1) best results are obtained when CP is given early in the course of disease, and (2) plasma containing high-titer neutralizing capacity is necessary to achieve optimal results. A variety of studies, including randomized controlled trials, have since illustrated the importance of both aspects, but this chapter will focus on the importance of adequate neutralizing antibody titers in both immunocompetent and immunocompromised individuals who are candidates for treatment using this approach (Casadevall et al. 2023; Franchini et al. 2024; Shibeeb et al. 2022).

Not every viral infection is associated with the development of high-titer antibodies, and CP has not been found to be useful in situations where there is a poor humoral antibody response or ones in which antibody titers routinely obtainable from donors do not appear to effectively interrupt disease progression, such as appears to be the case for infection with Ebola virus (van Griensven et al. 2016a, b). On the other hand, as noted above, several respiratory virus infections are associated with the development of high-titer antibody responses that potentially can be harnessed for prophylactic or therapeutic purposes in the form of CP (Ripoll et al. 2021).

To briefly review, CP had been used for several decades for the management of potentially life-threatening influenza infection.[1] In fact, the first reports of the use of convalescent sera for the treatment of influenza date back nearly a hundred years to the Spanish influenza pandemic of 1918 (Luke et al. 2006). In more modern times, randomized controlled trials have been challenging to conduct in this setting, and available data has shown conflicting results, with some studies showing no effect, and others a beneficial effect. A study conducted during the 2009 H1N1 pandemic using CP documented to have a neutralizing antibody titer of at least 1:160

documented a statistically significant reduction in mortality (20.0% vs 54.8%; $P = 0.01$) (Hung et al. 2011). However, another study conducted during non-pandemic times using CP documented to have a hemagglutination inhibition antibody titer of at least 1:80 documented no significant clinical benefit (Beigel et al. 2019). A meta-analysis of studies on influenza concluded that those studies that administered higher titer plasma early were more likely to be effective (Xu et al. 2020).

Although the use of CP was considered for SARS-CoV-1 and MERS, the limited nature of those outbreaks precluded the definitive investigation of CP for these infections (Mo and Fisher 2016). As another example, however, CP has been demonstrated to provide benefits in the management of Argentine Hemorrhagic Fever, caused by the Junin virus that is spread to humans by the inhalation of aerosolized rodent feces or by direct contact with infected animals (Ruggiero et al. 1986; Maiztegui et al. 1979). Of note, the importance of the early administration of CP with high neutralizing antibody titers for Argentine Hemorrhagic Fever was recognized by the early 1980s. Thus, based on experience with different infectious agents, including influenza and Junin virus, the basic principles for the effective application of convalescent plasma to the treatment of viral illnesses have been known for decades and its prophylactic and therapeutic use is reviewed in detail in the chapter by Dr. Nigel Paneth and colleagues.

2 Assessing Convalescent Plasma Titers During the COVID-19 Pandemic

Some studies conducted early in the COVID-19 pandemic in China in the first months of 2020 used early generation antibody assays to assess whether collected plasma had lower or higher titer neutralizing capacity and attempted to use CP units with higher neutralizing antibody titers (Ye et al. 2020; Li et al. 2020). These small studies, which were by no means definitive, reported encouraging results that supported consideration of the use of CP as part of an Expanded Access Program (EAP) in the United States, once the magnitude of the spread of SARS-CoV-2 and the morbidity and mortality associated with COVID-19 became apparent.

Ideally, an assay to assess neutralizing antibody levels or an assay that closely correlated with such levels would have been deployed initially at the outset of the EAP program. However, despite the early development of diagnostic PCR-based assays and the subsequent availability of lateral flow antibody tests, these methods did not allow for the assessment of the potential for viral neutralizing capacity. In fact, technical and logistic challenges encountered early on after declaration of the pandemic practically precluded the implementation of such testing of donated CP.

Specifically, no validated high-throughput neutralizing antibody test was available initially, and the practical logistics of having samples sent from across the country to a central laboratory to do any such testing during a time of tremendous societal and logistical disruption that was present in March and April of 2020 were insurmountable at the time. Instead, based on the limited neutralizing antibody data available from global reports that appeared early on, it was estimated that about half of the units collected in the United States following SARS-CoV-2 infection might qualify to be of higher titer. Thus, the administration of two units of CP was thought to help increase the chance of administration of at least one high-titer unit, and this was sometimes implemented in practice (Joyner et al. 2020).

Ultimately, the magnitude of the COVID-19 pandemic along with the initial lack of effective therapeutic alternatives, combined with the relative safety of the approach of administration of CP, led to the expansion of enrollment into the EAP beyond all initial expectations (Senefeld et al. 2021). Although titers were not available at the time that units were administered as part of the EAP, with the foresight that a form of randomization by the necessity of individuals to receive CP that would ultimately be shown to be of higher or lower titer was occurring, samples from each unit administered were retained so that testing could be performed retrospectively and could then be correlated with clinical outcomes. Thus, the accurate determination of the neutralizing capacity of each CP unit became important to understanding the potential clinical benefit of this therapy.

However, choosing which assay to employ in Spring 2020 was less than straightforward. At that time there were a variety of assays of research grade that had become available to assess antibodies to COVID-19: these included enzyme-linked immunoassays (ELISAs) for SARS-CoV-2 IgG and IgM as well as pseudo-virus and live-virus neutralization assays (Belov et al. 2022). Each of these assays had its unique advantages and disadvantages, and a decision was made to work through the assays to select the most accurate candidate that could be efficiently implemented.

The two major categories of titer tests implemented were ELISAs and viral neutralization assays, and each of these can be broken down further into subcategories, such as IgM, IgA, and IgG ELISAs and pseudo-viral or live-viral neutralization tests (Neerukonda et al. 2022). A sample bank was initially developed using about 20,000 samples from the EAP and aliquots of several thousand samples were sent to various laboratories for evaluation. A comparison of several different SARS-CoV-2 ELISAs, none of which had been thoroughly validated at that point several months after the start of the pandemic, revealed that there was not a clear and compelling strong correlation between the various assays being assessed. Ultimately, from comparison to a live-viral neutralization assay conducted manually in a BSL-3 laboratory, it became clear that the "gold standard" was an automated live-viral neutralization assay that made use of a fluorescent readout (Zhang et al. 2020).

An analysis performed in 13,794 individuals from the EAP for which clinical outcomes were known following the administration of single units of COVID-19 CP between the period of April and August 2020 indicated that higher titer COVID-19 CP was associated with a modest reduction in absolute mortality (Fig. 1). Because outcomes were collected in the hospital, the most data are available for survival at 7 days. Intubated individuals did not appear to benefit from

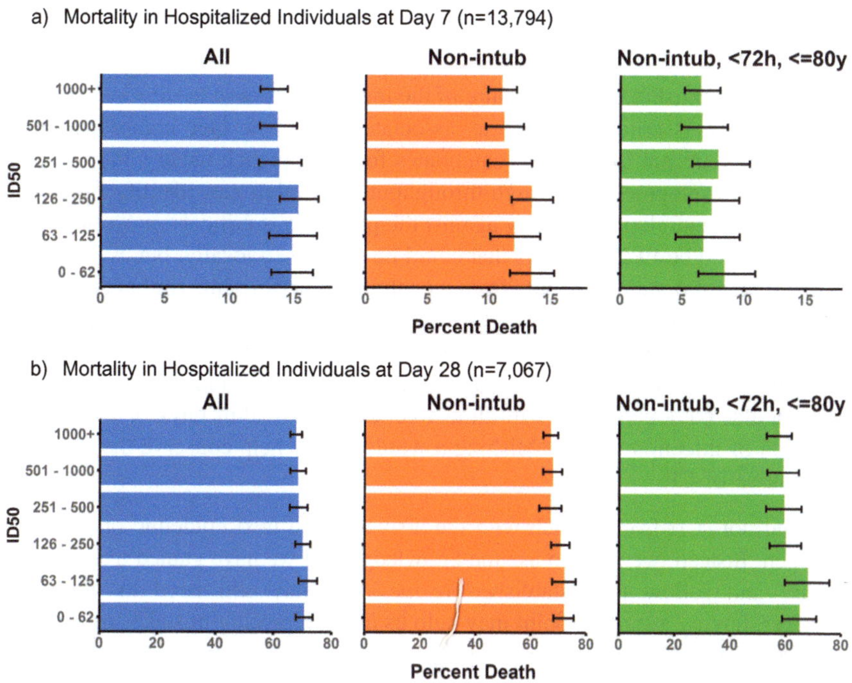

Fig. 1 Higher neutralizing antibody titers determined using a live-virus neutralization assay had some correlation with a reduction in mortality in non-intubated individuals receiving COVID-19 convalescent plasma early on during the COVID-19 pandemic. No reduction in mortality was observed in intubated individuals. In the U.S. COVID-19 Convalescent Plasma Expanded Access Program, follow-up was only available for continuously hospitalized individuals

the administration of CP. However, those who were not intubated (n = 9933) showed a 1.95% reduction in death at that timepoint from 13.12 to 11.19%. Of note, the benefit in terms of absolute risk reduction for those who were not intubated for whom 28-day follow-up was available (n = 4392) was about twice that which was seen at 7 days (4.27%). Although a threshold was ultimately used to classify units into “high” and “low” titer categories, the modest “dose–response” observed with higher neutralization titers trending toward improved outcomes is supportive of the importance of the use of high-titer CP in therapeutic applications (Belov et al. 2022).

3 Key Takeaways from the COVID-19 Experience

Although initial use of CP following the identification of a new epidemic or pandemic pathogen may end up proceeding empirically at first, particularly if there are no other therapeutic options readily available, key take-aways from the COVID-19 pandemic for the use of CP for any future emerging infectious disease outbreaks are the following:

1. Prompt establishment of the nature of the immune response to the viral pathogen in question is critical. Pathogens associated with low titer humoral immune responses are not likely to be candidates for an approach using CP.
2. Timely development of a high-throughput assay that can identify units with high-titer neutralizing capacity against the pathogen is critical—most preferably these would be ELISA assays that could be performed locally, but as a fallback, a system for obtaining results from a central site performing viral neutralization could be acceptable.
3. Once high-titer units can be identified, they should be deployed in clinical trials to most efficiently define the optimal window for CP administration for that particular pathogen. This will almost certainly be within days of the identification of infection, but establishing a window may help conserve the important resource of CP.

We need not start from scratch with the arrival of the next epidemic or pandemic pathogen. Instead, we need to use the knowledge that we have gained over nearly a century and follow the established recipe. We may still need to improvise. However, having a starting point may help us avoid repeating mistakes made in the past. COVID-19 CP has now become for all intents and purposes a standard of care for the treatment of immunocompromised individuals with hematologic malignancies who have persistent viral infection with SARS-CoV-2, and it also may play a role in the early treatment of inpatients and outpatients as future waves of subvariants become prevalent. Additionally, having this class of tool in the toolbox may be very handy as new pathogens emerge in the future.

Declarations The author has no conflicts of interest to declare, and no funding was obtained for this work. This chapter reflects the views of the author and should not be construed to represent FDA's views or policies.

References

Beigel JH, Aga E, Elie-Turenne MC et al (2019) Anti-influenza immune plasma for the treatment of patients with severe influenza A: a randomised, double-blind, phase 3 trial. Lancet Respir Med 7(11):941–950. https://doi.org/10.1016/S2213-2600(19)30199-7

Belov A, Huang Y, Villa CH et al (2022) Early administration of COVID-19 convalescent plasma with high titer antibody content by live viral neutralization assay is associated with modest clinical efficacy. Am J Hematol 97(6):770–779. https://doi.org/10.1002/ajh.26531

Casadevall A, Joyner MJ, Pirofski LA et al (2023) Convalescent plasma therapy in COVID-19: unravelling the data using the principles of antibody therapy. Expert Rev Respir Med 17 (5):381–395. https://doi.org/10.1080/17476348.2023.2208349

Franchini M, Cruciani M, Casadevall A et al (2024) Safety of COVID-19 convalescent plasma: a definitive systematic review and meta-analysis of randomized controlled trials. Transfusion 64 (2):388–399. https://doi.org/10.1111/trf.17701

Hung IF, To KK, Lee CK et al (2011) Convalescent plasma treatment reduced mortality in patients with severe pandemic influenza A (H1N1) 2009 virus infection. Clin Infect Dis 52(4):447–456. https://doi.org/10.1093/cid/ciq106
Joyner MJ, Wright RS, Fairweather D et al (2020) Early safety indicators of COVID-19 convalescent plasma in 5000 patients. J Clin Invest 130(9):4791–4797. https://doi.org/10.1172/JCI140200
Li L, Yang R, Wang J et al (2020) Feasibility of a pilot program for COVID-19 convalescent plasma collection in Wuhan, China. Transfusion 60(8):1773–1777. https://doi.org/10.1111/trf.15921
Luke TC, Kilbane EM, Jackson JL, Hoffman SL (2006) Meta-analysis: convalescent blood products for Spanish influenza pneumonia: a future H5N1 treatment? Ann Intern Med 145(8):599–609. https://doi.org/10.7326/0003-4819-145-8-200610170-00139
Maiztegui JI, Fernandez NJ, de Damilano AJ (1979) Efficacy of immune plasma in treatment of Argentine haemorrhagic fever and association between treatment and a late neurological syndrome. Lancet 2(8154):1216–1217. https://doi.org/10.1016/s0140-6736(79)92335-3
Mo Y, Fisher D (2016) A review of treatment modalities for Middle East respiratory syndrome. J Antimicrob Chemother 71(12):3340–3350. https://doi.org/10.1093/jac/dkw338
Neerukonda SN, Vassell R, Weiss CD, Wang W (2022) Measuring neutralizing antibodies to SARS-CoV-2 using lentiviral spike-pseudoviruses. Methods Mol Biol 2452:305–314. https://doi.org/10.1007/978-1-0716-2111-0_18
Ripoll JG, van Helmond N, Senefeld JW et al (2021) Convalescent plasma for infectious diseases: historical framework and use in COVID-19. Clin Microbiol Newsl 43(4):23–32. https://doi.org/10.1016/j.clinmicnews.2021.02.001
Ruggiero HA, Pérez Isquierdo F et al. (1986) Traitement de la fièvre hémorragique argentine par le plasma de convalescent. 4,433 cas [Treatment of Argentine hemorrhagic fever with convalescent's plasma. 4433 cases]. Presse Med 15(45):2239–2242. French. PMID: 2949253
Senefeld JW, Johnson PW, Kunze KL et al (2021) Access to and safety of COVID-19 convalescent plasma in the United States expanded access program: a national registry study. PLoS Med 18(12):e1003872. https://doi.org/10.1371/journal.pmed.1003872
Shibeeb S, Ajaj I, Al-Jighefee H, Abdallah AM (2022) Effectiveness of convalescent plasma therapy in COVID-19 patients with hematological malignancies: a systematic review. Hematol Rep 14(4):377–388. https://doi.org/10.3390/hematolrep14040052
van Griensven J, Edwards T, de Lamballerie X et al (2016a) Evaluation of convalescent plasma for Ebola virus disease in Guinea. N Engl J Med 374(1):33–42. https://doi.org/10.1056/NEJMoa1511812
van Griensven J, Edwards T, Baize S (2016b) Ebola-Tx consortium. Efficacy of convalescent plasma in relation to dose of Ebola virus antibodies. N Engl J Med 375(23):2307–2309. https://doi.org/10.1056/NEJMc1609116
Xu Z, Zhou J, Huang Y et al (2020) Efficacy of convalescent plasma for the treatment of severe influenza. Crit Care 24(1):469. https://doi.org/10.1186/s13054-020-03189-7
Ye M, Fu D, Ren Y et al (2020) Treatment with convalescent plasma for COVID-19 patients in Wuhan, China. J Med Virol. 92(10):1890–1901. https://doi.org/10.1002/jmv.25882
Zhang S, Ma P, Orzechowski M et al (2023) High-throughput neutralization and serology assays reveal correlated but highly variable humoral immune responses in a large population of individuals infected with SARS-CoV-2 in the US between March and August 2020. 14(2): e0352322. https://doi.org/10.1128/mbio.03523-22

Peter W. Marks received his graduate degree in cell and molecular biology and his medical degree at New York University and completed Internal Medicine residency and Hematology/ Medical Oncology training at Brigham and Women's Hospital in Boston. He has worked in academic settings teaching and caring for patients and in industry on drug development and is an author or co-author of over 125 publications. He joined the FDA in 2012 as Deputy Center Director for CBER and became Center Director in 2016 and is a Fellow of the American College of Physicians and a member of the National Academy of Medicine.

Convalescent Plasma and the US Expanded Access Program: A Personal Narrative

Michael J. Joyner

Contents

Abstract Between early April 2020 and late August 2020, nearly 100,000 patients hospitalized with SARS-CoV2 infections were treated with COVID-19 convalescent plasma (CCP) in the US under the auspices of an FDA-authorized Expanded Access Program (EAP) housed at the Mayo Clinic. Clinicians wishing to provide CCP to their patients during that 5-month period early in the COVID pandemic had to register their patients and provide clinical information to the EAP program. This program was utilized by some 2,200 US hospitals located in every state ranging from academic medical centers to small rural hospitals and facilitated the treatment of an ethnically and socio-economically diverse cross section of patients. Within 6 weeks of program initiation, the first signals of safety were found in 5,000 recipients of CCP, supported by a later analysis of 20,000 recipients (Joyner et al. in J Clin Invest 130:4791–4797, 2020a; Joyner et al. in Mayo Clin Proc 95:1888–

M. J. Joyner (✉)
Department of Anesthesiology and Perioperative Medicine, Mayo Clinic, Rochester, MN, USA
e-mail: Joyner.Michael@mayo.edu

Current Topics in Microbiology and Immunology (2025) 443: 63–78
https://doi.org/10.1007/82_2024_269

Published Online: 15 June 2024

1897, 2020b). By mid-summer of 2020, strong evidence was produced showing that high-titer CCP given early in the course of hospitalization could lower mortality by as much as a third (Joyner et al. in N Engl J Med 384:1015–1027, 2021; Senefeld et al. in PLoS Med 18, 2021a). These data were used by the FDA in its August decision to grant Emergency Use Authorization for CCP use in hospitals. This chapter provides a personal narrative by the principal investigator of the EAP that describes the events leading up to the program, some of its key outcomes, and some lessons learned that may be applicable to the next pandemic. This vast effort was a complete team response to a crisis and included an exceptional level of collaboration both inside and outside of the Mayo Clinic. Writing just 4 years after the initiation of the EAP, this intense professional effort, comprising many moving parts, remains hard to completely understand or fully explain in this brief narrative. As Nelson Mandela said of the perception of time during his decades in prison, "*the days seemed like years, and the years seemed like days*."

Glossary of Abbreviations and Definitions

BARDA Biomedical Advanced Research and Development Authority (Biomedical Advanced Research and Development Authority (BARDA) [Internet] 2021). This federal agency provides support for the development of vaccines, drugs, therapies, and diagnostic tools for public health medical emergencies such as chemical, biological, radiological, and nuclear accidents, incidents and attacks, pandemic influenza, and emerging infectious diseases

CP Convalescent Plasma, a blood product rich in disease fighting antibodies. It is harvested from donors after they have recovered from an infectious disease. CP can be used to passively transfer these antibodies to healthy people to prevent infection or given to infected people to treat infection. It has a history dating back to the nineteenth century. The first Nobel Prize in Physiology and Medicine was awarded for the use of a CP-like product derived from animals to treat diphtheria (The Nobel Prize 2024)

CCP COVID-19 Convalescent Plasma, derived from patients recovering from COVID-19 and presumed to contain (or proven to contain) antibodies to COVID-19

CCCP-19 The National COVID-19 Convalescent Plasma Project. A nationwide organization studying and later promoting use and access to CCP (National COVID-19 Convalescent Plasma Project [Internet] 2024)

EAP Expanded Access Program (FDA 2022). EAPs are mechanisms used by the US Food and Drug Administration (FDA) to allow widespread use of experimental therapies for life-threatening diseases that may be effective when there are limited treatment options. Patient consent and data collection are required. They are used most frequently for rare "incurable" diseases

eIND	Emergency Investigational New Drug (FDA 2020). Authorization by the FDA to use a new drug in a single patient
EUA	Emergency Use Authorization (FDA 2024). A step above an EAP that allows easier access to novel treatments that have shown promise. Patient consent and data collection are not required
FDA	The Food and Drug Administration. The US Agency responsible for formally authorizing use of drugs and biologic products by the health care system

1 My COVID Marathon

I first became aware of CP from reading a Wall Street Journal editorial by my friend and colleague Arturo Casadevall (Casadevall 2020). This was in late February 2020, and I was returning from the US Olympic Marathon Trials in Atlanta where I gave a talk on the physiological limits of fast marathon running—research I had been conducting since the early 1990s when I published a paper suggesting that humans might run a marathon in less than 2 hours (Joyner 1991). The editorial, which had been tweeted by the writer David Epstein, a mutual friend, described the use of CP to prevent a measles outbreak at a boy's school near Philadelphia in the 1930s. I immediately emailed Arturo and explained that CP could be used widely in the US to address the COVID pandemic because of our vast blood banking capacity. I had learned about this capacity in the course of my work on artificial blood and other transfusion-related research relevant to my anesthesiology practice.

Early in March 2020, I alerted Mayo Clinic Leadership to the potential value of CP in COVID. Though I received no response from institutional leadership, I began to repurpose my lab to conduct small clinical studies of CP in collaboration with Johns Hopkins colleagues. Arturo's team would focus on outpatient prophylactic studies at Hopkins, while we would undertake inpatient studies at Mayo. Protocols were drafted, and it is important to note that the Mayo Institutional Review Board (IRB), under the leadership of chairman, R. Scott Wright MD, was a key facilitator of all of our efforts. Dr. Wright recognized that while patient safety and confidentiality are paramount, a pandemic is uncharted territory requiring good judgment and regulatory flexibility to speed treatments to patients.

By mid-March 2020, the SARS-CoV2 situation had become desperate both nationally and internationally, largely due to the absence of effective treatments other than supportive care. Recognizing the historical track record of CP and related products in many infectious diseases in the pre-antibiotic era, an ad hoc national demand for CP emerged. This was characterized largely by one-at-a-time emergency requests to the FDA for permission to use CP in a single patient using the eIND mechanism. A gray market of patient families searching for CP for sick relatives on social media began to emerge, and a conference call with the FDA in mid-March was arranged by a group of stakeholders.

It was about this time that a group of academic physicians interested in CP formed an organization—CCP-19 (National COVID-19 Convalescent Plasma Project [Internet] 2024)—that included Arturo, myself, Nigel Paneth from Michigan State University, Shmuel Shoham from Johns Hopkins, Liise-anne Pirofski from Albert Einstein, and Jeffrey Henderson and Brenda Grossman from Washington University in St. Louis. Several of us were part of a pre-existing network of precision medicine skeptics. Arturo, Jeff, Shmuel, and Liise-anne are infectious disease specialists. Nigel is an epidemiologist. Brenda is a transfusion medicine specialist. Because interest in CP was exploding, we had a nightly call **for several months** to coordinate efforts. After a few months we went to calls two or three times per week. Four years later, we still hold biweekly evening calls. CCP-19 quickly developed a website housed at Michigan State University (with resources provided gratis by Amazon Web Services) that disseminated information about all things CP—research studies, papers, locations of CCP providers, and much else. We recruited many hundreds of physicians nationwide who met weekly by ZOOM for several months, and who participated as signatories to consensus statements and open letters to organizations such as NIH.

Some background: In 2015 Nigel emailed me—we had not yet met—after I published a controversial OpEd in the New York Times about the limits of the precision medicine narrative (Joyner 2015). This OpEd troubled the Mayo authorities who were quite invested in pursuing that particular vision of medicine. Nigel felt, as did I, that the precision medicine narrative neglected pressing public health needs—among other things—preparing for pandemics. Drs. Casadevall and Henderson joined us in forming an email group of some 30 precision medicine skeptics who began writing pieces questioning the narrative, eventuating in an entire issue of the journal *Perspectives in Medicine and Biology* in 2018 entitled the Precision Medicine Bubble (Paneth and Joyner 2018).

As part of our precision medicine skepticism, Nigel and I wrote an editorial in the Journal of Clinical Investigation (JCI) in early 2019 (solicited by Arturo, one of the journal's editors) about the need—among other things—to prepare for the next flu pandemic (Joyner and Paneth 2019). That these fears have been "vindicated" by the COVID pandemic is of little consolation. For what it is worth, Francis Collins, whose tenure as long-time NIH Director was characterized by powerful support for the precision medicine agenda, and particularly its links to the human genome project, is the only director of NIH to have life expectancy in the US fall on his watch, a fall that preceded the onset of the COVID pandemic.

By mid-March I was working a minimum of 12 h per day on CP-related issues, without a day off from the topic since that time. The hospital and my laboratory had been locked down and our usual human physiology studies had been suspended. Throughout this entire time my lab staff and especially my post-doctoral fellows and trainees stepped up to work on all things related to CP. Notwithstanding that my post-doctoral fellows are typically focused on exercise physiology, they put their work on hold for up to 3 years, risking career advancement. But the skills they learned and experience they gained working on the EAP and other CP-related activities have far outweighed what they lost by having their normal work and career progress interrupted. Importantly, all of them have found academic positions after their fellowships.

Two key lessons emerged from the events of March 2020. First, the concept that history matters was reinforced (Ripoll et al. 2021). We revisited the history of CP and related antibody products used to treat outbreaks of infectious diseases and learned of their considerable potential. Second, we showed that a network of collaborators from a variety of backgrounds in different institutions, with a range of perspectives but working together with a common goal, can be especially effective in addressing novel situations (Senefeld et al. 2021a).

2 Birth of the Expanded Access Program

In late March 2020, I received an email from Dr. Peter Marks, Head of Biologics at the FDA and an Architect of Operation Warp Speed (OWS) (Marks 2022; Cohen 2020), a term he coined, that brought COVID vaccines rapidly to the world in 2021. The FDA was then being inundated by time-consuming eINDs and in a follow-up early morning phone call he described to me the concept of expanded access, which not only permits eINDs, but also, under certain circumstances, permits the use of as yet unapproved therapeutics to groups of patients outside of the trial mechanism (FDA 2022). Dr. Marks asked whether I could organize an Expanded Access Program (EAP) for CCP, in which Mayo would collect data on the safety and efficacy of CCP as it was being used in the field. I immediately contacted Dr. Wright to see if Mayo could serve as the central coordinating IRB. He checked with Mayo's CEO, and we gave Dr. Marks the thumbs up within 24 h. I have no idea why Dr. Marks asked me to lead the EAP. I believe he was told by someone at NIH that I could get things done.

It is also important to note that beyond providing access to CCP and assessing its safety, we embodied a prospective, testable hypothesis about efficacy in the EAP. The hypothesis was that a dose–response relationship showing a relationship between survival and the level of antibody in the plasma provided to the patient would constitute powerful evidence that CP was an effective treatment. This concept was especially useful because we had no way of prospectively knowing the level of antibody in infused plasma. In effect, the varying levels of antibody in plasma were distributed randomly. This hypothesis was later confirmed—mortality in recipients of high antibody plasma was lower if the plasma was provided prior to the onset of mechanical ventilation (Joyner et al. 2021).

We then set about putting a team together at Mayo. Many talented people were available because much of Mayo was locked down. From loading dock workers to repurposed basic scientists, more than 100 Mayo employees at all three Mayo sites in Rochester, Minnesota; Jacksonville, Florida; and Scottsdale, Arizona participated in the effort. Compliance experts, administrators, operations managers, biostatisticians, clinicians, and many others undertook a variety of roles to support the EAP. My lab team of about 15 people including my research fellows anchored a number of activities. Some individuals were assigned to us, but most found out about the EAP via informal word-of-mouth networks and volunteered. We had no federal funding at this time.

I would be remiss if I did not name the core group who worked on the EAP. These included the aforementioned Scott Wright as well as Kelly Avery, Jack Senefeld, Rickey Carter, DeLisa Fairweather, Steve Klassen, Chad Wiggins, Katelyn Bruno, Patrick Johnson, Katie Kunze, Emily Whelan, Kris Tree, Starr Guzman, and Lisa Muenkel. These stalwart scientists and administrators stood side by side with me, allowing me to be, in effect, the Executive Producer of the EAP, mixing and matching the impressive skill sets that team members brought to our program. This core group and others met daily, then several times a week, then weekly for about 2 years. Our many partners and collaborators from outside Mayo complemented these efforts.

We enrolled our first patient via an entirely automated web-based system within a week on April 7, 2020. Our original mandate was to facilitate the use of CCP in up to 5,000 patients across the country. We were asked to assess safety, and the hope was to generate some insights about efficacy, perhaps sometime in 2021. We also had to coordinate many activities with the national blood banking "system" which is in fact a diffuse network of not-for-profit blood collectors large and small. I learned that the American Red Cross—which is the only blood bank recognized by most of the public—generates only about 40% of the nation's blood supply.

Because we had no funding except Mayo support, I began to independently solicit philanthropic support. I was able to get funds from UnitedHealth Group (UHG) and initially suggested a $1 million donation to the program, but after I pointed out that Michael Bloomberg had given the Hopkins team $4 million, they were kind enough to provide $5 million. After a critical article in New York Times, they may have been sensitized to show they were engaged in addressing pandemic-related problems (Abelson 2020). Several 100 k donations from pharma companies arrived and even a substantial donation from the NBA. At some point Mayo's CEO allocated $600,000 from Eric Schmidt of Google to us.

In mid-April, we applied for funding to BARDA for $25 million (Biomedical Advanced Research and Development Authority (BARDA) [Internet] 2021). Drs. Mary Homer and Corey Hoffman at BARDA were true partners in this whole process, which included BARDA providing funds for the collection of CP by blood banks. Ultimately, we had a total of $70 million to work with for this nationwide program.

In the early days of the EAP, I had many calls with Peter Marks of the FDA who must have been under tremendous strain. These calls were as early as 4am Rochester, Minnesota time. Parenthetically, Peter Marks is the best federal scientist-administrator I have ever worked with since I began conducting federally funded research in the early 1990s. His work during the pandemic, including his envisioning the rapid development of vaccines via Operation Warp Speed, will be viewed by history as on a par with the work of General Leslie Groves, Administrator of the Manhattan Project and James Webb, Developer of NASA's moon landing program.

3 The Pence Visit

By mid-April 2020, national interest in the program was gathering steam. Hundreds of patients were being treated with CCP daily and in the following 5 months I made roughly 100 media appearances, with sources as far away as the BBC. CCP was a bright spot in the chaotic therapeutic landscape of a spreading pandemic.

On about April 20, I received a call on my cell phone while working from home. The call was from Vice-President Pence's office in the White House. Olivia Troye was on the phone (Wikipedia. Olivia Troye [Internet]. Available from: https://en.wikipedia.org/wiki/Olivia_Troye). Later a sharp critic of the Trump administration, she was then working for the Department of Homeland Security on the Vice-President's team. She asked me whether Mr. Pence could visit Mayo in the following week to point to the early success of the EAP and to announce the award of the BARDA grant to Mayo.

I immediately contacted Jeff Bolton, the Chief Administrator of Mayo, and from that point on the visit was handled by the Mayo Public Affairs Department. The visit was scheduled for late April and the Mayo Public Affairs unit, and I met the day before in Mayo's blood bank (where the visit would be centered) to go over the plan with a political advance team from Washington. The need for VP Pence to wear a mask while at Mayo was clearly agreed upon.

Unfortunately, Mr. Pence chose not to ***wear a mask*** on Mayo property, catching us all by surprise (Liptak 2020). The several media photos of me in close proximity to the unmasked VP generated much interest, leading to many texts and emails from friends and colleagues and a slew of negative comments on social media. I offered to publicly state that Pence's team had agreed that he would wear a mask as had been agreed with our institutional leadership. Mayo initially noted the mask agreement on social media, but then took the post down (Lapin 2020).

At about the same time, I had a conversation with Dr. Cliff Lane, Deputy Director of the National Institute of Allergy and Infectious Disease, the agency then directed by Dr. Anthony Fauci. Criticism of the EAP in medical circles focused on the fact that the EAP was not a randomized clinical trial (RCT) of CCP, but a case series of treated individuals. As noted earlier, we later used a comparison of mortality by antibody levels in received plasma—levels which were randomly distributed—to show effectiveness, obviating much of this criticism. While I proposed carving out an RCT from the EAP, Dr. Lane indicated that his agency was more interested developing a refined plasma product—Intravenous Immunoglobulin (IVIG)—that contained concentrated convalescent antibody to COVID-19 (NIH 2022). IVIG from pooled donors is used for many purposes in medicine and is a potentially useful product but is subject to a 6-month or more lag time from the collection of plasma to the release of the refined product.

Microsoft Research also focused on IVIG, encouraging donations to commercial plasma companies to make a refined IVIG product. With lockdown, however, paid donations to the plasma companies had declined. The NIH and Gates Foundation were also advocates of IVIG. However, IVIG specific to COVID did not come to be developed at scale, and the single US RCT of COVID-10 IVIG did not show effectiveness (cite paper). The NIH committee on therapeutics recommends against use of pooled IVIG in COVID.

4 The Website

Everything we did was facilitated by a special website created for the EAP by Mayo. We used this site to enroll and consent patients and record patient and hospital data. The Mayo IRB designed a consent form to be used for all enrollees in the EAP across the nation, with Mayo serving as a central IRB. Frequently asked questions were included, and a survey-based "drop-down" menu for outcomes reporting was developed by Drs. Delisa Fairweather and Katelyn Bruno that fed directly to the analytics team led by Dr. Rickey Carter.

Before COVID-19, Drs. Fairweather and Bruno's interest in viral myocarditis led them to develop a patient questionnaire to understand more about the medical conditions of the patients whose biospecimens they were studying in their basic science lab at Mayo Jacksonville. Dr. Wright was familiar with this effort and felt it could be repurposed and adapted for use by the EAP. This repurposing is an example of another key element of what made the EAP a success: repurpose and adapt what you can. The Library of Congress ultimately archived our website as significant for the pandemic response (Schanilec 2020).

5 CP is Safe!

By **early May 2020**, we had follow-up data on the first 5,000 people treated with CCP. The hard work of Jack Senefeld and our statistical team led by Rickey Carter along with our team of clinicians, who reviewed serious adverse events daily, allowed us to rapidly summarize any safety issues associated with CP in the hours and days after a patient transfusion. Few safety concerns were noted and there were no signals that we were doing harm. The team rapidly generated a manuscript for submission to the JCI, a prominent medical journal and we posted a preprint to get the information out as soon as possible (Joyner et al. 2020a, c). In fact, our safety publication was on the preprint server within 6 weeks of the EAP's start. When the pandemic started, preprints were relatively new in medical research and little used, but they soon became an essential tool for communicating the latest findings on COVID, allowing them to be shared rapidly.

6 Community Partners, Survivor Corp and the Fight is in US

As the demand for CCP grew, two community groups were especially effective in driving donations of CCP. The Orthodox Jewish community centered around New York City was affected early in the pandemic with many infections. A key member of that community, Chaim Lebovits, had already served as a patient advocate, and knew several key physicians, including Dr. Shmuel Shoham of Johns Hopkins, a leader in our CCP efforts. As described in Chap. 10, Chaim made use of the logistical skills developed in his manufacturing career and his connections in the medical community to help blood bankers increase the range and efficiency of the donation process. Later in the pandemic he developed novel community-based ways to distribute and administer monoclonal antibodies.

Survivor Corps, an organization founded by Diana Berrent Guthe, developed a grass-roots website and communication plan that boosted plasma donations around the country. As noted above, Microsoft Research was interested in using plasma for creating IVIG, and to increase plasma donations they created a donation-promoting initiative branded as The Fight Is In Us (TFIIU), which, after a time, began to focus on CCP donations for immediate use (Microsoft News Center 2020).

The EAP was fortunate to have such partners, who, in addition to spurring the collection of CCP, provided us with a direct link to patient and family concerns. Chaim and Diana and their teams demonstrated the remarkable power of community engagement, and their leadership remains one of the highlights of the EAP. Their efforts serve as a model for involving the community in patient-oriented research.

7 Does It Work?

As hundreds of patients were being treated daily across the country at hospitals, large and small, rural and urban, and academic and non-academic, pressure mounted on the FDA, BARDA, and our team to somehow determine whether CCP worked. Blood banks retain a small sample of the plasma they distribute, largely so that if complications ensue after the transfusion, the sample can be checked for the presence of unscreened infectious agents or toxins. Thus, in late May of 2020, a plan was developed to ship samples of the CCP that was being transfused all over the country to Mayo, so that the antibody levels in the plasma could be measured.

Plasma donations identified as convalescent required evidence of recent COVID infection either by PCR test or antibody profile. However, it was discovered that antibody levels varied considerably among recovering patients, and thus some CCP units were far more antibody rich than others. We hypothesized that we might find a dose–response relationship between antibody level in the plasma and the outcome of treatment, and finding such a relationship would provide impressive evidence of effectiveness. To execute this plan, logistics experts, working with our program who are part of Mayo Medical Labs, arranged for thousands of samples of plasma that had been transfused to patients in our registry to be shipped to the Mayo lab medicine group from all over the country. This started at breakneck speed on **Memorial Day weekend of 2020**. A total of about 20,000 samples were shipped over the next few weeks.

This sort of real-world data analysis, relying on a natural experiment—the essentially random assignment of antibody content to patients—followed in the footsteps of other natural experiments where a treatment was assigned randomly not by design, but in effect. Perhaps the most famous example is Semmelweis's analysis of deaths from puerperal sepsis in Vienna in the 1840s. He showed that mortality was much higher in mothers delivered on the medical service (whose practitioners autopsied women with puerperal sepsis and then delivered babies without washing their hands) than in mothers delivered on the midwife service. Recognizing that those two cohorts were formed by a hospital policy of alternate day assignment of mothers to each service means in effect, like our dose response data, it was randomized. That finding was widely accepted as valid, then as now, notwithstanding the de-facto, rather than planned, random assignment.

The bottom line is that we learned that if CCP with a lot of antibodies was administered early in the course of hospitalization—prior to ICU admission—hospital mortality could be reduced by about a quarter to a third. These findings were consistent with what was known about antibody therapy before World War II —it must be given early to work. The findings were published in preprint form and then in the NEJM in March 2021 (Joyner et al. 2021).

Matched control studies from Mt. Sinai in New York and Houston Methodist Hospital also showed CCP worked if given early in the course of disease, and the same was shown when less ill participants were examined in an early trial from China (Liu et al. 2020; Salazar et al. 2020). Case reports of CCP being linked to dramatic improvements in immunocompromised patients unable to make antibodies added to the evidence base (Senefeld et al. 2021b).

Our efficacy information was made available to the United States Government by **late June of 2020**. It was vetted by mid-July, but skeptics in the NIH (2020) led by the director Francis Collins delayed release of the data until it could be confirmed via additional antibody testing by a group in Boston. Consistent with Dr. Lane's lack of enthusiasm in April, the NIH continued to be skeptical of the value of CCP.

I learned later that the NIH had a functioning COVID-19 antibody assay but would not work with the FDA to analyze the CCP samples. When the EAP started, standard assays for levels of antibody to the COVID-19 virus did not yet exist. To test the plasma units, we needed an easily deployed testing system that could be used by blood banks, and one was developed at Mayo during the EAP in conjunction with a local start-up with some support from our EAP program. A newly developed commercial assay system was also used to test the CP. This episode is one of the several examples I have noted during the pandemic of less-than-ideal coordination between different elements of the US Government. Better coordination inside the government can only improve the federal response to any future pandemic.

While all this was going on we were having weekly Friday calls with hundreds of clinicians from across the country advocating for early use of CCP, and, when assays were developed, to try to also use plasma with high antibody titers. This message got across and by **July 2020** or so there was much less salvage therapy being attempted using convalescent plasma in the ICU, and early use, often in the emergency room, upon the decision to admit the patient, became the norm. Outpatient treatment with CCP, it should be noted, had yet to be authorized by the FDA outside of clinical trials. The increasing number of physicians dialing in to our weekly calls required us to increase the capacity of our dial-in service.

8 The Big Reveal

By **late July 2020** the NIH had stopped obstructing the release of our data showing a relationship between transfused antibody level and survival in early treated patients. Rickey Carter's Mayo statistical group in Florida, ably assisted by my post-doctoral fellow Jack Senefeld, worked feverishly in what I named *the data engine room* to generate a paper to be placed on the preprint server, co-authored by our federal government collaborators that required, by contract, 30 days for BARDA to review (Joyner et al. 2020). We also worked with Mayo Continuing Education to generate a webinar describing our findings on the first Saturday of August 2020 that had over 500 attendees including members of the press.

In July, Mayo was contacted by Adam Boehler, an associate of Jared Kushner in the Trump White House (Wikipedia. Adam Boehler [Internet]. Wikipedia: Available from: https://en.wikipedia.org/wiki/Adam_Boehler). Boehler's background was in business, and he seemed to want to make sure the data were valid, to avoid the communications problems that surrounded the Administration's earlier backing of hydroxychloroquine. I coordinated our response and ensured that our government partners were on the call to avoid crosstalk and confusion and we were able to address Boehler's concerns. The assignment of Boehler to this meeting reflected the White House's policy of constantly shifting who was in charge of the response to this national pandemic, and how the reporting and communication lines were structured. This lack of organization led to repetitive cycles of re-explaining critically important items of information to new individuals who were now "in charge."

In light of the above, it is perhaps not surprising that, during an August press conference on CCP that focused on our findings, then FDA commissioner Stephen Hahn reported that CCP saved the lives of 35 of every 100 who received the treatment. In fact, what we reported to the White House (and showed in the first Mayo preprint that came out shortly after the news conference) was that treatment in the first 3 days of hospitalization reduced mortality in the first 7 days of hospitalization from 11.9 to 8.7%, which represents a 37% reduction in mortality (35% after statistical adjustment) and thus might save some 3 to 4 lives per 100 people treated (Harris 2020). Not a trivial finding, but nowhere near the claim made by Hahn.

In spite of these difficulties, the webinar went well, media coverage was positive, and the FDA Center for Biologics Evaluation and Research (CBER) was now in position to provide an Emergency Use Authorization (EUA) for CCP, especially after they confirmed our findings using their own antibody tests and classifying patients differently (O'Shaughnessy 2021). Richard Server who leads the MedRxiv preprint server agreed to post our preprint showing the finding that Hahn had misunderstood. Within a few weeks, billing codes for CCP and other administrative details were ironed out. We also had to continuously increase the enrollment cap approved by our IRB, who eventually authorized treatment of 125,000 patients under the EAP which formally closed at the end of August.

After anticipating an enrollment of 5,000 patients, we enrolled 105,717 patients, 90% of whom actually received CCP. More than a third (2,200) of the approximately 6,000 US hospitals participated, as did some 13,000 physicians. Plasma was administered in every state and virtually every congressional district, most of it in non-academic community hospitals. By every metric the EAP exceeded—in the racial, ethnic, rural/urban status, and economic diversity of the patients treated—every expectation we had when the program started.

We had hoped to learn something about safety by the end of 2020, but enrollment was so rapid that we knew the product was safe by May 2020. With some luck we thought we might understand something about efficacy late in 2020 or in 2021, but we found clear signals of efficacy within 12 weeks of program onset and first published these findings on August 12, 2020. These early signals of efficacy have since been amply confirmed in studies of several sorts since that time (Casadevall et al. 2021; Senefeld et al. 2023a, b).

In the summer of 2021, the Mayo clinic was presented with BARDA Medal for Service to the US During the Pandemic by Dr. Mary Homer, our BARDA partner. She came to Mayo to make the presentation. While the FDA was the most visible government agency we worked with during the EAP, BARDA, and its tireless, innovative, and flexible leaders—who include Dr. Corey Hoffman in addition to Dr. Homer—fully supported and funded our efforts.

9 After the EAP

On August 26, 2020, the EAP was not needed; it had served its function, producing sufficiently convincing information about CCP effectiveness that the FDA could comfortably issue an Emergency Use Authorization (EUA) for use of this product in hospitalized patients. This authorization led to the use of at least a half million units of CCP over the next 6 months in the US. Since the EUA required antibody testing of infused plasma, most patients received high-titer plasma, and our educational efforts led to much administration of plasma early in the hospital course.

Several large randomized controlled trials (RCTs) of CCP were reported in 2021. Unfortunately, these trials—which tested CP late in hospitalization, and at times used plasma with low antibody titers—were generally negative. Notwithstanding that, trends for efficacy were found in subgroups of patients who received high-titer CP early in the course of hospitalization, these negative trials, plus the increasing availability of small chemical antivirals and monoclonal antibodies, as well as the rollout of vaccines, led to a substantial reduction in the use of CCP.

Members of our CCP leadership team linked together Centers for Disease Control and Prevention (CDC) data on hospitalizations and mortality and information from America's Blood Centers on the number of units provided to hospitals per week. This allowed us to document that as many as 40% of hospitalized patients were receiving CCP in the fall of 2020, but that fraction had dropped to 10% by the spring of 2021. The research also found a strong inverse correlation between the percent of patients receiving plasma in any week and the hospital mortality rate (Casadevall et al. 2021). This analysis indicated that tens of thousands of lives had been saved by CCP, but also showed, tragically, that the reduction in CCP use in 2021 is likely to have been accompanied by loss of life that might have been prevented.

Additional RCTs testing the early use case, including the use of CCP in outpatient settings, began to emerge in late 2021 and 2022 and showed positive results. More recently meta-analysis of all randomized controlled trials and well-conducted matched control studies clearly validates the findings of the EAP and shows that when high-titer CCP is administered early in the course of hospitalization mortality is reduced (Senefeld et al. 2023a, b). Additionally, CCP—especially CCP collected from donors who have both had infections and been vaccinated, which is very high titer—has special value in immunocompromised patients who are unable to generate their own antibodies to either vaccine or infection (Ripoll et al. 2022). This special use case has remained strong throughout the pandemic and is now even more important with the steady loss of effectiveness of monoclonals and the continued evolution of the virus.

In the final analysis, the lessons I summarized with my colleagues Jack Senefeld, Arturo Casadevall, and Peter Marks in an editorial published in 2023 are worth repeating (Senefeld et al. 2023).

1. The focus of treatment with CP in any infectious disease emergency should be early after symptoms arise—preferably in outpatients or immediately on admission to the hospital.
2. Regulators and funding agencies should be involved in the pre-planning and development of *simple* CP administration protocols and their oversight. They should also work to expedite the development and deployment of assays to determine the neutralization capacity or antibody content of CP.
3. Mechanisms—including funding—should be in place so that the blood collection and the blood banking system can obtain CP from qualified recovered donors as soon as they become available. The experience with CP for COVID-19 suggests that local plasma is more likely to be effective than distantly sourced plasma (see Chap. 20 by Kunze). Samples of the CP sent to hospitals and clinics should be stored in a central biobank for later analysis as assays come online.
4. A diverse network of sites should be available for the administration of CP with pre-established ground rules for a central IRB, patient consent, and the reporting of fundamental patient outcomes.
5. A data management system and analytics team should be in place to facilitate rapid reporting of results.
6. Simulation exercises should be performed to test the system.
7. All elements of the program should focus on simplicity and the "*enemy of good is perfect…*" axiom of crisis-based medical care must be rigorously adhered to.

Conflict Declaration
Disclaimer: All opinions expressed in this chapter are those of the author and do not necessarily represent the views of his collaborators, the Mayo clinic or the US government.

References

Abelson R (2020) Major U.S. health insurers report big profits, benefiting from the pandemic [Internet]. Available from: https://www.nytimes.com/2020/08/05/health/covid-insurance-profits.html

Biomedical Advanced Research and Development Authority (BARDA) [Internet] (2021) Digital Communications Division (DCD): U.S. Department of Health and Human Services. Available from: https://www.hhs.gov/about/agencies/orgchart/aspr/barda/index.html

Casadevall A (2020) How a boy's blood stopped an outbreak [Internet]. Available from: https://www.wsj.com/articles/how-a-boys-blood-stopped-an-outbreak-11582847330

Casadevall A, Dragotakes Q, Johnson PW et al (2021) Convalescent plasma use in the USA was inversely correlated with COVID-19 mortality. Elife 10:e69866

Cohen J (2020) Unveiling 'Warp Speed,' the White House's America-first push for a coronavirus vaccine [Internet]. Science. Available from: https://www.science.org/content/article/unveiling-warp-speed-white-house-s-america-first-push-coronavirus-vaccine

Expanded Access to Convalescent Plasma [Internet]. US Covid Plasma; Available from: https://www.uscovidplasma.org/

FDA (2020) Emergency investigational new drug (EIND) applications for antiviral products [Internet]. Available from: https://www.fda.gov/drugs/investigational-new-drug-ind-application/emergency-investigational-new-drug-eind-applications-antiviral-products
FDA (2022) Expanded access [Internet]. Available from: https://www.fda.gov/news-events/public-health-focus/expanded-access
FDA (2024) Emergency use authorization [Internet]. Available from: https://www.fda.gov/emergency-preparedness-and-response/mcm-legal-regulatory-and-policy-framework/emergency-use-authorization
Harris R (2020) FDA's Hahn apologizes for overselling plasma's benefits as a COVID-19 treatment [Internet]. Available from: https://www.npr.org/sections/health-shots/2020/08/25/905792261/fdas-hahn-apologizes-for-overselling-plasmas-benefits-as-a-covid-19-treatment
Joyner MJ (1991) Modeling: optimal marathon performance on the basis of physiological factors. J Appl Physiol 70(2):683–687
Joyner MJ, Paneth N (2019) Promises, promises, and precision medicine. J Clin Invest 129 (3):946–948
Joyner MJ, Wright RS, Fairweather D et al (2020a) Early safety indicators of COVID-19 convalescent plasma in 5000 patients. J Clin Invest 130(9):4791–4797
Joyner MJ, Bruno KA, Klassen SA et al (2020b) Safety update. Mayo Clin Proc 95(9):1888–1897
Joyner MJ, Carter RE, Senefeld JW et al (2021) Convalescent plasma antibody levels and the risk of death from Covid-19. N Engl J Med 384(11):1015–1027
Joyner MJ (2015) 'Moonshot' medicine will let us down [Internet]. Available from: https://www.nytimes.com/2015/01/29/opinion/moonshot-medicine-will-let-us-down.html
Joyner MJ, Senefeld JW, Klassen SA et al (2020) Effect of convalescent plasma on mortality among hospitalized patients with COVID-19: initial three-month experience. medRxiv
Joyner M, Wright RS, Fairweather D et al (2020c) Early safety indicators of COVID-19 convalescent plasma in 5,000 patients [Internet]. Infect Dis (except HIV/AIDS) [cited 2024 Feb 7]. Available from: https://doi.org/10.1101/2020.05.12.20099879
Lapin T (2020) Mike Pence criticized for not wearing coronavirus mask on Mayo Clinic tour [Internet]. Available from: https://nypost.com/2020/04/28/mike-pence-blasted-for-not-wearing-coronavirus-mask-on-mayo-clinic-tour/
Liptak K (2020) Pence flouts Mayo Clinic policy on masks—which is to wear one [Internet]. Available from: https://www.cnn.com/2020/04/28/politics/mike-pence-mayo-clinic-mask/index.html
Liu STH, Lin H-M, Baine I et al (2020) Convalescent plasma treatment of severe COVID-19: a propensity score-matched control study. Nat Med 26(11):1708–1713
Marks P (2022) [Internet]. Director - Center for biologics evaluation and research (CBER): FDA. 2022. Available from: https://www.fda.gov/about-fda/fda-organization/peter-marks
Microsoft News Center (2020) Racing against time, medical researchers, life science companies and COVID-19 survivors launch national campaign to drive blood plasma donation [Internet]. Available from: https://news.microsoft.com/2020/05/26/racing-against-time-medical-researchers-life-science-companies-and-covid-19-survivors-launch-national-campaign-to-drive-blood-plasma-donation/
National COVID-19 Convalescent Plasma Project [Internet] (2024). Available from: https://ccpp19.org/who-are-we/index.html
NIH (2020) Warns data on effectiveness of convalescent plasma in treating COVID-19 is weak [Internet]. Available from: https://www.npr.org/2020/08/19/903982549/nih-warns-data-on-effectiveness-of-convalescent-plasma-in-treating-covid-19-is-w
NIH (2022) Hyperimmune intravenous immunoglobulin does not improve outcomes for adults hospitalized with COVID-19 [Internet]. NIH. Available from: https://www.nih.gov/news-events/news-releases/hyperimmune-intravenous-immunoglobulin-does-not-improve-outcomes-adults-hospitalized-covid-19
O'Shaughnessy J (2021) Convalescent plasma EUA letter of authorization. FDA
Paneth N, Joyner MJ (2018) Perspectives in biology and medicine. Special issue on the precision medicine bubble. John Hopkins University Press

Ripoll JG, van Helmond N, Senefeld JW et al (2021) Convalescent plasma for infectious diseases: historical framework and use in COVID-19. Clin Microbiol Newsl 43(4):23–32

Ripoll JG, Gorman EK, Juskewitch JE et al (2022) Vaccine-boosted convalescent plasma therapy for patients with immunosuppression and COVID-19. Blood Adv 6(23):5951–5955

Salazar E, Christensen PA, Graviss EA et al (2020) Treatment of coronavirus disease 2019 patients with convalescent plasma reveals a signal of significantly decreased mortality. Am J Pathol 190 (11):2290–2303

Schanilec K (2020) Library of congress selects Mayo's convalescent plasma website for Coronavirus Web Archive [Internet]. Available from: https://newsnetwork.mayoclinic.org/discussion/library-of-congress-selects-mayos-convalescent-plasma-website-for-coronavirus-web-archive/

Senefeld JW, Johnson PW, Kunze KL et al (2021a) Access to and safety of COVID-19 convalescent plasma in the United States expanded access program: a national registry study. PLoS Med 18(12):e1003872

Senefeld JW, Klassen SA, Ford SK et al (2021b) Use of convalescent plasma in COVID-19 patients with immunosuppression. Transfusion 61(8):2503–2511

Senefeld JW, Franchini M, Mengoli C et al (2023a) COVID-19 convalescent plasma for the treatment of immunocompromised patients: a systematic review and meta-analysis. JAMA Netw Open 6(1):e2250647

Senefeld JW, Gorman EK, Johnson PW et al (2023b) Rates among hospitalized patients with COVID-19 treated with convalescent plasma: a systematic review and meta-analysis. Mayo Clin Proc Innov Qual Outcomes 7(5):499–513

Senefeld JW, Marks P, Casadevall A, Joyner MJ (2023) The value of observational registry studies for the next infectious disease emergency. mBio 14(6):e0256523

The Nobel Prize (2024) Emil von Behring [Internet]. [cited 2024 Feb 7]. Available from: https://www.nobelprize.org/prizes/medicine/1901/behring/facts/

Wikipedia. Adam Boehler [Internet]. Wikipedia: Available from: https://en.wikipedia.org/wiki/Adam_Boehler

Wikipedia. Olivia Troye [Internet]. Available from: https://en.wikipedia.org/wiki/Olivia_Troye

The Safety Profile of COVID-19 Convalescent Plasma

Michael J. Joyner, Rickey E. Carter, R. Scott Wright and Jonathon W. Senefeld

Contents

Abstract Despite concerns about potential side effects, based both on historical experience with plasma products and more recent concerns about contemporary use of plasma, COVID-19 convalescent plasma has been shown to be a very safe product. Research early in the COVID-19 pandemic documented—among the very large population of convalescent plasma recipients in the US Convalescent Plasma Study component of the FDA-authorized Expanded Access Program—that the overall risk profile was no different than that seen for fresh frozen plasma, a product used routinely in medical practice. The safety of CCP was further demonstrated using real-world evidence, pragmatic trials, and formal randomized trials. The rates

M. J. Joyner (✉)
Department of Anesthesiology and Perioperative Medicine, Mayo Clinic, Rochester, MN, USA
e-mail: joyner.michael@mayo.edu

R. E. Carter
Department of Quantitative Health Sciences, Mayo Clinic, Jacksonville, FL, USA
e-mail: carter.rickey@mayo.edu

R. S. Wright
Department of Cardiovascular Medicine, Mayo Clinic, Rochester, MN, USA
e-mail: wright.scott@mayo.edu

Human Research Protection Program, Mayo Clinic, Rochester, MN, USA

J. W. Senefeld
Department of Health and Kinesiology, University of Illinois Urbana-Champaign, Urbana, IL, USA
e-mail: senefeld@illinois.edu

Current Topics in Microbiology and Immunology (2025) 443: 79–88
https://doi.org/10.1007/82_2024_282

Published Online: 19 December 2024

of all serious adverse events were very low, an especially impressive finding in light of the fact that nearly all safety data came from the use of COVID-19 convalescent plasma in patients who were hospitalized, were older, and/or had significant co-morbid cardiopulmonary and metabolic disorders. The well-known complications of blood and plasma transfusions—transfusion-associated circulatory overload and transfusion-related acute lung injury—were found with no higher incidence than with standard use of blood and plasma, nor was there evidence for antibody-dependent enhancement or increased incidence of thromboembolic events. The comprehensive safety profile derived from studies enrolling hundreds of thousands of recipients of COVID-19 convalescent plasma across the world should allay safety fears about the rapid deployment of convalescent plasma in future pandemics.

From the earliest days of convalescent plasma and serum use in clinical medicine, the problem of safety and adverse effects were prominently discussed—including both known and theoretical risks (Casadevall and Pirofski 2020). These concerns were especially pronounced with the widespread use of serum prepared in animals, especially horses. Serum sickness, an immune-complex-mediated hypersensitivity reaction presenting as a combination of fever, rashes, and joint pain, was found in up to 70% of recipients of some animal-based passive antibody preparations (Lawley et al. 1984). Even though it was generally self-limited, these reactions did much to deter passive antibody use. The problem did not disappear as a concern until the development of refined immunoglobulin products. Serum sickness is, however, quite rare when human-based convalescent plasma is used, as in the COVID-19 pandemic, and thus was not a major concern when human convalescent plasma began to be used in the COVID-19 epidemic (Casadevall and Pirofski 2020; Luke et al. 2006).

However, the transfusion of serum or plasma of human origin is not completely free of potential complications. These adverse reactions can be categorized into four main categories with outstanding "textbook" summaries available from several key publications (Children's Hospital of Philadelphia; Suddock and Crookston 2024; Cho et al. 2024). For the sake of brevity and clarity, these references are quoted and/or paraphrased and *italicized* in the following brief definitions and descriptions of pathophysiology.

(1) *Transfusion-Related Lung Injury (TRALI)*: Transfusion-related acute lung injury (TRALI) is a clinical syndrome in which there is an acute, noncardiogenic pulmonary edema associated with hypoxia that occurs during or after a transfusion. The pathophysiology of TRALI can be summarized as: "Acute lung injury due to antibodies in the donor product (human leukocyte antigen or human neutrophil antigen) reacting with antigens in the recipient. The recipient's immune system responds and causes the release of mediators that lead to pulmonary edema. Possibly contributing to this are clinical conditions that predispose the patient including infection, recent surgery, or inflammation."

(2) *Transfusion Associated Cardiac Overload (TACO)*: This occurs when the volume of the transfused component causes hypervolemia (volume overload).
(3) *Antibody Dependent Enhancement (ADE)*: ADE occurs when the antibodies generated during an immune response recognize and bind to a pathogen, but they are unable to prevent infection. Instead, these antibodies act as a "Trojan horse"… In this situation, the antibodies bind to the virus and help it more easily get into cells than it would on its own.
(4) *Other Transfusion Reactions (TR)*: Other TRs of interest in the context of the use of CCP to treat SARS-CoV-2 infections include the following:

- Mild allergic: Attributed to hypersensitivity to a foreign protein in the donor product.
- Anaphylactic: Similar to a mild allergic reaction, however resulting in a more severe reaction. Sometimes this can occur in a patient with IgA deficiency who makes alloantibodies against IgA and then receives blood products containing IgA.
- Febrile non-hemolytic: Generally thought to be caused by cytokines released from blood donor leukocytes (white blood cells).
- Septic: Caused by bacteria or bacterial byproducts (such as endotoxin) which may contaminate blood.
- Acute hemolytic transfusion reactions: Can result in intravascular or extravascular hemolysis, depending on the specific etiology (cause). Immune-mediated reactions are often a result of recipient antibodies present to blood donor antigens. Non-immune reactions are possible and occur when red blood cells are damaged before transfusion (e.g., by heat or incorrect osmotic conditions).

In Chapter 5, the narrative history of the US Expanded Access Program (EAP) for CCP was reviewed. The EAP encompassed multiple emergency-use single center IND approvals, the large, robust US Convalescent Plasma Study (CPS), and the post-EUA approval of plasma use as high neutralizing-titer convalescent plasma. We here describe findings from the US Convalescent Plasma Study, which used an expanded access regulatory mechanism coupled with a pragmatic nationwide, multi-center, single-arm trial design to evaluate the safety and efficacy of convalescent plasma in critically ill hospitalized patients with COVID-19. This program, as noted in Chapter 5, recorded the details of treatment of over 100,000 COVID-19 patients (Senefeld et al. 2021). It was recognized early that a major value of the detailed patient registry created by the CPS was the assessment of complications of CCP transfusion. Thus, one of the initial goals of the CPS was to assess the safety of CCP. Due to the unanticipated scale of the CPS, sufficient patients had been enrolled by May 2020 to report on the first 5000 transfused patients (Joyner et al. 2020a). This analysis was made possible because data were collected on the immediate (four hours or less) responses to the administration of CCP. All serious adverse event reports were independently adjudicated over the course of the study by the IND sponsor and trained designees using the National Healthcare Safety Network Biovigilance Component Hemovigilance Module Surveillance Protocol as a framework (Control C for D, Prevention (CDC) 2023). In frequent meetings (several times per week or more), a panel of experienced clinicians reviewed clinical notes about serious adverse events and their potential relationship to the administration of CCP. These meetings were facilitated by the nearly real-time data reporting and analysis of key outcome variables. We inferred that any major transfusion reactions, including TRALI and TACO, would be detected during this timeframe. Because this was a single-arm trial that lacked a control group not transfused with convalescent plasma, we compared our outcomes to standard rates for these complications.

It is also important to note that the FDA Expanded Access Program was the regulatory mechanism that allowed CCP to be administered at scale in the US prior to its formal approval by the FDA. The CPS developed a sophisticated medical registry that included investigator registration; extensive data collection on outcomes, mortality, and complications; and the ability to link and geocode the collection locations of CCP. This registry allowed far broader access to CCP than could be achieved by the case-by-case approval mechanism that preceded it. The CPS served as a real-world/pragmatic observational cohort registry study that permitted Mayo clinical investigators and colleagues not only to provide expanded access but also to perform sophisticated analyses of real-world data on very large numbers of patients early in the course of the challenging conditions of a pandemic.

When safety data from the first 5,000 patients was evaluated, we observed (Joyner et al. 2020a): "*The incidence of all serious adverse events (SAEs), including mortality rate (0.3%), in the first 4 h after transfusion was < 1%* (Fig. 1). *Of the 36 reported SAEs, there were 25 reported incidences of related SAEs, including mortality (n = 4), transfusion-associated circulatory overload (n = 7), transfusion-related acute lung injury (n = 11), and severe allergic transfusion reactions (n = 3). However, only 2 of 36 SAEs were judged as definitely related to the convalescent plasma transfusion by the treating physician.*"

These low rates of complications were confirmed a few months later when we reported our results in 20,000 patients (Joyner et al. 2020b): "*The incidence of all serious adverse events was low; these included transfusion reactions (n = 78; <1%), thromboembolic or thrombotic events (n = 113; <1%), and cardiac events (n = 677, ~ 3%). Notably, the vast majority of the thromboembolic or thrombotic events (n = 75) and cardiac events (n = 597) were judged to be unrelated to the plasma transfusion* per se".

Additionally, we stratified our analysis of serious adverse events by a 6-level ordinal scale (obtained at the onset of the transfusion) that represented the severity of COVID-19, to better understand whether thrombotic or thromboembolic events were commoner among patients with more severe COVID-19 (Joyner et al 2020a). The rates of serious adverse events were similarly low (<1%) among all patients, including patients in the most severe category of COVID-19 (Fig. 2). The low rates of thrombotic and thromboembolic events provide strong support for the safety profile of CCP, even among hospitalized patients with severe COVID-19.

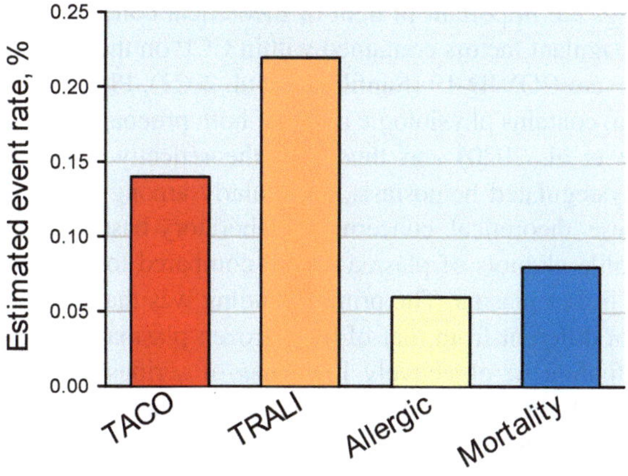

Fig. 1 Serious adverse events among 5,000 patients transfused with COVID-19 convalescent plasma. Vertical bar charts displaying the rates of serious adverse events (including possibly, probably, and definitely related events) relative to the denominator of 5,000 transfusions. Wilson confidence interval calculation for binomial proportions was used to estimate the 95% confidence intervals for each point estimate. Adapted, with permission, from the US CPS (Joyner et al. 2020a)

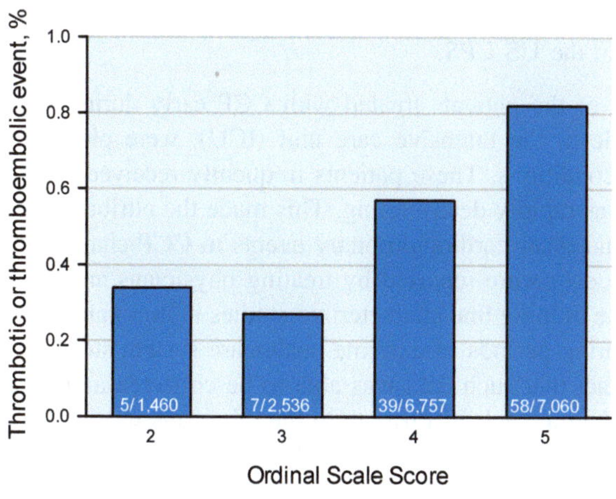

Fig. 2 The rate of thrombotic and thromboembolic events stratified using a 6-level ordinal scale to assess the clinical course of COVID-19. Scores on the ordinal scale were defined as follows: a score of 1 indicated not hospitalized; 2, hospitalized and not receiving supplemental oxygen; 3, hospitalized and receiving supplemental oxygen; 4, hospitalized and receiving oxygen supplementation administered by a high-flow nasal cannula or noninvasive ventilation; 5, hospitalized and receiving mechanical ventilation; and 6, death. This ordinal scale represents the condition of the patient at the start of CCP transfusion. In this way, a score of 6 (death) is not represented. Wilson confidence interval calculation for binomial proportions was used to estimate the 95% confidence intervals for each point estimate. Adapted, with permission, from the US CPS (Klompas et al. 2021)

These findings are important in light of theoretical concerns about the potential impact of procoagulant factors contained within CCP on the coagulation cascade in patients with severe COVID-19 (Sanfilippo et al. 2021). Plasma, both convalescent and fresh frozen, contains physiologic ratios of both procoagulant and anticoagulant proteins (Gratz et al. 2020) and thus it is theoretically possible for plasma to contribute to dysregulated hemostasis, particularly among prothrombotic patients. To address these theoretical concerns, a laboratory-based study evaluated the coagulation profile aliquots of plasma—CCP compared to both pathogen-reduced CCP and fresh frozen plasma. The primary finding was that the coagulation profile of CCP was not different from that of fresh frozen plasma (Klompas et al. 2022).

Our initial finding of objectively low rates of serious adverse events among 5,000 and 20,000 transfused patients was confirmed with the entire cohort of ~95,000 patients (Senefeld et al. 2021). Among these ~95,000 patients, 597 serious adverse events were classified as related to convalescent plasma transfusion, fewer than 1% of all transfusions (Fig. 3). After adjudication, TACO, TRALI, and severe allergic events were often deemed "probably" or "definitely" related to CCP transfused, whereas hypotensive events were deemed as "possibly" related in most cases. In this way, among the 597 serious adverse events related to convalescent plasma, 107 events were deemed "definitely" related, 267 "probably" related, and 223 "possibly" related.

Five important considerations should be kept in mind in interpreting the early safety data from the US CPS.

1. First, many of the patients treated with CCP early during the pandemic were treated while in the intensive care unit (ICU), were older, and had multiple co-morbid conditions. These patients frequently received CCP as their clinical condition was rapidly deteriorating. This made the attribution of worsening gas exchange and other cardiopulmonary events to CCP challenging.
2. Second, the data were reported by treating physicians and their teams without the usual site training that characterizes studies in non-pandemic times and were recorded during periods of extreme healthcare system stress.
3. Third, the fact that such data was able to be collected and adjudicated in generally less than a week is a tribute to the clinicians using CCP in the field and shows what is possible with streamlined electronic reporting and analytics during the most trying circumstances.
4. Fourth, the use of preprints allowed these data to be reported rapidly and to inform clinical practice during a time when CCP was one of a very limited number of effective therapies for SARS-CoV-2 infections.
5. The fifth point is that we saw no evidence of ADE. The positive signals that emanated from the US CPS data showing that when high titer plasma was administered early it was associated with reduced mortality generally put that concern to rest. Together, the data highlighted above informed the lifesaving use of CCP in the United States at Scale through early 2021 (Casadevall et al. 2021a, b; Dragotakes et al. 2024).

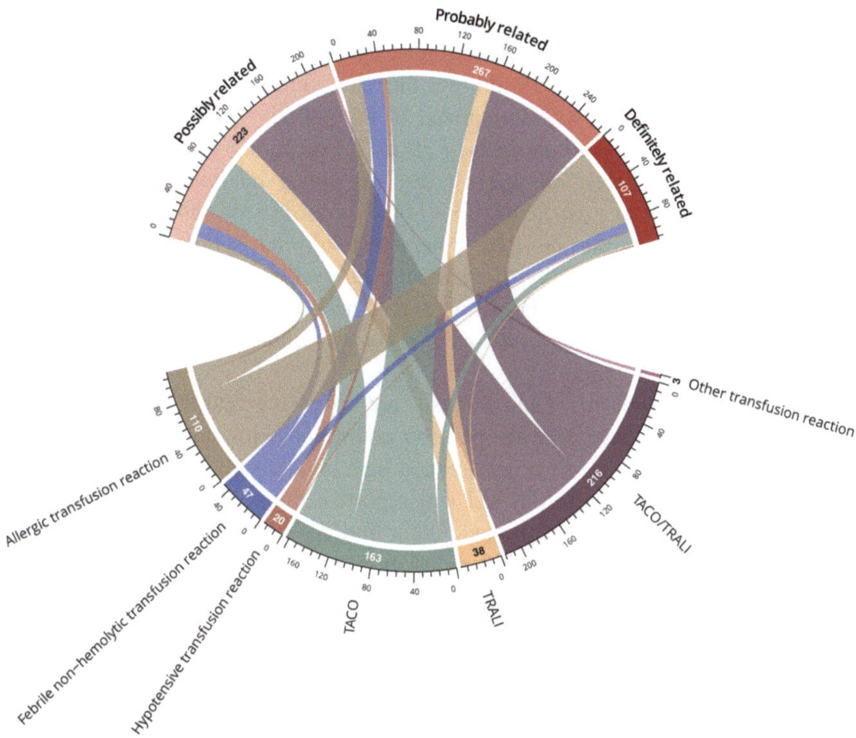

Fig. 3 Chord diagram of serious transfusion reaction associated attributions. Attribution to a category of relatedness is depicted by lines connecting each serious transfusion reaction type. The width of each line and the circumferential axis indicate the number of patients in each combined serious transfusion reaction and relatedness category group. Other transfusion reactions which were adjudicated to be possibly related to the transfusion ($n = 3$) included leukemoid reaction ($n = 1$), pulmonary embolism ($n = 1$), and red cell dilution ($n = 1$). Adapted, with permission, from the US CPS (Senefeld et al. 2021)

Our early findings that CCP was safe and that its safety profile was similar to that of other blood products were confirmed by the fact that no trials including the large ($N > 500$ patients) randomized controlled trials were stopped early for safety concerns (Bégin et al. 2021; RECOVERY Collaborative Group 2021; Troxel et al. 2022; Estcourt et al. 2021). In all trials with detailed safety reporting, the safety profile of CCP was similar to the safety profile of blood products generally, and any differences between treatment and control groups were minimal. Additionally, the vast majority of the safety data have been reported in supplements, attesting to the fact that no major unanticipated safety signals were seen.

To assess the overall safety of CCP, we performed two comprehensive meta-analyses of all safety data reported for CCP. The first meta-analysis was of all inpatient and outpatient RCTs during the first three years of the pandemic (Franchini et al. 2024b). As summarized in the paper: "*...we have performed a systematic review and meta-analysis of 39 RCTs, including nearly 24,000 participants. We have explored all safety issues, including all and treatment-related adverse reactions, with a particular focus on serious and thromboembolic adverse events. Our meta-analysis, the most complete published to date, confirmed on a larger number of studies and participants the safety data from previous systematic reviews* (Iannizzi et al. 2023; Franchini and Cruciani 2021). *No significant differences emerged between cases and controls from the pooled analysis for any grade of or serious adverse reactions*".

The findings from this meta-analysis are extensive and it is important to remember that not all trials were placebo controlled with standard fresh frozen plasma. The incidence of treatment-related serious adverse events was very low (0.5%) and only one case of TRALI was reported. Importantly, no differences in thromboembolic events were reported. Other findings from the meta-analysis show no evidence of transmission of infectious disease, no evidence for ADE, and no evidence that the endogenous immune response to SARS-CoV-2 was blunted by CCP.

In the second meta-analysis, the safety and efficacy of CCP combined with other therapeutic agents were evaluated (Franchini et al. 2024a). While this analysis showed that CCP in combination with remdesivir (but not corticosteroids) reduced mortality, there were insufficient data to draw conclusions about safety issues with combined therapy. However, no major safety signals associated with combined therapies were obviously apparent.

In future pandemics, convalescent plasma should be used as a therapeutic agent as soon as the first patients have recovered from the infectious disease and are able to donate plasma. The theoretical risks of convalescent plasma such as ADE, cytokine storms, exacerbation of coagulation derangement, and transfusion-related events such as TACO and TRALI are very rare and can be minimized by following standard transfusion medicine procedures. Additionally, real-time reporting of serious adverse events allows the detection of unexpected adverse events should they occur. This should also allow a determination as to whether the safety profile of CP for a new, non-COVID-19 pathogen differs from SARS-CoV2 and evokes higher rates of transfusion-related or unusual adverse events.

Conflict of Interest The authors report no conflicts of interest, financial or otherwise.

References

Bégin P, Callum J, Jamula E, Cook R, Heddle NM, Tinmouth A et al (2021) Convalescent plasma for hospitalized patients with COVID-19: an open-label, randomized controlled trial. Nat Med 27(11):2012–2024

Casadevall A, Pirofski LA (2020) The convalescent sera option for containing COVID-19. J Clin Invest 130(4):1545–1548

Casadevall A, Dragotakes Q, Johnson PW, Senefeld JW, Klassen SA, Wright RS et al (2021b) Convalescent plasma use in the USA was inversely correlated with COVID-19 mortality. Elife 4(10):e69866

Casadevall A, Dragotakes Q, Johnson PW, Senefeld JW, Klassen SA, Wright RS et al (2021a) Convalescent plasma use in the United States was inversely correlated with COVID-19 mortality: did plasma hesitancy cost lives? medRxiv 2021.04.07.21255089

Children's Hospital of Philadelphia. Antibody-dependent enhancement and vaccines [Internet]. Children's Hospital of Philadelphia. Available from: https://www.chop.edu/vaccine-education-center/vaccine-safety/other-vaccine-safety-concerns/antibody-dependent-enhancement-and-vaccines

Cho MS, Modi P, Sharma S (2024) Transfusion-related acute lung injury. In: StatPearls [Internet]. StatPearls Publishing, Treasure Island (FL). Available from: http://www.ncbi.nlm.nih.gov/books/NBK507846/

Control C for D, Prevention (CDC) (2023) National healthcare safety network (NHSN) biovigilance component: hemovigilance module protocol [Internet]. Centers for Disease Control and Prevention (CDC). Available from: https://www.cdc.gov/nhsn/pdfs/biovigilance/bv-hv-protocol-current.pdf

Dragotakes Q, Johnson PW, Buras MR, Carter RE, Joyner MJ, Bloch E et al (2024) Estimates of actual and potential lives saved in the United States from the use of COVID-19 convalescent plasma [Internet]. Available from: https://doi.org/10.1101/2024.05.16.24307505

Estcourt LJ, Turgeon AF, McQuilten ZK, McVerry BJ, Al-Beidh F, Writing Committee for the REMAP-CAP Investigators et al (2021) Effect of convalescent plasma on organ support-free days in critically ill patients with COVID-19: a randomized clinical trial. JAMA 326(17):1690–1702

Franchini M, Cruciani M (2021) How safe is COVID-19 convalescent plasma? Mayo Clin Proc 96 (8):2279–2281

Franchini M, Focosi D, Cruciani M, Joyner MJ, Pirofski LA, Senefeld JW et al (2024a) Safety and efficacy of convalescent plasma combined with other pharmaceutical agents for treatment of COVID-19 in hospitalized patients: a systematic review and meta-analysis. Diseases 12(3):41

Franchini M, Cruciani M, Casadevall A, Joyner MJ, Senefeld JW, Sullivan DJ et al (2024b) Safety of COVID-19 convalescent plasma: a definitive systematic review and meta-analysis of randomized controlled trials. Transfusion 64(2):388–399

Gratz J, Ponschab M, Iapichino GE, Schlimp CJ, Cadamuro J, Grottke O et al (2020) Comparison of fresh frozen plasma vs. coagulation factor concentrates for reconstitution of blood: an in vitro study. Eur J Anaesthesiol 37(10):879–888

Iannizzi C, Chai KL, Piechotta V, Valk SJ, Kimber C, Monsef I et al (2023) Convalescent plasma for people with COVID-19: a living systematic review. Cochrane Database Syst Rev 2(2): CD013600

Joyner MJ, Wright RS, Fairweather D, Senefeld JW, Bruno KA, Klassen SA et al (2020a) Early safety indicators of COVID-19 convalescent plasma in 5000 patients. J Clin Invest 130 (9):4791–4797

Joyner MJ, Bruno KA, Klassen SA, Kunze KL, Johnson PW, Lesser ER et al (2020b) Safety update: COVID-19 convalescent plasma in 20,000 hospitalized patients. Mayo Clin Proc 95 (9):1888–1897

Klompas AM, Klassen SA, Senefeld JW, Joyner MJ, Johnson PW, Carter RE et al (2021) In reply—micro-thrombosis, perfusion defects, and worsening oxygenation in COVID-19 patients: a word of caution on the use of convalescent plasma. Mayo Clin Proc 96(1):259–261

Klompas AM, van Helmond N, Juskewitch JE, Pruthi RK, Sexton MA, Soto JCD et al (2022) Coagulation profile of human COVID-19 convalescent plasma. Sci Rep 12(1):637

Lawley TJ, Bielory L, Gascon P, Yancey KB, Young NS, Frank MM (1984) A prospective clinical and immunologic analysis of patients with serum sickness. N Engl J Med 311(22):1407–1413

Luke TC, Kilbane EM, Jackson JL, Hoffman SL (2006) Meta-analysis: convalescent blood products for Spanish influenza pneumonia: a future H5N1 treatment? Ann Intern Med 145 (8):599–609

RECOVERY Collaborative Group (2021) Convalescent plasma in patients admitted to hospital with COVID-19 (RECOVERY): a randomised controlled, open-label, platform trial. Lancet 397(10289):2049–2059

Sanfilippo F, La Rosa V, Astuto M (2021) Micro-thrombosis, perfusion defects, and worsening oxygenation in COVID-19 patients: a word of caution on the use of convalescent plasma. Mayo Clin Proc 96(1):259

Senefeld JW, Johnson PW, Kunze KL, Bloch EM, van Helmond N, Golafshar MA et al (2021) Access to and safety of COVID-19 convalescent plasma in the United States expanded access program: a national registry study. PLoS Med 18(12):e1003872

Suddock JT, Crookston KP (2024) Transfusion reactions. In: StatPearls [Internet]. StatPearls Publishing, Treasure Island (FL). Available from: http://www.ncbi.nlm.nih.gov/books/NBK482202/

Troxel AB, Petkova E, Goldfeld K, Liu M, Tarpey T, Wu Y et al (2022) Association of convalescent plasma treatment with clinical status in patients hospitalized with COVID-19: a meta-analysis. JAMA Netw Open 5(1):e2147331

Michael J. Joyner MD is a physiologist and anesthesiologist at the Mayo Clinic. During the pandemic he repurposed his professional efforts and used a distributed leadership model to facilitate and study the use of convalescent plasma to treat SARS-CoV-2.

Rickey E. Carter PhD is Professor of Biostatistics and the Director of the Biostatistics, Epidemiology and Research Design resource within Mayo Clinic's Center for Clinical and Translational Science Activities. Dr. Carter helped assemble, mentor and empower the biostatistical team that supported the Mayo Clinic Convalescent Plasma clinical trial.

R. Scott Wright MD is a Professor of Medicine, a consulting cardiologist and research leader at Mayo Clinic. He chairs the Mayo Clinic multisite IRB and is Director of the Human Research Protection Program across all Mayo Clinic sites. He was honored to be involved and co- lead the Mayo Convalescent plasma trial during Covid-19.

Jonathon W. Senefeld PhD is an Assistant Professor in the Department of Health and Kinesiology at the University of Illinois Urbana-Champaign. Jack was a member of the Mayo Clinic team that led the Mayo Clinic COVID-19 Convalescent Plasma trial and has a special interest in passive antibody use in the immunocompromised.

The Importance of Geographic Proximity of Convalescent Plasma Donors

Katie L. Kunze and Patrick W. Johnson

Contents

Abstract Donor–recipient proximity emerged as an important factor influencing the efficacy of COVID-19 convalescent plasma (CCP) treatment during the early stages of the COVID-19 pandemic. This relationship was uncovered while analyzing data collected in the collaborative Expanded Access Program (EAP) for CCP at Mayo Clinic, a project aimed to establish protocols for CCP use amid the uncertainty of the novel disease. Analysis of data from nearly 28,000 patients revealed a significant reduction in risk of 30-day mortality for those receiving near-sourced plasma when compared to those receiving distantly sourced plasma [pooled relative risk, 0.73 (95% CI 0.67–0.80)], prompting adjustments in treatment protocols at selected institutions, and highlighting the importance of proximity in optimizing CCP outcomes. Despite its significance, subsequent studies of CCP effectiveness in COVID-19 have often overlooked donor–recipient proximity. Our

K. L. Kunze (✉)
Department of Quantitative Health Sciences, Mayo Clinic, Scottsdale, AZ, USA
e-mail: Kunze.Katie@mayo.edu

K. L. Kunze · P. W. Johnson
Digital Innovation Lab, Mayo Clinic, Jacksonville, FL, USA
e-mail: Johnson.Pat@mayo.edu

P. W. Johnson
Department of Quantitative Health Sciences, Mayo Clinic, Jacksonville, FL, USA

Current Topics in Microbiology and Immunology (2025) 443: 89–98
https://doi.org/10.1007/82_2024_270

Published Online: 9 August 2024

findings emphasize the importance of donor–recipient proximity in CCP treatment in the current pandemic, and we discuss potential methods for improving CCP efficacy in future pandemics. Our recommendations include prioritizing virus genotyping for vulnerable patients, establishing a robust testing infrastructure, and collecting additional donor data to enhance plasma selection. This chapter underscores the importance of comprehensive data collection and sharing to navigate the evolving landscape of newly emerging infectious diseases.

Glossary

BARDA	Biomedical Advanced Research and Development Authority
CCP	COVID-19 convalescent plasma
CMH	Cochran–Mantel–Haenszel
COVID-19	Novel Coronavirus Disease 2019, which is the result of the SARS-CoV-2 virus
EAP	Expanded Access Program for Convalescent Plasma
FDA	U.S. Food and Drug Administration
GBM	Gradient Boosting Machine
ICU	Intensive Care Unit

1 Background

In the early days of the SARS-CoV-2 novel coronavirus disease-2019 (COVID-19) pandemic, with limited knowledge of this novel disease, treatment options for hospitalized patients beyond symptomatic management were nonexistent. In an attempt to identify effective treatments, a consortium of researchers from across the U.S. partnered with the U.S. Food and Drug Administration (FDA) and Biomedical Advanced Research and Development Authority (BARDA) to establish protocols and data collection requirements for administering COVID-19 convalescent plasma (CCP) from donors who had recovered from a COVID-19 infection. This effort evolved into the Expanded Access Program for CCP in COVID-19 patients (EAP). Due to the urgent demand, plasma was sourced from a network of blood banks from across the U.S. and was often shipped long distances to support needs and to treat patients in a timely manner. Proximity of donor to recipient varied widely.

As the pandemic progressed and meaningful changes in disease expression were observed across newly identified variants, a highly relevant question presented itself: did timely near-sourced and distantly source plasma have the same treatment benefit? Data from the EAP, which ran from March 1, 2020 to August 31, 2020, were used to explore this potential effect. We found a lower rate of 30-day mortality in patients who had received near-sourced plasma compared to those who had received distantly sourced plasma (Kunze et al. 2021).

After this publication, some institutions altered treatment protocols to administer only near-sourced plasma in an attempt to provide patients with the plasma potentially most similar to the variant with which they were infected (Sullivan et al. 2022). Other institutions were either unable to make such a change or were not aware of the possible impact of sourcing plasma from nearby donors who had provided plasma recently. As a result, some studies of the efficacy of convalescent plasma for use in COVID-19 patients may not have captured its full treatment benefit.

In addition to the relevance of donor proximity of convalescent plasma for the treatment of COVID-19, this observation could be relevant to future pandemics or to the spread of other infectious diseases. Convalescent plasma has been used to treat infectious diseases for over one hundred years and likely will be used again (Casadevall and Scharff 1995). Knowing the impact of donor proximity could help establish future distribution networks and effective treatment protocols.

2 Proximity Research in the EAP

After initial research which demonstrated the safety (Joyner et al. 2020a, b) and efficacy (Joyner et al. 2021) of treatment with CCP for subsets of patients hospitalized with COVID-19, and in light of a quickly mutating virus, we turned our attention to reasons for differing outcomes among patients. We explored whether the distance between the location of the recipient and the location of CCP collection might influence the effectiveness of this form of treatment (Kunze et al. 2021). Combining information on the location of plasma collection and the location of the treating hospital, we assessed whether patients received near-sourced or distantly sourced CCP, defining near-sourced as ≤ 150 miles between the two locations and distantly sourced as above this threshold. This distance of 150 miles was used to represent a reasonable local area for treatment facilities and a likely threshold for commute time for patients and plasma. For some analyses, the continuous values of distance were used, and the relationship between distance and mortality was observed across the full range of distances. Early in the program, most patients were treated with locally sourced convalescent plasma. In April, locally sourced CCP accounted for 80% of all CCP, but by June this figure was below 40% (Fig. 1a). The average distance between source and recipient surpassed 150 miles in late May and remained above this cutoff for the rest of the program (Fig. 1b). Beginning in July, a downward trend in average distance was observed, which coincided with an increase in the number of blood banks providing CCP. Patterns of source–recipient distance varied by blood bank. For example, Carter BloodCare primarily provided near-sourced plasma throughout the entirety of the EAP, whereas the American Red Cross contributed an increasing percentage of units to distantly located hospitals starting in July. Figure 2 shows the trends in distance categories for the 12 blood banks that contributed the greatest number of units of plasma during the EAP.

We analyzed data from nearly 28,000 of the approximately 94,000 hospitalized patients in the EAP, whose plasma came from over 200 blood centers to see whether source–recipient distance affected 30-day mortality. Figure 3 shows (1) the number of units donated by state, (2) the number of units infused by hospital

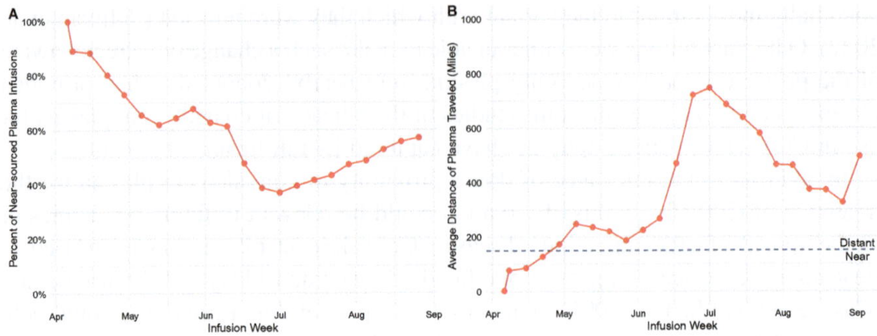

Fig. 1 Percent of local plasma and distance of plasma traveled over time. Panel A depicts the percent of near-sourced plasma infusions by calendar week of the EAP. Depicted here are only samples for which a donor identification number was provided. Panel B shows the average distance that plasma donations traveled in miles by calendar week of the EAP. A threshold of 150 miles was chosen to delineate near-sourced versus distantly sourced plasma. Depicted here are only samples for which a donor identification number was provided

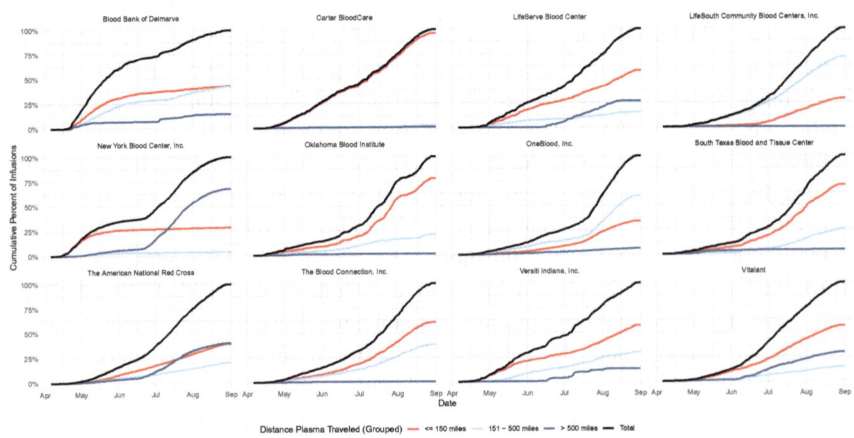

Fig. 2 The cumulative percent of infusions grouped by distance plasma traveled during the EAP for the top 12 participating blood banks (by total number of donations). This plot shows the cumulative percent of infusions over the duration of the EAP for each of the top 12 participating blood banks as defined by total units of plasma infused. The distance plasma traveled between collecting blood bank and the infusion hospital is categorized into ≤ 150 miles in red (i.e., near-sourced), 151–500 miles (light blue), greater than 500 miles (dark blue), and the total of all donations (black). Depicted here are only samples for which a donor identification number was provided

location, and (3) the number of units donated by blood bank location. Our analyses included regression models and Cochran–Mantel–Haenszel (CMH) techniques to estimate the adjusted relative risk of death within the 30 days following transfusion with CCP in relation to near or distantly sourced CCP. Additionally, a gradient

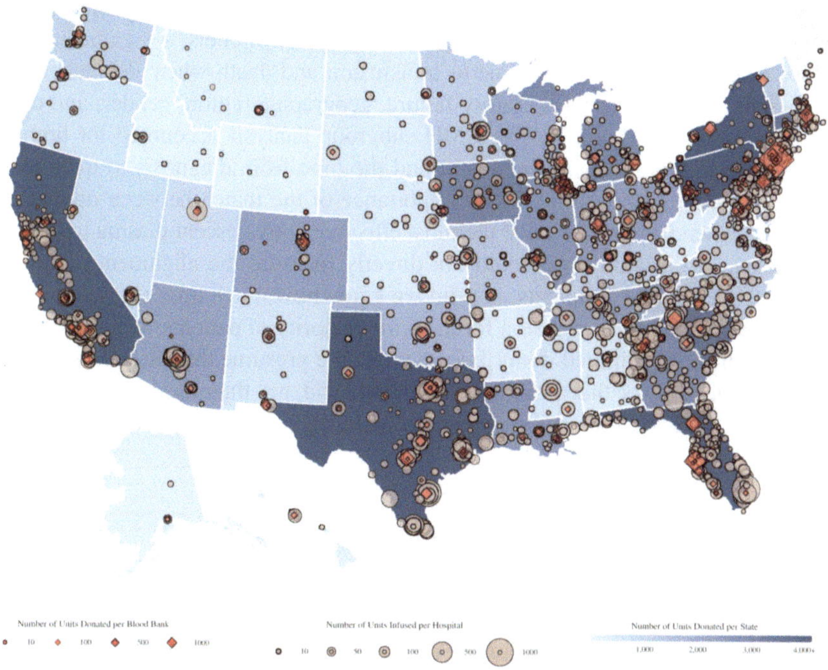

Fig. 3 Geographical representation of the number of units of plasma donated and infused during the EAP. This map shows the number of units donated by each state as depicted by the shading of each state with dark blue representing a higher number of units donated which were used in the EAP. Due to some states with outsized contributions data were winsorized at 4,000 units (i.e., any state with 4,000 or more units is represented by the same shade of dark blue). Each circle denotes a hospital that participated in the EAP. The size of the circle captures the number of units infused at a particular participating site. The participating U.S. territories of Guam, Puerto Rico, and the U.S. Virgin Islands are not included as plasma donations were not exported. Each diamond represents a participating blood bank with the size of the diamond capturing the number of units donated by that blood bank. Depicted here are only samples for which a donor identification number was provided.

boosting machine (GBM) model was used to explore the relationship between patient data and treatments and 30-day mortality. Patients requiring mechanical ventilation and patients over age 65 were excluded, because we did not find mortality benefits of CCP in this group of patients (Joyner et al. 2021; Yoon et al. 2021).

We found that patients receiving near-sourced convalescent plasma had a lower risk of 30-day mortality than patients who received distantly sourced plasma [CMH pooled relative risk, 0.73 (95% CI 0.67–0.80)]. In the GBM analysis, only advanced age, intensive care unit (ICU) admission, and respiratory failure had a stronger influence on mortality than proximity in the prediction of 30-day mortality.

Time to transfusion, defined as the time from patient diagnosis to treatment with CCP, also played an important but less impactful role in predicting 30-day mortality for these patients. In the GBM, time to transfusion was the eighth most important variable, and, similarly, in our relative risk regression models, a significant association was observed between time to transfusion and death when also accounting for patient age, ICU care, respiratory failure, geographic region, gender, and donor distance (Kunze et al. 2021). The CMH subgroup analysis accounted for both the time from symptom onset to treatment and the time from diagnosis to treatment.

These findings showed the likely importance of the match between donors and recipients when trying to provide the most effective convalescent plasma treatments to patients. In our study, we could not directly measure the alignment of disease variants in donor and recipients, nor did we know how much time elapsed between plasma collection and transfusion for the wide majority of patients with only certain labs providing very limited donor information. We presume that these factors were the principal reasons that proximity mattered, and we thus used proximity as a stand-in for these factors in the absence of detailed information on either.

3 Impact of Proximity on the Current COVID-19 Pandemic

In the years since the publication of our research from the EAP, we found that donor–recipient proximity was rarely mentioned in studies exploring the effectiveness of convalescent plasma for treating COVID-19. Most studies that took note of donor–recipient proximity invoked this factor only to explain results contrary to expectation (Alemany et al. 2022; Bar et al. 2021; Celikgil et al. 2023; Drews et al. 2022; Ortigoza et al. 2022; Qian et al. 2022; Song et al. 2022; Yu et al. 2022). Some studies expressed consideration for donor–recipient distances but separated near from far using different metrics from ours, often including distances of several hundred miles to be “nearby”, making the comparison to our findings difficult. Some studies prioritized using near-sourced plasma (Arnold Egloff et al. 2021; Herman et al. 2022; Root et al. 2022; Sanz et al. 2022), but no consistent effect on patient outcomes has been formally documented since our publication. In addition, some reviews, guidelines, and recommendations addressing both current and future pandemics emphasize the use of near-sourced plasma as an approach to improve the effectiveness of convalescent plasma treatment (Sullivan et al. 2022; Casadevall et al. 2023; Focosi et al. 2022a, b; Ostrovskaya et al. 2022; Denakker et al. 2023).

4 Further Implications for the Ongoing Pandemic

Given the current body of research, several questions remain to be explored regarding the role of proximity in the use and effectiveness of convalescent plasma during the initial phase of the COVID-19 pandemic. First, was proximity acting solely as a proxy for regional differences in the specific viral variant, or were other factors involved? The high cost and lack of ready availability of genotyping capacity of COVID-19 variants made it impossible to examine this factor directly in the early days of the COVID-19 pandemic.

It is plausible that travel by infected individuals may impact variant spread and patient disease. While it is most likely that both plasma recipients and plasma donors would be infected with the predominant viral variant in their region, travel may have resulted in either the donor or the recipient having acquired a different strain, particularly when, later in the pandemic, travel was less restricted. The capacity to rapidly genotype variants might, in the future, allow better matching of convalescent plasma regardless of its source, whether near or far from the recipient. Additional studies could help to identify the separate impacts of overlapping factors including time from plasma collection to treatment administration, time from infection/symptoms to treatment, and distance plasma traveled (which could influence these time elements).

Inasmuch as the COVID-19 virus continued to mutate throughout the pandemic, and our data were obtained during the first few months of 2020, we cannot assess the changing impact of donor–recipient distance during later waves of the pandemic (Focosi et al. 2022a). Given the timeline of the EAP, we can assume that the patients we studied contracted one of several early subvariants that circulated prior to the worst waves of infection as defined by caseloads and hospitalizations, produced by the Delta, Omicron, and Omicron-sublineage variants.

5 Recommendations and Future Considerations

With the emergent phase of the pandemic behind us, it is worth reflecting on processes that have been quickly built out which would be useful for future pandemics while paying special attention to the effective distribution of convalescent plasma. The most impactful of these processes may be the ability to quickly genotype variants in patients who test positive for the disease. In the COVID-19 pandemic, we did not have sufficient infrastructure to type many samples. With the ability to type most or all samples, we would be able to quickly match patients to what is likely to be the most effective donor plasma for their specific variant/subvariant and location. We recognize that a testing infrastructure of this nature would be costly to maintain on the chance that another virus presents itself. However, having agreed-upon procedures in place and a reserve of emergency funding would allow these processes to be stood up quickly. If resources are few, we recommend prioritizing genotyping of donor samples and patient variants for the most vulnerable patients such as those who are immunocompromised. Antibody testing (including variant typing and neutralizing antibody effectiveness) at the time of donation could contribute to this effort to obtain the best possible match. Needless to say, testing infrastructure of this nature would likely be useful for applications beyond simply creating the most effective match of plasma donor and recipient infection.

In the meantime, blood collection centers can do much to improve the match of donors and recipients. We recommend that blood collection centers collect the address of the donor, the location of the infection (if known), and any history of recent domestic travel (history of foreign travel is routinely obtained in blood donation). Having additional donor location data available to treatment centers and hospitals would assist clinicians in selecting the most appropriate donor plasma for patients.

In sum, donor–recipient proximity proved to be a key factor in CCP effectiveness, and we anticipate that it will be so in future pandemics as well. Building the capacity to collect, describe, and share important disease and donor data will be important in addressing patient needs in future pandemics. We hope that learning from this research and our shared recent past will aid us in our efforts to adequately prepare.

Disclosures The work to compose this book chapter was supported by a gift from UnitedHealth Group. The authors have no other conflicts of interest to disclose.

References

Alemany A, Millat-Martinez P, Corbacho-Monne M, Malchair P, Ouchi D, Ruiz-Comellas A et al (2022) High-titre methylene blue-treated convalescent plasma as an early treatment for outpatients with COVID-19: a randomised, placebo-controlled trial. Lancet Respir Med 10 (3):278–288

Ambrose N, Amin A, Anderson B, Bertagnolli M, Campion F, Chow D et al (2023) The influence of social determinants on receiving outpatient treatment with monoclonal antibodies, disease risk, and effectiveness for COVID-19. J Gen Intern Med

Arnold Egloff SA, Junglen A, Restivo JS, Wongskhaluang M, Martin C, Doshi P et al (2021) Convalescent plasma associates with reduced mortality and improved clinical trajectory in patients hospitalized with COVID-19. J Clin Invest 131(20)

Bar KJ, Shaw PA, Choi GH, Aqui N, Fesnak A, Yang JB et al (2021) A randomized controlled study of convalescent plasma for individuals hospitalized with COVID-19 pneumonia. J Clin Invest 131(24)

Casadevall A, Scharff MD (1995) Return to the past: the case for antibody-based therapies in infectious diseases. Clin Infect Dis 21(1):150–161

Casadevall A, Joyner MJ, Pirofski LA, Senefeld JW, Shoham S, Sullivan D et al (2023) Convalescent plasma therapy in COVID-19: unravelling the data using the principles of antibody therapy. Expert Rev Respir Med 17(5):381–395

Celikgil A, Massimi AB, Nakouzi A, Herrera NG, Morano NC, Lee JH et al (2023) SARS-CoV-2 multi-antigen protein microarray for detailed characterization of antibody responses in COVID-19 patients. PLoS ONE 18(2):e0276829

Drews SJ, Abe KT, Hu Q, Samson R, Gingras AC, Colwill K et al (2022) Resistance of SARS-CoV-2 beta and gamma variants to plasma collected from Canadian blood donors during the spring of 2020. Transfusion 62(1):37–43

Focosi D, Franchini M, Joyner MJ, Casadevall A (2022a) Are convalescent plasma stocks collected during former COVID-19 waves still effective against current SARS-CoV-2 variants? Vox Sang 117(5):641–646

Focosi D, Franchini M, Pirofski LA, Burnouf T, Paneth N, Joyner MJ et al (2022b) COVID-19 convalescent plasma and clinical trials: understanding conflicting outcomes. Clin Microbiol Rev 35(3):e0020021

Herman JD, Wang C, Burke JS, Zur Y, Compere H, Kang J et al (2022) Nucleocapsid-specific antibody function is associated with therapeutic benefits from COVID-19 convalescent plasma therapy. Cell Rep Med 3(11):100811

Joyner MJ, Wright RS, Fairweather D, Senefeld JW, Bruno KA, Klassen SA et al (2020a) Early safety indicators of COVID-19 convalescent plasma in 5000 patients. J Clin Invest 130 (9):4791–4797

Joyner MJ, Bruno KA, Klassen SA, Kunze KL, Johnson PW, Lesser ER et al (2020b) Safety update: COVID-19 convalescent plasma in 20,000 hospitalized patients. Mayo Clin Proc 95 (9):1888–1897

Joyner MJ, Carter RE, Senefeld JW, Klassen SA, Mills JR, Johnson PW et al (2021) Convalescent plasma antibody levels and the risk of death from Covid-19. N Engl J Med 384(11):1015–1027

Kunze KL, Johnson PW, van Helmond N, Senefeld JW, Petersen MM, Klassen SA et al (2021) Mortality in individuals treated with COVID-19 convalescent plasma varies with the geographic provenance of donors. Nat Commun 12(1):4864

McNeely CL, Schintler LA, Stabile B (2020) Social determinants and COVID-19 disparities: differential pandemic effects and dynamics. World Med Health Policy 12(3):206–217

Organization WH (2008) Closing the gap in a generation: health equity through action on the social determinants of health. WHO, Geneva

Ortigoza MB, Yoon H, Goldfeld KS, Troxel AB, Daily JP, Wu Y et al (2022) Efficacy and safety of COVID-19 convalescent plasma in hospitalized patients: a randomized clinical trial. JAMA Intern Med 182(2):115–126

Ostrovskaya EA, Kostin AI, Bulanov AY (2022) Efficiency and safety of convalescent plasma therapy in patients with COVID-19: a systematic review. Ann Crit Care 2:108–120

Prevention CfDCa (2013) CDC health disparities and inequalities report—United States. Morbidity Mortality Weekly Rep 62(3)

Qian Z, Zhang Z, Ma H, Shao S, Kang H, Tong Z (2022) The efficiency of convalescent plasma in COVID-19 patients: a systematic review and meta-analysis of randomized controlled clinical trials. Front Immunol 13:964398

Root HB, Gilleskie M, Lu CH, Gilmore A, Evans M, Nelson BG et al (2022) Evaluation of a COVID-19 convalescent plasma program at a U.S. academic medical center. PLoS One 17(12): e0277707

Sanz C, Nomdedeu M, Pereira A, Sauleda S, Alonso R, Bes M et al (2022) Efficacy of early transfusion of convalescent plasma with high-titer SARS-CoV-2 neutralizing antibodies in hospitalized patients with COVID-19. Transfusion 62(5):974–981

Song ATW, Rocha V, Mendrone-Junior A, Calado RT, De Santis GC, Benites BD et al (2022) Treatment of severe COVID-19 patients with either low- or high-volume of convalescent plasma versus standard of care: a multicenter Bayesian randomized open-label clinical trial (COOP-COVID-19-MCTI). Lancet Reg Health Am 10:100216

Sullivan DJ, Gebo KA, Shoham S, Bloch EM, Lau B, Shenoy AG et al (2022) Early outpatient treatment for Covid-19 with convalescent plasma. N Engl J Med 386(18):1700–1711

Turner-Musa J, Ajayi O, Kemp L (2020) Examining social determinants of health, stigma, and COVID-19 disparities. Healthcare (Basel) 8(2)

Upshaw TL, Brown C, Smith R, Perri M, Ziegler C, Pinto AD (2021) Social determinants of COVID-19 incidence and outcomes: a rapid review. PLoS ONE 16(3):e0248336

Van Denakker TA, Al-Riyami AZ, Feghali R, Gammon R, So-Osman C, Crowe EP et al (2023) Managing blood supplies during natural disasters, humanitarian emergencies, and pandemics: lessons learned from COVID-19. Expert Rev Hematol 16(7):501–514

Yoon HA, Bartash R, Gendlina I, Rivera J, Nakouzi A, Bortz RH et al (2021) Treatment of severe COVID-19 with convalescent plasma in Bronx, NYC. JCI Insight 6(4)
Yu D, Li YF, Liang H, Wu JZ, Hu Y, Peng Y et al (2022) Potent anti-SARS-CoV-2 efficacy of COVID-19 hyperimmune globulin from vaccine-immunized plasma. Adv Sci (weinh) 9(14): e2104333

Katie L. Kunze is a Principal Biostatistician, Assistant Professor of Biostatistics, and Associate in the Department of Quantitative Health Sciences at Mayo Clinic in Arizona. Dr. Kunze received her M.A. and Ph.D. in Measurement Statistics and Methodological Studies from Arizona State University and joined the Mayo Clinic in 2017. In her current position, she supports research faculty within her department, clinicians from across the Mayo Clinic, and trainees in a variety of fields. She serves as lead statistician for a family of studies examining the benefits of genetic testing, is lead statistician for the Department of Critical Care Medicine at Mayo Clinic Florida, and served as one of the lead statisticians for the Expanded Access Program for Convalescent Plasma in COVID-19 patients led by Mayo Clinic. In addition to her consulting work, Dr. Kunze conducts methodological research in survey development, patient-reported outcomes, and quality-of-life measures.

Patrick W. Johnson is a Senior Data Science Analyst in the Department of Quantitative Health Sciences at Mayo Clinic in Florida. He received BS degrees in Biology and Applied Mathematics at the University of Oregon and an MS in Computer Science from the University of Minnesota. As a member of the Digital Innovation Lab, he works to create, validate, and implement machine learning and AI solutions within the healthcare space. He was lead data science analyst for the Expanded Access Program for Convalescent Plasma in COVID-19 patients and helped create data pipelines for Mayo Clinic's COVID-19 predictive models. More recently, Pat has contributed to several high-impact studies including the design, implementation, and regulatory approval of multiple novel AI algorithms for use with electrocardiograms and pulmonary function tests as alternatives to more invasive procedures, and modeling of real-time patient telemetry data.

Evidence for the Efficacy of COVID-19 Convalescent Plasma

Stephen A. Klassen and Jonathon W. Senefeld

Contents

S. A. Klassen (✉)
Department of Kinesiology, Faculty of Applied Health Sciences, Brock University, St. Catharines, Canada
e-mail: sklassen@brocku.ca

J. W. Senefeld
Department of Health and Kinesiology, University of Illinois Urbana-Champaign, Urbana, IL, USA
e-mail: senefeld@illinois.edu

Current Topics in Microbiology and Immunology (2025) 443: 99–117
https://doi.org/10.1007/82_2024_280

Published Online: 28 August 2024

Abstract During the global health emergency caused by the coronavirus disease 2019 (COVID-19), evidence relating to the efficacy of convalescent plasma therapy —evidence critically needed for both public policy and clinical practice—came from multiple levels of the epistemic hierarchy. The challenges of conducting clinical research during a pandemic, combined with the biological complexities of convalescent plasma treatment, required the use of observational data to fully assess the impact of convalescent plasma therapy on COVID symptomatology, hospitalization rates, and mortality rates. Observational studies showing the mortality benefits of convalescent plasma emerged early during the COVID-19 pandemic from multiple continents and were substantiated by real-time pragmatic meta-analyses. Although many randomized clinical trials (RCTs) were initiated at the onset of the pandemic and were designed to provide high-quality evidence, the relative inflexibility in the design of clinical trials meant that findings generally lagged behind other forms of emerging information and ultimately provided inconsistent results on the efficacy of COVID-19 convalescent plasma. In the pandemic framework, it is necessary to emphasize more flexible analytic strategies in clinical trials, including secondary, subgroup, and exploratory analyses. We conclude that in totality, observational studies and clinical trials taken together provide strong evidence of a mortality benefit conferred by COVID-19 convalescent plasma, while acknowledging that some randomized clinical trials examined suboptimal uses of convalescent plasma.

1 The Hierarchy of Evidence

Evidence-based medicine focuses on understanding the efficacy of treatments and has become the dominant epistemic framework for medical practice and clinical research since its inception in the early 1990s (Bingeman 2016). While there are many versions of the "hierarchy of evidence", which assigns to each type of evidence a ranking or confidence in the quality of evidence, there is broad-based agreement on the general framework as presented in Fig. 1. The highest quality of evidence represents synthesized information (clinical practice guidelines or meta-analyses). Among primary studies, randomized controlled trials (RCTs) represent the highest quality of evidence, followed by cohort studies, case-control studies, and case series and reports. The RCT has thus become the "gold standard" for comparison with other experimental approaches (Bothwell et al. 2016) and has become so dominant that evidence *not* derived from RCTs is viewed by some as unusable for clinical or policy decisions (Paneth et al. 2022). However, RCTs do not have a monopoly on medical knowledge production, particularly in the context of novel, life-threatening pathogens, and can mislead if the dose and timing of the intervention, and the study population, do not conform to the biology of the condition or the intervention (Hellman and Hellman 1991). Observational study

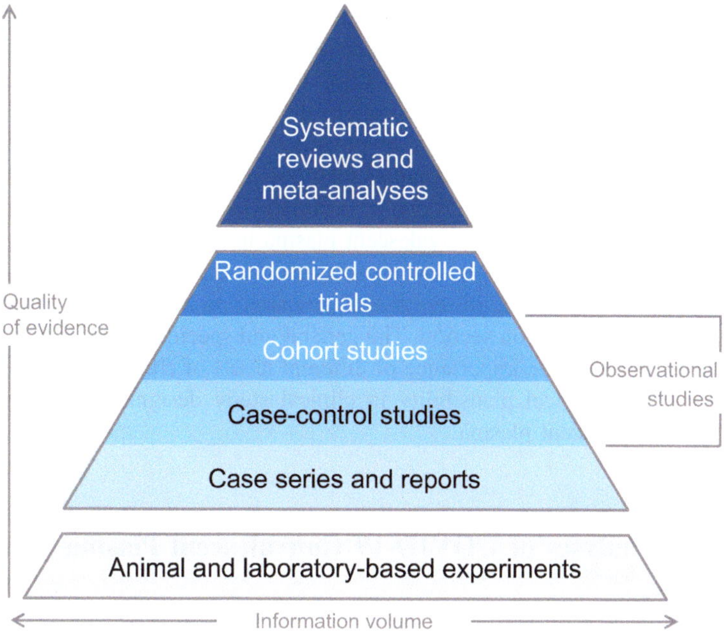

Fig. 1 Conceptualization of the hierarchy of evidence for clinical research. A conceptual model displaying different types of experimental approaches based on data synthesis approach, allocation of an intervention, availability of data for a comparator group, and more. Although randomized controlled trials (RCTs) are generally considered to be the highest quality of evidence and least susceptible to bias, observational studies may be adequate to assess the efficacy of therapeutic interventions and inform the appropriate design of subsequent RCTs

designs can provide crucial information early during epidemics and pandemics for three reasons:

(1) their speed and ease of implementation across settings,
(2) their relatively low cost, and
(3) their flexibility in study design (Ricotta et al. 2023; Paneth and Joyner 2021).

Although observational studies are more likely to suffer from methodological shortcomings, especially the potential for confounding, and the findings may be more difficult to interpret by the public compared to RCTs, properly done observational studies provide critical information, particularly during a public health crisis. Additionally, observational studies often produce more generalizable results, not having many of the enrollment constraints faced by RCTs. This chapter considers the totality of available clinical evidence—including that generated by RCTs and observational studies—for the analysis of the convalescent plasma therapeutic strategy used during the COVID-19 pandemic.

The range of studies conducted during the COVID-19 pandemic included RCTs and several types of observational studies including propensity-matched cohorts, real-world large-scale studies using electronic medical records or vital data, uncontrolled case series, dose–response natural experiments, and evidence emerging from animal studies. While in RCTs, treatment is assigned randomly, in observational studies of clinical interventions the decision to treat or not is made by clinicians based on the perception of potential risks and benefits. The general strategy of presentation for this chapter includes eleven sections, beginning with this introduction followed by six sections categorically describing the evidence-base of COVID-19 convalescent plasma in the approximate order of the hierarchy of evidence. In the following sections, we discuss historical precedents, biological plausibility, and observational research on COVID-19 convalescent plasma, before a conclusion section. The general and specific concepts discussed in this chapter highlight the importance of different levels of clinical research and the importance of biological plausibility in clinical study design, particularly in the context of convalescent plasma.

2 Meta-Analyses of COVID-19 Convalescent Plasma Efficacy

Data from twelve studies of COVID efficacy were aggregated as early as July 2020 (Joyner et al. 2020) and published via a preprint server to inform the design of RCTs and to draw inferences about convalescent plasma efficacy, inform contemporaneous public policy, and support clinical treatments. This synthesis of emerging clinical data aimed to provide the first iteration of a living systematic review and to provide a pragmatic aggregation of available data during an ongoing public health crisis. Publications from this continuously updated meta-analysis expanded to include 32 studies in May 2021 (Klassen et al. 2021) and 109 studies in October 2023 (Senefeld et al. 2023) highlighting the value of preprints (Nelson et al. 2022) and iterations of meta-analyses during a public health crisis.

Our most recent meta-analysis (Senefeld et al. 2023) identified 39 RCTs and 70 matched cohorts or other observational studies investigating the clinical benefits of treating patients with COVID-19 convalescent plasma. The 39 RCTs (11,303 treated and 10,226 untreated patients) revealed a 13% reduction in mortality risk conferred by treatment with convalescent plasma (Fig. 2), while the 70 observational studies that included mortality as a primary outcome (14,500 treated and over 35,000 untreated) found a 23% reduction in mortality risk. Thus, a mortality benefit of COVID-19 convalescent plasma emerged from both observational studies and RCTs.

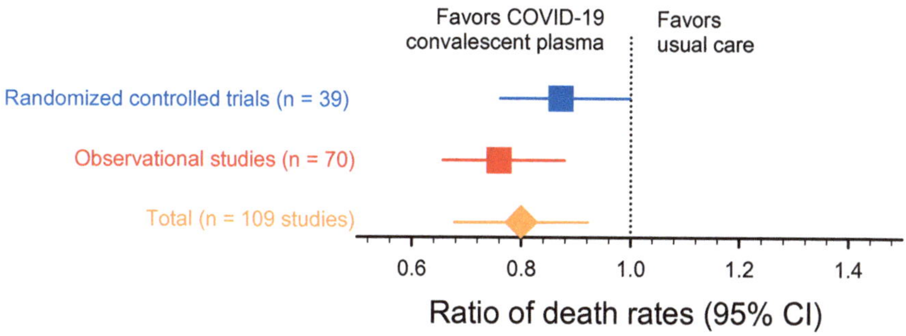

Fig. 2 Summarized forest plot of mortality among clinical studies investigating COVID-19 convalescent plasma efficacy. Symbols represent pooled estimates of randomized controlled trials (RCTs; blue square), observational studies (red square), and pooled estimate of both study types (orange diamond). Horizontal lines indicate the 95% confidence interval of odds ratio (OR). Adapted, with permission, from a contemporary meta-analysis (Senefeld et al. 2023)

3 Randomized Controlled Trials of COVID-19 Convalescent Plasma

Enormous efforts were made across the world to understand the potential efficacy of COVID-19 convalescent plasma, as illustrated by the 39 published RCTs conducted less than three years from the onset of the pandemic (Senefeld et al. 2023; Casadevall et al. 2023). Findings from individual RCTs were heterogenous and many RCTs revealed no statistically significant improvement in mortality rates among patients treated with convalescent plasma—including several of the large and highly influential RCTs such as the CONCOR (Bégin et al. 2021), RECOVERY (RECOVERY Collaborative Group 2021), REMAP-CAP (Estcourt et al. 2021), and TSUNAMI (Menichetti et al. 2021) trials. As one example, the largest RCT—RECOVERY (RECOVERY Collaborative Group 2021)—observed similar mortality rates of 24% in the convalescent plasma group (1,399 of 5,795 patients) and 24% in the usual care group (1,408 of 5,763 patients) and concluded that convalescent plasma did not improve survival among patients hospitalized with COVID-19. Careful examination of the RECOVERY trial reveals that convalescent plasma conferred a mortality benefit if transfused within 7 days of symptom onset (rate ratio, 0.92; 95% confidence interval, 0.82 to 1.03) and among patients who were not receiving supplemental oxygen therapy (rate ratio, 0.83; 95% confidence interval, 0.58 to 1.18). Despite these encouraging efficacy signals, on average, only 7% of transfused patients were not receiving supplemental oxygen therapy (442 of 5795 patients) and just 38% of patients were transfused within 7 days of symptom onset (2226 of 5790 patients). Thus, the largest trial did not treat participants early in the disease course and consequently did not provide optimal treatment according to the principles of antibody therapy. By contrast, a multicenter RCT from Spain (Avendaño-Solá et al. 2021) enrolled only patients who could be treated early in the disease course—within 7 days of symptom onset and not receiving high-flow oxygen support or mechanical ventilation. This trial treated 350 patients (179 in convalescent plasma group and 171 in usual care group) and concluded that transfusion of convalescent plasma can improve survival (hazard ratio, 0.46; 95% confidence interval, 0.19 to 1.14) and mitigate escalation of oxygen support (a proxy for disease progression).

In aggregate, the available RCT evidence-base supports that COVID-19 convalescent plasma confers a mortality benefit. However, clinical use of convalescent plasma dropped after reports of negative RCTs, particularly the large, influential trials (Casadevall et al. 2023). This reduced use of convalescent plasma was not based on the best available evidence-base and may have been even associated with excess COVID-19 deaths (Casadevall et al. 2021). Although most RCTs did not conduct prespecified hypothesis-testing analyses of subgroups based on categorical factors (e.g., early vs. late treatment, higher vs. lower titer plasma), data associated with these subgroups are often presented in data tables. Careful examination of optimal use cases embedded within these large RCTs reveals that the mortality benefit of convalescent plasma is most apparent in patients who received plasma transfusions containing high levels of antibodies early in the disease course. In this context, RCTs provide evidence supporting a mortality benefit of convalescent plasma among COVID-19 patients, particularly when plasma is transfused in accordance with optimized biological plausibility by following the principles of antibody therapy.

4 Propensity-Matched Cohort Studies of COVID-19 Convalescent Plasma Efficacy

Perhaps the strongest observational evidence regarding COVID-19 convalescent plasma arises from studies that employed propensity-score matching to evaluate the clinical effectiveness among patients transfused with convalescent plasma. We will highlight two such large studies from the United States.

The first example is work by Musser, Salazar, and colleagues (Salazar et al. 2021) at the Houston Methodist hospital system (Salazar et al. 2020a, b, 2021). To adjust for confounding factors, Musser and colleagues (Salazar et al. 2021) used a rigorous propensity analysis that matched transfused patients to non-transfused patients on age, sex, body mass index, presence of chronic diseases (e.g., diabetes, hypertension, etc.), ventilation status, and use of other COVID-19 therapies (e.g., steroids, remdesivir, hydroxychloroquine, etc.). Sixty-day mortality was reduced in the 341 transfused patients compared to 594 propensity-score matched controls (6.2% vs. 12.4%) (Fig. 3) (Salazar et al. 2021). A large mortality benefit of COVID-19 convalescent plasma was observed in patients who were transfused with high-titer plasma (anti-receptor binding domain IgG titer $\geq$ 1:1350) within 72 h of admission (convalescent plasma 60-day mortality: 5.7% vs. matched controls

60-day mortality: 10.7%, a reduction of 47%) (Salazar et al. 2021). Conversely, convalescent plasma transfusion was not associated with 60-day mortality reductions in intubated patients, in patients receiving low-titer plasma, and in patients transfused after 72 h of hospital admission (Salazar et al. 2021). Thus, these data highlight the importance of transfusing high-titer COVID-19 convalescent plasma according to the accepted principles of passive antibody therapy.

The second example, from Liu, Bouvier, and colleagues (Liu et al. 2020) at Mount Sinai Hospital in New York City, examined the efficacy of convalescent plasma very early in the COVID-19 pandemic, among patients admitted by early April 2020. Liu et al. (2020) used propensity-score matching (similar to that used by Musser and colleagues) to investigate the effect of convalescent plasma on oxygen requirements at day 14 after transfusion and 30-day survival in 39 transfused patients and 156 matched controls. Fourteen days after transfusion, patients transfused with convalescent plasma were less likely to require oxygen compared to controls (18% vs. 28%, respectively) and experienced lower 30-day mortality rates (13% vs. 24%, respectively; Adjusted hazard ratio: 0.34, 95% confidence interval 0.13–0.89). As in the study by Salazar et al. trends emerged pointing to the effective use of convalescent plasma with the strongest effects noted in non-intubated patients and recipients of transfusion within 7 days of symptom onset.

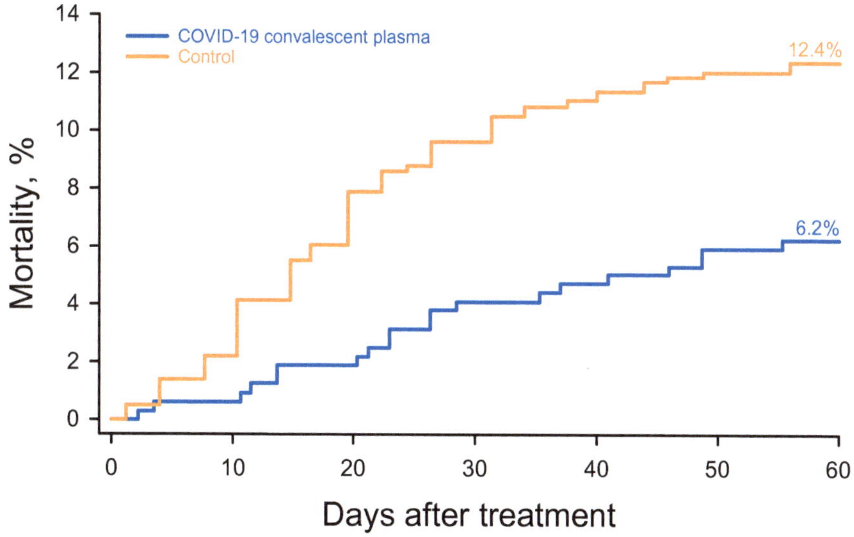

Fig. 3 Kaplan–Meier curves of 60-day mortality among patients transfused with COVID-19 convalescent plasma compared to controls. Line charts displaying the cumulative all-cause mortality up to 60 days after treatment or hospital admission among patients transfused with COVID-19 convalescent plasma (blue lines) and controls (orange lines). The cumulative 60-day mortality rate was greater among controls (12.4%) compared to patients transfused with COVID-19 convalescent plasma (6.2%) Adapted, with permission, from Salazar and colleagues (Salazar et al. 2021)

5 Real-World Large-Scale Studies Using Electronic Medical Records or Vital Data to Investigate COVID-19 Convalescent Plasma Effectiveness

Several analyses leveraged large population databases and healthcare records to draw inferences regarding the mortality and symptomatology benefit of COVID-19 convalescent plasma in humans. This section will focus on two such analyses.

To investigate real-world signals of efficacy, Casadevall and colleagues (Casadevall et al. 2021) relied on correlations between time-varying convalescent plasma distribution by blood banking organizations and hospital mortality data in the first years of the pandemic. The overall premise was that a real-world signal of efficacy could be derived based on linking data on hospital admissions, plasma use, and mortality. Using data from blood banking organizations, Casadevall et al. identified that convalescent plasma usage per hospital admission peaked in Fall 2020, with more than 40% of inpatients estimated to have received convalescent plasma between late September and early November 2020 (Casadevall et al. 2021). However, after RCT data emerged suggesting no mortality benefit, convalescent plasma usage declined steadily to less than about 10% of admissions by March 2021. Casadevall et al. found a strong inverse correlation ($r = -0.52$, $P = 0.002$) between weekly convalescent plasma usage per hospital admission and deaths in hospitalized patients occurring two weeks after admission (Fig. 4) (Casadevall et al. 2021). Changes in the number of hospital admissions, SARS-CoV-2 variants, and age of patients could not explain these findings. Based on their analyses, it is postulated that the retreat from convalescent plasma may have been associated with 29,000 or more excess deaths from mid-November 2020 to February 2021.

Egloff and colleagues investigated the association between convalescent plasma use and clinical outcomes among patients hospitalized for COVID-19 at Hospital Corporations of American (HCA) Healthcare-affiliated community hospitals in the United States (Arnold Egloff et al. 2021). This study matched COVID-19 patients

Fig. 4 Correlation between all-cause mortality (deaths per admission) and COVID-19 convalescent plasma doses per admitted patients. The blue line represents a linear model regression with Pearson's correlation coefficient of −0.518 ($p = 0.0024$). Adapted, with permission, from Casadevall and colleagues (Casadevall et al. 2021)

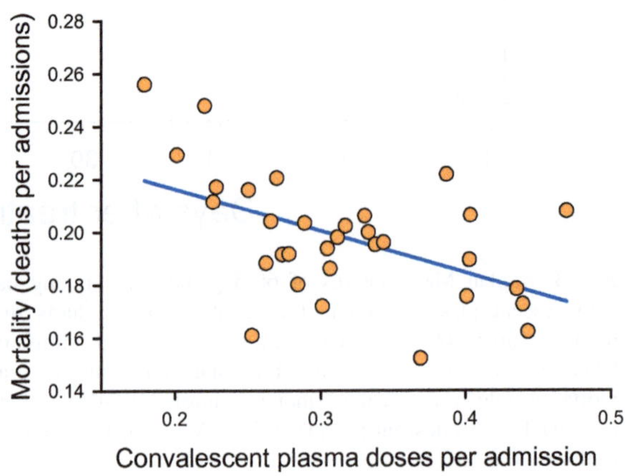

treated with convalescent plasma to non-transfused patients in a multicenter registry using both a daily real-time risk model for COVID-19 risk trajectories and the COVID-19 World Health Organization progression scale. The registry consisted of 44,770 patients admitted to any of the 176 HCA Healthcare-affiliated community hospitals where convalescent plasma was used. Patient matching using either the real-time COVID-19 risk model or the World Health Organization Progression Scale demonstrated a 25–29% lower risk for mortality among patients transfused with convalescent plasma compared to controls. This mortality reduction was observed regardless of intubation status at the time of convalescent plasma transfusion. Moreover, based on patient risk trajectories, patients who received convalescent plasma recovered from COVID-19 more quickly than controls. Time from admission to transfusion of convalescent plasma was significantly associated with risk of mortality, such that convalescent plasma transfusion within 3 days of admission, as contrasted to 4–7 days, was associated with 47% lower risk of mortality risk relative to the comparator group.

Overall, these two very large studies of real-world data from health systems, blood banks, and CDC-derived hospitalization and mortality data support the meta-analytical evidence indicating a mortality benefit of COVID-19 convalescent plasma. Importantly, these real-world data highlight the deleterious population health consequences of convalescent plasma disuse in the United States that were prompted by findings from randomized clinical studies focusing on later treatment of patients with COVID-19.

6 Uncontrolled Case Studies of COVID-19 Convalescent Plasma Effectiveness

Unlike some of the other observational studies described above, case studies included only patients transfused with convalescent plasma, did not match these patients with non-transfused patients, and thus could not control for potential confounding factors in analyses. In many countries including the United States, the regulatory avenues available during the COVID-19 pandemic such as emergency use investigational new drug (eIND) applications, expanded access program (EAP), or emergency use authorization (EUA) enabled continuous and broad access to COVID-19 convalescent plasma. The interested reader may find many single-arm studies (over 300 case reports or case series) that evaluated the risk of death from COVID-19 after transfusion with convalescent plasma (Senefeld et al. 2023). To date, case series and case reports have documented convalescent plasma transfusions to over 135,000 patients with mortality occurring in 33,771 participants ($\sim$25% mortality rate). Evidently, the uncontrolled nature and strong heterogeneity of case series and reports represent important caveats when interpreting these mortality data. One particularly useful study of convalescent plasma recipients alone is the analysis of antibody levels by Joyner et al. (Focosi et al. 2022) described below and in Chapter 7.

7 Animal Challenge Studies of COVID-19 Convalescent Plasma

Invaluable data on the effectiveness of convalescent plasma therapy have emerged from several observational studies of preclinical models including mice (Golden et al. 2021; Zheng et al. 2021), hamsters (Haagmans et al. 2021; Imai et al. 2020), and macaques (McMahan et al. 2021). Convalescent serum from a group of SARS-CoV-2 infected hamsters exerted an antiviral effect when administered to newly infected hamsters (Imai et al. 2020) and also was effective in preventing disease (Haagmans et al. 2021). Likewise, in green monkeys infected with COVID-19, transfusion of convalescent plasma from the same species improved lung pathology, reduced viral shedding, and reduced markers of inflammation (Cross et al. 2021). When human convalescent plasma was transfused in mice expressing the human ACE2 receptor it prevented SARS-CoV-2 infection (Golden et al. 2021; Zheng et al. 2021). Overall, observational studies in animal models confirmed that viral neutralization was likely the primary biological mechanism of action of convalescent plasma. These data emerged early in the COVID-19 pandemic and highlighted that convalescent plasma contained SARS-CoV-2 virus-neutralizing antibodies that could protect against infection, slow disease progression, and improve outcomes in animal models. These animal experiments provided key insights into the biological plausibility of COVID-19 convalescent plasma treatment and augmented the rationale for supporting ongoing efforts to maintain broad access to convalescent plasma throughout the pandemic.

8 The Historical Precedent for Convalescent Plasma Was Formed by Observational Data

The extensive use of convalescent plasma in the COVID-19 pandemic, including over 500,000 units of convalescent plasma transfused in the United States alone, was supported by strong signals of efficacy and well-understood principles of passive immune therapy that arose from earlier research. As reviewed in Chapter 5 by Paneth et al., convalescent plasma therapy has been used since the late 1800s to treat infectious diseases, with a sharp decline in use related to the advent of antibiotics in the 1930s (Kaufmann 2017). For the discovery of antitoxin, von Behring was awarded the first Nobel Prize in Physiology or Medicine in 1901 (Kaufmann 2017). Since those early days, convalescent plasma therapy has been studied in an observational fashion for an array of bacterial and viral diseases (e.g., H1N1 influenza, H5N1 influenza, SARS-CoV-1, MERS-CoV, Ebola virus) (Ripoll et al. 2021). For example, strong observational evidence for convalescent plasma therapy arose from the H5N1 influenza pandemic of 1918–1920. As detailed by Luke and colleagues, in seven observational studies including over 1,700 patients, convalescent plasma was associated with lower H5N1 influenza mortality compared to non-transfused patients (16% vs. 37% mortality, respectively), with early treatment, within 3 days of illness onset, conferring a strong survival benefit (Luke et al. 2006). In addition, as reviewed by Paneth et al. (Chapter 5), antiserums and antitoxins against diseases as varied as diphtheria, plague, pneumococcal pneumonia, and Argentine Hemorrhagic Fever were assessed, and found effective, through randomized or alternate assignment trials. These fundamental data informed academic research, clinical practice, and policy for COVID-19 convalescent plasma at the onset of the pandemic (Casadevall and Pirofski 2020).

9 COVID-19 Convalescent Plasma Efficacy and Principles of Antibody Therapy

While the forest plot in Fig. 2 depicts a pooled estimate indicating a mortality benefit of COVID-19 convalescent plasma therapy, individual studies contributing to this forest plot were very heterogeneous. Under pandemic circumstances, several features of the therapeutic approach for convalescent plasma transfusion were not standardized, leading to substantial biological heterogeneity in several studies, especially in three key areas:

(1) ***SARS-CoV-2 variant match***, i.e., the relative match between the prevailing SARS-CoV-2 variant and the anti-SARS-CoV-2 antibody within the transfused plasma,
(2) ***Antibody titers***, i.e., the anti-SARS-CoV-2 antibody levels within transfused plasma, and
(3) ***Transfusing timing***, i.e., the timing of the plasma transfusion relative to disease course (COVID-19 onset, progression, and symptomatology) (Casadevall et al. 2023).

These fundamental principles of passive immune therapy—match, dosage, and timing—were affected by numerous factors during the COVID-19 pandemic, including but not limited to, the availability of antibody assays, the geographical location of available convalescent plasma donors (Kunze et al. 2021), and the capacity and focus of the health system during the study period.

Variant match. Contextual challenges associated with the COVID-19 pandemic complicated the study of the relation between SARS-CoV-2 variant match and the efficacy of convalescent plasma. Adaptive mutations of the SARS-CoV-2 viral genome altered the virus's pathogenic potential, and the genetic evolution of SARS-CoV-2 generated several variants of concern (VOCs) in different times and locations throughout the COVID-19 pandemic (Carabelli et al. 2023). Variants of concern independently emerged (e.g., Alpha, Beta, Gamma, Delta, and Omicron VOCs) and there were changes in intrinsic functional properties of the virus which conferred changes to virus antigenicity and ability to evade immune responses. Because COVID-19 diagnostic tests were designed to be produced at-scale with low costs, these clinical tests generally can detect all known SARS-CoV-2 variants but cannot identify the specific variant. Like COVID-19 diagnostic testing, serologic testing of convalescent plasma generally quantified antibodies without identification of the SARS-CoV-2 variant. In this framework, there were limited studies that assessed the relation between SARS-CoV-2 variant and the efficacy of convalescent plasma. Although this principle of antibody therapy (variant match) was often disregarded in study design during the COVID-19 pandemic, there is a complex interplay between virus biology and human immunity, and this principle should not be forgotten. Importantly, the polyclonal nature of convalescent plasma represents an important strength over monoclonal antibody therapies. Indeed, due to the rapid evolution of the SARS-CoV-2 and mutations associated with the Omicron VOC, by January 2023 all FDA-approved monoclonal antibodies had lost their therapeutic efficacy (Focosi et al. 2022).

Antibody Titers. During the COVID-19 pandemic, several groups found strong evidence for a dose–response effect conferred by convalescent plasma, whereby higher antibody levels reduced mortality, but lower antibody levels did not (Joyner et al. 2021; Belov et al. 2022; Libster et al. 2021; Maor et al. 2020; Senefeld et al. 2021). Generally, COVID-19 convalescent plasma units were qualified for clinical use based on broad cut points of anti-SARS-CoV-2 antibody titers, but the precise level of antibody titers was unknown until high-resolution testing on small aliquots of the infused plasma after patient treatment. Several dose–response studies leveraged this in-effect random allocation of convalescent plasma by antibody level to hospitalized patients with COVID-19. Joyner et al. examined the dose–response relationship between antibody titer level and 30-day mortality among 3,082 transfused patients via the United States Expanded Access Program for convalescent plasma (Joyner et al. 2021). In non-intubated patients, transfusion of higher titer plasma was associated with lower 30-day mortality than transfusion with lower antibody titers (22 vs. 30%) (Joyner et al. 2021). Indeed, the fifteen observational studies (over 29,000 patients) that explored the link between antibody levels in transfused plasma and patient mortality found that transfusion of plasma with higher antibody levels was associated with 15% lower mortality than when plasma had lower antibody levels (risk ratio, 0.85; 95% confidence interval, 0.73 to 0.99) (Fig. 5) (Senefeld et al. 2023).

Therapeutic Timing. As described in the multicenter Spanish RCT (Avendaño-Solá et al. 2021), the mortality benefit of convalescent plasma is generally more apparent among patients treated earlier in the disease course. The United States Expanded Access Program concluded that convalescent plasma conferred a survival benefit only in patients transfused earlier in the disease course, as this association was not observed in patients who were mechanically ventilated at the time of transfusion (Joyner et al. 2021). Similarly, in the Argentine Expanded Access Program experience (González et al. 2022), death within 28 days after COVID-19 convalescent plasma transfusion occurred in 563 of 3,113 patients (18.1%) in patients transfused within 3 days of hospitalization and in 88 of 226 patients (38.9%) transfused more than 7 days after hospitalization. It is of interest, however, that a multicenter RCT

Source	Deaths/patients treated (%) High Antibody	Deaths/patients treated (%) Low Antibody	Ratio of death rates, OR (95% CI)	Weight
Maor et al, 2020	7/30 (23.3%)	2/19 (11%)	2.31 (0.53-10.03)	1.07%
Fazeli et al, 2021	351/1147 (31%)	244/850 (29%)	1.10 (0.90-1.33)	20.26%
Cain et al, 2022	11/48 (23%)	11/48 (23%)	1.00 (0.39-2.58)	2.46%
González et al, 2022	344/1536 (23%)	218/915 (24%)	0.92 (0.76-1.12)	20.21%
Belov et al, 2022	3123/13636 (23%)	2314/9482 (24%)	0.92 (0.87-0.98)	28.31%
Gachoud et al, 2022	1/19 (5%)	1/17 (6%)	0.89 (0.05-14.91)	0.30%
Khan et al, 2021	5/33 (13%)	3/17 (18%)	0.83 (0.17-4.05)	0.93%
Joyner et al, 2021	115/515 (22%)	166/561 (30%)	0.69 (0.52-0.90)	15.50%
Sanz et al, 2022	7/127 (5%)	12/134 (9%)	0.60 (0.24-1.53)	2.53%
Bartelt et al, 2022	2/14 (14%)	11/41 (27%)	0.51 (0.12-2.08)	1.15%
Franchini et al, 2022	42/281 (15%)	9/124 (7%)	0.50 (0.26-0.94)	4.95%
Romon et al, 2021	2/21 (10%)	4/20 (20%)	0.44 (0.08-2.44)	0.80%
Mendoza et al, 2021	7/19 (38%)	7/9 (78%)	0.21 (0.04-0.98)	0.96%
Hemsinliogu et al, 2021	0/7 (0%)	5/9 (56%)	0.09 (0.01-0.70)	0.55%
Total	3984/17280 (23.1%)	3040/12411 (24.5%)	0.85 (0.73-0.99)	

Heterogeneity: $\tau^2 = 0.021$; $\chi^2 = 23.26$ (df=13; P=.04); I^2=44%

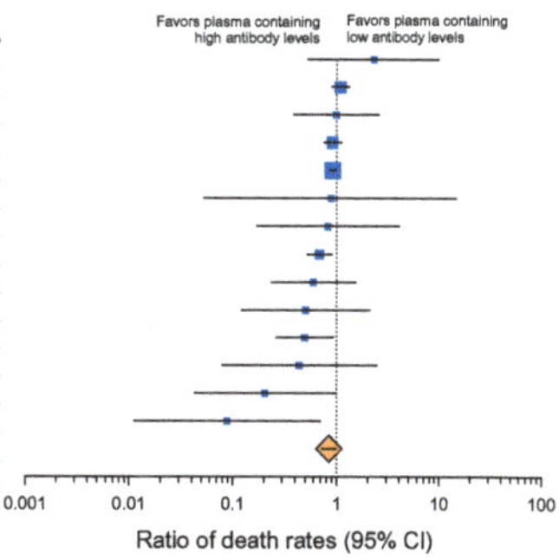

Fig. 5 Forest plot of mortality among clinical studies investigating COVID-19 convalescent plasma containing higher compared with lower anti-SARS-CoV-2 antibodies. Blue squares represent individual studies and symbol size represents relative weights used in meta-analysis proportional to both study size and study variance. Orange diamond represents the pooled estimate from the meta-analysis. Horizontal lines indicate the 95% confidence interval of odds ratio (OR) estimate. Adapted, with permission, from a contemporary meta-analysis (Senefeld et al. 2023)

from Belgium (Misset et al. 2023) found a mortality benefit of convalescent plasma with high levels of neutralizing antibodies among patients with COVID-19-induced acute respiratory distress syndrome (ARDS) who were receiving mechanical ventilation for less than five days—35.4% mortality in convalescent plasma group and 45.0% mortality in usual care group. This unexpected finding with late treatment suggests that some patient populations may benefit from later convalescent plasma treatment (Senefeld et al. 2024) and calls for a better understanding of the mechanisms of efficacy of convalescent plasma (Casadevall et al. 2024).

Contemporary meta-analyses explored the impact of the timing of COVID-19 convalescent plasma on hospitalized patient mortality (Klassen et al. 2021; Senefeld et al. 2023). Timing of convalescent plasma transfusion relative to COVID-19 disease course was assessed in 26 studies with over 18,000 participants, 25 of which were observational in design. Earlier use of plasma in the disease course was associated with 37% lower mortality than later use (risk ratio: 0.63, 95% confidence interval: 0.48–0.82) (Fig. 6) (Senefeld et al. 2023).

One caveat to the interpretation of these meta-analyses and subgroup analyses is that they relied on the study-level categorization of convalescent plasma therapy—higher vs. lower antibody levels and earlier vs. later transfusion timing. Throughout the COVID-19 pandemic, there were several iterations of regulatory guidance on assay systems used to measure specific SARS-CoV-2 antibodies which led to the use of many at-times discordant assay systems (Larkey et al. 2022). Although quantifying the timing of convalescent plasma transfusion seems more straightforward, studies show no clear consensus or standardization of the precise meaning of "day 0" in the disease course. Examples of "day 0" include the date of viral exposure, date of first positive test, date of hospital admission, and date of intensive

Source	Deaths/patients treated (%) Early group	Late group	Ratio of death rates, OR (95% CI)	Weight
Hueso et al, 2022	16/29 (55%)	23/83 (28%)	3.32 (1.37-8.03)	3.86%
Balcells et al, 2021	5/28 (18%)	2/30 (7%)	2.82 (0.59-13.51)	2.02%
RECOVERY, 2021	606/2226 (27%)	790/3564 (22%)	1.32 (1.17-1.49)	6.60%
Sajmi et al, 2021	10/22 (45%)	6/15 (40%)	1.24 (0.34-4.89)	2.56%
Kocayigit et al, 2021	43/84 (51%)	29/57 (51%)	1.01 (0.52-1.98)	4.70%
Akay Cizmecioglu et al, 2021	6/22 (19%)	4/18 (22%)	0.81 (0.19-3.38)	2.29%
Briggs et al, 2021	30/80 (38%)	32/71 (45%)	0.73 (0.38-1.40)	4.80%
Salazar E et al, 2021	14/244 (6%)	6/77 (8%)	0.70 (0.24-2.03)	3.26%
Cain et al, 2022	14/67 (21%)	8/29 (28%)	0.69 (0.25-1.93)	3.34%
Lattanzio et al, 2021	18/92 (20%)	28/105 (27%)	0.67 (0.35-1.30)	4.75%
Mahapatra et al, 2021	214/1039 (21%)	42/150 (28%)	0.65 (0.43-0.98)	5.76%
Fazeli et al, 2022	386/1484 (26%)	410/1135 (36%)	0.62 (0.52-0.73)	6.52%
Gazitua et al, 2020	18/129 (14%)	13/63 (21%)	0.61 (0.27-1.39)	4.11%
Bajpai et al, 2022	20/115 (17%)	22/85 (26%)	0.60 (0.30-1.19)	4.64%
Moniuszko-Malinowska et al, 2020	6/55 (11%)	4/23 (17%)	0.56 (0.13-2.40)	2.25%
Janaka et al, 2021	5/35 (14%)	2/9 (22%)	0.56 (0.08-4.06)	1.43%
Fodor et al, 2021	24/101 (24%)	34/88 (39%)	0.50 (0.27-0.93)	4.92%
Romon et al, 2021	2/20 (10%)	4/21 (19%)	0.49 (0.09-2.73)	1.78%
Asem etal, 2022	27/61 (44%)	21/33 (64%)	0.46 (0.20-1.08)	4.01%
González et al, 2022	563/3113 (18%)	508/1606 (32%)	0.46 (0.40-0.53)	6.57%
De Silvestro et al, 2022	122/1112 (11%)	85/387 (22%)	0.40 (0.28-0.56)	6.06%
Ateş et al, 2021	2/61 (3%)	5/50 (10%)	0.32 (0.07-1.50)	2.08%
Franchini et al, 2022	5/131 (4%)	46/274 (17%)	0.31 (0.17-0.58)	4.89%
Greenbaum et al, 2021	1/13 (8%)	11/31 (35%)	0.25 (0.06-1.07)	2.27%
Jeyaraman et al, 2021	9/25 (36%)	6/8 (75%)	0.22 (0.05-1.05)	2.01%
Ibrahim et al, 2020	2/16 (13%)	12/22 (55%)	0.17 (0.05-0.64)	2.54%
Total	2168/10414 (20.8%)	2153/8034 (26.8%)	0.63 (0.48-0.82)	
Heterogeneity: $\tau^2 = 0.276$; $\chi^2 = 179.57$ (df=25; $P<.01$); $I^2=86\%$				

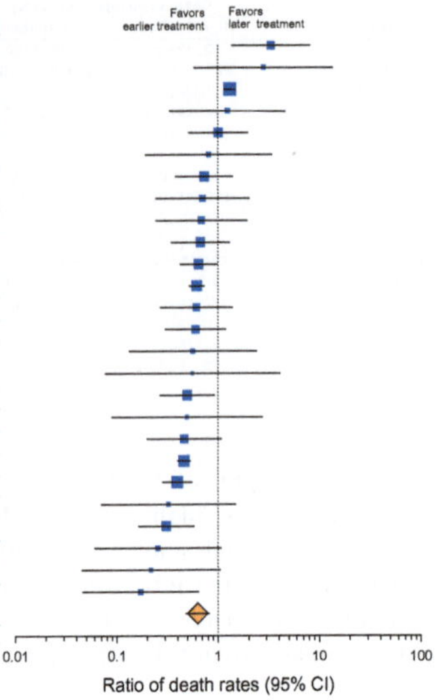

Fig. 6 Forest plot of mortality among clinical studies investigating earlier compared with later transfusion of COVID-19 convalescent plasma. Blue squares represent individual studies and symbol size represents relative weights used in meta-analysis proportional to both study size and study variance. Orange diamond represents the pooled estimate from the meta-analysis. Horizontal lines indicate the 95% confidence interval of odds ratio (OR) estimate. Adapted, with permission, from a contemporary meta-analysis (Senefeld et al. 2023).

care unit (ICU) admission, making it difficult to assess the timing of plasma transfusion. Nonetheless, the data highlight two core principles of antibody therapy—convalescent plasma must contain sufficiently high antibody levels and be transfused early in the disease course (Klassen et al. 2021).

10 Strengths, Limitations, and Caveats of Observational Research in the Context of COVID-19 Convalescent Plasma

In the context of convalescent plasma therapy for COVID-19, there are several key strengths of observational studies that warrant discussion. ***First***, observational studies are fundamentally flexible. Although overall there was an observed mortality benefit associated with COVID-19 convalescent plasma when all RCTs were amalgamated (Senefeld et al. 2023), a review of results from several large and influential RCTs (such as RECOVERY, REMAP-CAP, and CONCOR) showed no overall reduction in mortality among patients transfused with convalescent plasma compared to controls (Bégin et al. 2021; RECOVERY Collaborative Group 2021; Estcourt et al. 2021; Menichetti et al. 2021; Agarwal et al. 2020; Simonovich et al.

2021; Hamilton et al. 2021; Franchini et al. 2022; Ortigoza et al. 2022; Casadevall and Henderson 2022; Self et al. 2022; Senefeld et al. 2022). As reviewed elsewhere (Casadevall et al. 2023), the rigid design and difficult logistics associated with RCTs contributed to suboptimal treatment strategies in many instances, specific examples include studies in which:

(1) One-third of transfused convalescent plasma units contained no neutralizing antibody (Agarwal et al. 2020),
(2) Nine percent of patients allocated to the convalescent plasma group did not actually receive convalescent plasma (RECOVERY Collaborative Group 2021), and
(3) One-third of patients were already on mechanical ventilation (Estcourt et al. 2021).

While some RCTs found no effect of convalescent plasma in suboptimal use conditions and the rigidity of RCT reporting typically does not enable exploratory analyses, secondary analysis of data from RCTs was used to test important hypotheses about early transfusion of high-titer COVID-19 convalescent plasma (analyses which were not part of registered study outcomes in ongoing randomized clinical trials). ***Second***, observational studies can be executed rapidly. Case series, case reports, and small matched studies of convalescent plasma therapy emerged early in the pandemic, providing important signals of safety and efficacy. Flexibility and rapid study implementation were important factors in situations when there were new developments in convalescent plasma that required study, such as convalescent plasma from vaccinated donors. ***Third***, carefully constructed observational studies can inform the design of future randomized controlled trials. ***Fourth***, the criticism of large registries such as the convalescent plasma Expanded Access Program and Emergency Use Authorization is that they may have interfered with the enrollment of RCTs of convalescent plasma during the pandemic. However, the COVID-19 experience has shown that large RCTs and observational studies could be successfully completed in parallel and that many RCTs were conducted after these authorizations (Joyner et al. 2023).

There are some caveats and considerations that inform the interpretation of findings from observational studies of convalescent plasma. Observational studies are ranked lower than RCTs on the epistemological hierarchy due to the potential for several types of bias or confounding associated with the non-random distribution of confounders between groups (or lack of comparison group altogether) and selection bias associated with group construction, among other risks. Indeed, careful inspection of the 70 observational studies evaluating the efficacy of COVID-19 convalescent plasma reveals substantial heterogeneity in the mortality benefit associated with convalescent plasma (Senefeld et al. 2023). All studies, including both RCTs and observational studies, must be examined carefully before drawing definitive conclusions about causal inference, and signals of efficacy for convalescent plasma arising from observational research should not be dismissed as, unfortunately, it often was during the pandemic.

11 Conclusions

The totality of evidence—including an amalgam of 39 RCTs, 70 observational studies, and several hundred single-arm studies—reveals that convalescent plasma confers a clinically meaningful mortality benefit among hospitalized patients with COVID-19. Importantly, given the large diversity of the included trials and heterogeneity of both methodology and findings, the results must be analyzed and interpreted critically. Lessons learned during the COVID-19 pandemic may inform the optimal use of convalescent plasma in future major infectious disease emergencies and the design of clinical research (both RCTs and observational studies). Although much was learned during COVID-19 pandemic, two key lessons discussed in this chapter include: (1) clinical research should be designed to assess the efficacy of convalescent plasma with the optimal use case following basic principles of antibody therapy (earlier treatment with high levels of pathogen-specific antibodies), and (2) a broad spectrum of study designs may be implemented to inform study design early on during an infectious disease emergency, and RCTs should be more widely implemented after the optimal use case is clearly defined. Importantly, the findings from the COVID-19 pandemic support the deployment of convalescent plasma during future infectious disease emergencies until better therapies are available.

Acknowledgements SA Klassen is supported by a Natural Sciences and Engineering Research Council of Canada (NSERC) Discovery Grant (No. 05293), a NSERC Discovery Launch Supplement (No 00320), and Brock University New Investigator Funding.

Conflict of Interest The authors report no conflicts of interest, financial or otherwise.

References

Agarwal A, Mukherjee A, Kumar G, Chatterjee P, Bhatnagar T, Malhotra P et al (2020) Convalescent plasma in the management of moderate covid-19 in adults in India: open label phase II multicentre randomised controlled trial (PLACID Trial). BMJ 22(371):m3939

Arnold Egloff SA, Junglen A, Restivo JS, Wongskhaluang M, Martin C, Doshi P et al (2021) Convalescent plasma associates with reduced mortality and improved clinical trajectory in patients hospitalized with COVID-19. J Clin Invest 131(20):e151788

Avendaño-Solá C, Ramos-Martínez A, Muñez-Rubio E, Ruiz-Antorán B, Malo De Molina R, Torres F et al (2021) A multicenter randomized open-label clinical trial for convalescent plasma in patients hospitalized with COVID-19 pneumonia. J Clin Invest 131(20):e152740

Bégin P, Callum J, Jamula E, Cook R, Heddle NM, Tinmouth A et al (2021) Convalescent plasma for hospitalized patients with COVID-19: an open-label, randomized controlled trial. Nat Med 27(11):2012–2024

Belov A, Huang Y, Villa CH, Whitaker BI, Forshee R, Anderson SA et al (2022) Early administration of COVID-19 convalescent plasma with high titer antibody content by live viral neutralization assay is associated with modest clinical efficacy. Am J Hematol 97(6):770–779

Bingeman E (2016) Evaluating normative epistemic frameworks in medicine: EBM and casuistic medicine. J Eval Clin Pract 22(4):490–495

Bothwell LE, Greene JA, Podolsky SH, Jones DS (2016) Assessing the gold standard-lessons from the history of RCTs. N Engl J Med 374(22):2175–2181
Carabelli AM, Peacock TP, Thorne LG, Harvey WT, Hughes J, COVID-19 Genomics UK Consortium et al (2023) SARS-CoV-2 variant biology: immune escape, transmission and fitness. Nat Rev Microbiol [Internet]. 2023 Jan 18 [cited 2024 Jul 4]; Available from: https://www.nature.com/articles/s41579-022-00841-7
Casadevall A, Henderson JP (2022) COVID-19 convalescent plasma and concomitant therapies in PassITON. Chest 162(5):e284–e285
Casadevall A, Pirofski LA (2020) The convalescent sera option for containing COVID-19. J Clin Invest 130(4):1545–1548
Casadevall A, Joyner MJ, Pirofski LA, Senefeld JW, Shoham S, Sullivan D et al (2023) Convalescent plasma therapy in COVID-19: Unravelling the data using the principles of antibody therapy. Expert Rev Respir Med 17(5):381–395
Casadevall A, Shoham S, Pirofski LA (2024) Convalescent plasma for Covid-19-induced ARDS. N Engl J Med 390(4):385–386
Casadevall A, Dragotakes Q, Johnson PW, Senefeld JW, Klassen SA, Wright RS et al (2021) Convalescent plasma use in the USA was inversely correlated with COVID-19 mortality. eLife 10:e69866
Cross RW, Prasad AN, Borisevich V, Woolsey C, Agans KN, Deer DJ et al (2021) Use of convalescent serum reduces severity of COVID-19 in nonhuman primates. Cell Rep 34 (10):108837
Estcourt LJ, Turgeon AF, McQuilten ZK, McVerry BJ, Al-Beidh F, Writing Committee for the REMAP-CAP Investigators et al (2021) Effect of convalescent plasma on organ support-free days in critically ill patients with COVID-19: a randomized clinical trial. JAMA 326(17):1690–1702
Focosi D, McConnell S, Casadevall A, Cappello E, Valdiserra G, Tuccori M (2022) Monoclonal antibody therapies against SARS-CoV-2. Lancet Infect Dis 22(11):e311–e326
Franchini M, Focosi D, Casadevall A, Joyner MJ, Perotti C (2022) Convalescent plasma for COVID-19. TSUNAMI is not the final word. Eur J Intern Med 97:116–118
Golden JW, Zeng X, Cline CR, Garrison AR, White LE, Fitzpatrick CJ et al (2021) Human convalescent plasma protects K18-hACE2 mice against severe respiratory disease. J Gen Virol 102(5):001599
González SE, Regairaz L, Salazar MR, Ferrando NS, González Martínez VV, Carrera Ramos PM et al (2022) Timing of convalescent plasma administration and 28-day mortality in COVID-19 pneumonia. J Investig Med off Publ Am Fed Clin Res. 70(5):1258–1264
Haagmans BL, Noack D, Okba NMA, Li W, Wang C, Bestebroer T et al (2021) SARS-CoV-2 neutralizing human antibodies protect against lower respiratory tract disease in a hamster model. J Infect Dis 223(12):2020–2028
Hamilton FW, Lee T, Arnold DT, Lilford R, Hemming K (2021) Is convalescent plasma futile in COVID-19? A Bayesian re-analysis of the RECOVERY randomized controlled trial. Int J Infect Dis IJID off Publ Int Soc Infect Dis 109:114–117
Hellman S, Hellman DS (1991) Of mice but not men. Problems of the randomized clinical trial. N Engl J Med 324(22):1585–1589
Imai M, Iwatsuki-Horimoto K, Hatta M, Loeber S, Halfmann PJ, Nakajima N et al (2020) Syrian hamsters as a small animal model for SARS-CoV-2 infection and countermeasure development. Proc Natl Acad Sci U S A 117(28):16587–16595
Joyner MJ, Carter RE, Senefeld JW, Klassen SA, Mills JR, Johnson PW et al (2021) Convalescent plasma antibody levels and the risk of death from Covid-19. N Engl J Med 384(11):1015–1027
Joyner MJ, Carter RE, Fairweather D, Wright RS (2023) Convalescent plasma and COVID-19: time for a second-second look? Transfus Med Oxf Engl. 33(1):16–20
Joyner MJ, Klassen SA, Senefeld JW, Johnson PW, Carter RE, Wiggins CC et al (2020) Evidence favouring the efficacy of convalescent plasma for COVID-19 therapy [Internet]. [cited 2024 May 14]. Available from: https://doi.org/10.1101/2020.07.29.20162917

Kaufmann SHE (2017) Remembering Emil von Behring: from tetanus treatment to antibody cooperation with phagocytes. mBio 8(1):e00117–17

Klassen SA, Senefeld JW, Johnson PW, Carter RE, Wiggins CC, Shoham S et al (2021) The effect of convalescent plasma therapy on mortality among patients with COVID-19: systematic review and meta-analysis. Mayo Clin Proc 96(5):1262–1275

Klassen SA, Senefeld JW, Senese KA, Johnson PW, Wiggins CC, Baker SE et al (2021) Convalescent plasma therapy for COVID-19: a graphical mosaic of the worldwide evidence. Front Med 8:684151

Kunze KL, Johnson PW, van Helmond N, Senefeld JW, Petersen MM, Klassen SA et al (2021) Mortality in individuals treated with COVID-19 convalescent plasma varies with the geographic provenance of donors. Nat Commun 12(1):4864

Larkey NE, Ewaisha R, Lasho MA, Roforth MM, Granger D, Jerde CR et al (2022) Limited correlation between SARS-CoV-2 serologic assays for identification of high-titer COVID-19 convalescent plasma using FDA thresholds. Microbiol Spectr 10(4):e0115422

Libster R, Pérez Marc G, Wappner D, Coviello S, Bianchi A, Braem V et al (2021) Early high-titer plasma therapy to prevent severe Covid-19 in older adults. N Engl J Med 384(7):610–618

Liu STH, Lin HM, Baine I, Wajnberg A, Gumprecht JP, Rahman F et al (2020) Convalescent plasma treatment of severe COVID-19: a propensity score-matched control study. Nat Med 26 (11):1708–1713

Luke TC, Kilbane EM, Jackson JL, Hoffman SL (2006) Meta-analysis: convalescent blood products for Spanish influenza pneumonia: a future H5N1 treatment? Ann Intern Med 145 (8):599–609

Maor Y, Cohen D, Paran N, Israely T, Ezra V, Axelrod O et al (2020) Compassionate use of convalescent plasma for treatment of moderate and severe pneumonia in COVID-19 patients and association with IgG antibody levels in donated plasma. E Clin Med 26:100525

McMahan K, Yu J, Mercado NB, Loos C, Tostanoski LH, Chandrashekar A et al (2021) Correlates of protection against SARS-CoV-2 in rhesus macaques. Nature 590(7847):630–634

Menichetti F, Popoli P, Puopolo M, Spila Alegiani S, Tiseo G, Bartoloni A et al (2021) Effect of high-titer convalescent plasma on progression to severe respiratory failure or death in hospitalized patients with COVID-19 pneumonia: a randomized clinical trial. JAMA Netw Open 4(11):e2136246

Misset B, Piagnerelli M, Hoste E, Dardenne N, Grimaldi D, Michaux I et al (2023) Convalescent plasma for Covid-19-induced ARDS in mechanically ventilated patients. N Engl J Med 389 (17):1590–1600

Nelson L, Ye H, Schwenn A, Lee S, Arabi S, Hutchins BI (2022) Robustness of evidence reported in preprints during peer review. Lancet Glob Health 10(11):e1684–e1687

Ortigoza MB, Yoon H, Goldfeld KS, Troxel AB, Daily JP, Wu Y et al (2022) Efficacy and safety of COVID-19 convalescent plasma in hospitalized patients: a randomized clinical trial. JAMA Intern Med 182(2):115–126

Paneth NS, Joyner MJ, Casadevall A (2022) The fossilization of randomized clinical trials. J Clin Invest 132(4):e158499

Paneth N, Joyner M (2021) The use of observational research to inform clinical practice. J Clin Invest 131(2):e146392, 146392

RECOVERY Collaborative Group (2021) Convalescent plasma in patients admitted to hospital with COVID-19 (RECOVERY): a randomised controlled, open-label, platform trial. Lancet Lond Engl. 397(10289):2049–2059

Ricotta EE, Rid A, Cohen IG, Evans NG (2023) Observational studies must be reformed before the next pandemic. Nat Med 29(8):1903–1905

Ripoll JG, van Helmond N, Senefeld JW, Wiggins CC, Klassen SA, Baker SE et al (2021) Convalescent plasma for infectious diseases: historical framework and use in COVID-19. Clin Microbiol Newsl 43(4):23–32

Salazar E, Christensen PA, Graviss EA, Nguyen DT, Castillo B, Chen J et al (2020a) Treatment of coronavirus disease 2019 patients with convalescent plasma reveals a signal of significantly decreased mortality. Am J Pathol 190(11):2290–2303

Salazar E, Perez KK, Ashraf M, Chen J, Castillo B, Christensen PA et al (2020b) Treatment of coronavirus disease 2019 (COVID-19) patients with convalescent plasma. Am J Pathol 190 (8):1680–1690

Salazar E, Christensen PA, Graviss EA, Nguyen DT, Castillo B, Chen J et al (2021) Significantly decreased mortality in a large cohort of coronavirus disease 2019 (COVID-19) patients transfused early with convalescent plasma containing high-titer anti-severe acute respiratory syndrome coronavirus 2 (SARS-CoV-2) spike protein IgG. Am J Pathol 191(1):90–107

Self WH, Wheeler AP, Stewart TG, Schrager H, Mallada J, Thomas CB et al (2022) Neutralizing COVID-19 convalescent plasma in adults hospitalized with COVID-19: a blinded, randomized. Placebo-Controlled Trial Chest 162(5):982–994

Senefeld JW, Johnson PW, Kunze KL, Bloch EM, van Helmond N, Golafshar MA et al (2021) Access to and safety of COVID-19 convalescent plasma in the United States expanded access program: a national registry study. PLoS Med 18(12):e1003872

Senefeld JW, Paneth NS, Carter RE, Wright RS, Fairweather D, Bruno KA et al (2022) Late treatment for COVID-19 with convalescent plasma. Chest 162(5):e283–e284

Senefeld JW, Gorman EK, Johnson PW, Moir ME, Klassen SA, Carter RE et al (2023) Mortality rates among hospitalized patients with COVID-19 treated with convalescent plasma: a systematic review and meta-analysis. Mayo Clin Proc Innov Qual Outcomes 7(5):499–513

Senefeld JW, Henderson JP, Joyner MJ (2024) Convalescent plasma for Covid-19-induced ARDS. N Engl J Med 390(4):385

Simonovich VA, Burgos Pratx LD, Scibona P, Beruto MV, Vallone MG, Vázquez C et al (2021) A randomized trial of convalescent plasma in Covid-19 severe pneumonia. N Engl J Med 384 (7):619–629

Zheng J, Wong LYR, Li K, Verma AK, Ortiz ME, Wohlford-Lenane C et al (2021) COVID-19 treatments and pathogenesis including anosmia in K18-hACE2 mice. Nature 589(7843):603–607

Stephen A. Klassen Ph.D is an Assistant Professor in the Department of Kinesiology at Brock University in Canada. He performed his Doctoral training at Western University in Canada and was a postdoctoral fellow at the Mayo Clinic. A human physiologist by training, his primary research interests relate to neural control of the human cardiovascular system. During the COVID-19 pandemic, Stephen was a member of the Mayo Clinic team that led the US Expanded Access Program for COVID-19 convalescent plasma and maintains research interests in passive immune therapies.

Jonathon W. Senefeld Ph.D. is an Assistant Professor in the Department of Health and Kinesiology at the University of Illinois Urbana-Champaign in the United States. He performed his doctoral training at Marquette University before completing a postdoctoral fellowship at the Mayo Clinic. His research focuses on the limits of human performance with specific interests in fatigue of skeletal muscle, oxygen transport, and sex differences in physiological regulation. Jack was a member of the Mayo Clinic team that led the US Expanded Access Program for COVID-19 for COVID-19 convalescent plasma and has a special interest in passive antibody use in the immunocompromised.

Convalescent Plasma for Immunocompromised Patients

Shmuel Shoham

Contents

Abstract COVID-19 convalescent plasma (CCP) is an important therapeutic option for immunocompromised patients with COVID-19. Such patients are at increased risk for serious complications of infection and may also develop a unique syndrome of persistent infection. This article reviews the rationale for CCP utilization in immunocompromised patients and the evidence for its value in immunosuppressed patients with both acute and persistent COVID-19. Both historical precedence and understanding of the mechanisms of action of antibody treatment support this use, as do several lines of evidence derived from case series, comparative studies, randomized trials, and systematic reviews of the literature. A summary of recommendations from multiple practice guidelines is also provided.

S. Shoham (✉)
Department of Medicine, Johns Hopkins School of Medicine, 1830 East Monument St., Room 447, Baltimore, MD 21205, USA
e-mail: sshoham1@jhmi.edu

Current Topics in Microbiology and Immunology (2025) 443: 119–130
https://doi.org/10.1007/82_2024_272

Published Online: 9 August 2024

Glossary of Abbreviations

AABB	Association for the Advancement of Blood & Biotherapies
CAR-T	Chimeric antigen receptor T cell therapy
CCP	COVID-19 convalescent plasma
CT	Computed tomography
CVID	Common variable immunodeficiency
ECIL	The European Conference on Infections in Leukemia
FDA	United States Food and Drug Administration
IDSA	Infectious Diseases Society of America
mAb	Monoclonal antibody
NCCN	National Comprehensive Cancer Network
PCR	Polymerase chain reaction
SOT	Solid Organ Transplant

1 Background

Immunocompromised patients with COVID-19 have higher rates of hospitalization, longer stays in the hospital, more frequent need for intensive care, and increased mortality (Bahremand et al. 2023; Singson et al. 2022). Additionally, some immunocompromised patients suffer from a syndrome of persistent infection with symptoms of infection and viral replication that can last many weeks to months (Dioverti et al. 2022).

Since the start of the pandemic, COVID-19 convalescent plasma (CCP) has been used as treatment for COVID-19 in immunocompromised patients. That use is supported by multiple lines of evidence including:

(1) Historical precedence—passive immunotherapy has often been used for the prevention and treatment of viral infections in immunocompromised patients (Ghosh et al. 2000; Alexander et al. 2010),
(2) The mechanistic rationale:
    a. naturally occurring neutralizing antibodies are critical for viral clearance and recovery from infection,
    b. treatment with CCP can provide such antibodies via passive transfer, and
    c. such transfer can protect against progression of infection in animal models (Klein et al. 2020; Dispinseri et al. 2021; Shoham et al. 2023a; Imai et al. 2020; Sun et al. 2020),
(3) Observational data from individual patients and small case series providing evidence for the effectiveness of CCP in the immunocompromised patient population (Senefeld et al. 2021),

(4) Comparative efficacy data from non-randomized trials (Thompson et al. 2021),
(5) Results of randomized trials (Denkinger et al. 2022; Lacombe et al. 2022), and
(6) Pooled data from systematic reviews and meta-analyses of clinical trials (Senefeld et al. 2023; Estcourt et al. 2022).

2 Immunocompromised Patients and COVID-19

Immunocompromised patients are a diverse population that makes up about 2–3% of the US population (Harpaz et al. 2016). During the first two years of the pandemic, an estimated 12% of people hospitalized in the US for COVID-19 were immunocompromised, suggesting a four-to-six fold elevation in risk of hospitalization for the immunocompromised (Singson et al. 2022). Conditions such as transplantation, advanced HIV infection, active hematologic malignancies, receipt of cancer therapy, primary immunodeficiencies, and treatment with immunosuppressive medications for autoimmune disease impair the capacity to efficiently clear the SARS-CoV-2 (Shoham et al. 2023b). Additionally, immunocompromised people tend to be older and have additional and sometimes multiple comorbid conditions, which place them at additional risk for complications of COVID-19.

B cell immunosuppression with potentially impaired ability to produce anti-SARS-CoV-2 neutralizing antibodies is found in solid organ transplant recipients, people with hematologic malignancies such as leukemia, lymphoma, and multiple myeloma who in the past year had received active treatment (e.g. chemotherapy, targeted therapies, CAR-T, and immunotherapy), hematopoietic stem cell transplant recipients who are less than two years from transplant or are still receiving immunosuppressive medications for graft vs host disease, people who have received anti-CD20 or B cell depleting agents in the past two years, and people with other severe immune deficiency syndromes affecting lymphocyte function and number (Antinori and Bausch-Jurken 2023).

Vaccine responses among people with immunocompromising conditions tend to be weaker than in non-immunocompromised people, and this has also been the case with COVID-19 vaccines. Multiple studies have demonstrated reduced vaccine efficacy in the immunocompromised population (Yek et al. 2022; Boyarsky et al. 2021; Werbel et al. 2021; Sun et al. 2022; Kwon et al. 2022; Galmiche et al. 2022). Certain immunosuppressive conditions (e.g. solid organ transplantation, hematologic malignancies, receipt of B cell depleting drugs) severely impair immune responses to vaccination (Boyarsky et al. 2021). For example, among SOT recipients, COVID-19 mRNA vaccine effectiveness was 29% for 2 doses and 77% for 3 doses, while vaccine effectiveness in non-immunocompromised people was 88% for 2 doses and 96% for 3 doses (Kwon et al. 2022). In a systematic review of vaccine responses in immunocompromised populations, the proportion of non-responders varied by reason for immune compromise. Poor vaccine responses were somewhat more common among solid organ transplant recipients (range 18–

100%) and patients with hematological malignancy (range 14–61%), than in patients with solid cancers (range 2–36%) and patients on dialysis (range 2–30%) (Galmiche et al. 2022). In that review, additional factors for non-response included older age, use of corticosteroids and receipt of immunosuppressive or anti-CD20 agents.

Even with vaccinations, immunocompromised patients as a whole, and particularly those belonging to the highest-risk groups have a substantially increased risk for hospitalization, admission to intensive care, and death (Bahremand et al. 2023; Singson et al. 2022). The rates of breakthrough infections despite vaccination vary by group. For example, the risk of breakthrough infection for solid organ recipients is nearly twice that of persons immunocompromised by rheumatoid arthritis and its treatments (Sun et al. 2022). The sum of these factors has rendered certain immunocompromised patients extremely vulnerable to complications of COVID-19.

A subset of immunocompromised patients with abnormal lymphocyte number and/or function may develop persistent COVID-19. This group includes solid organ transplant recipients, patients with lymphoid malignancies, recipients of B cell depleting agents such as rituximab, and patients with low immunoglobulin levels as in common variable immunodeficiency (CVID). These patients respond poorly to vaccination and when infected, can develop severe infections with prolonged lengths of hospitalization and persistent illnesses that may not clear for weeks or months (Dioverti et al. 2022; Duléry et al. 2021). In a study of patients receiving anti-CD20 therapy (with subsequent B cell depletion), the median time for the clearance of the virus was 22 days, and 17% of patients developed the relapsed clinical disease after an initial improvement (Calderón-Parra et al. 2022). In a study from Korea, patients with solid organ transplantation or hematologic malignancies shed infectious virus for a median of 4 weeks (Kang et al. 2023).

The clinical significance of ongoing viral shedding varies. For some patients, such as those with persistently positive but very low viral levels (e.g. cycle thresholds >35 on SARS-CoV-2 PCR tests) the ongoing positive tests may represent harmless laboratory-testing artifacts. In others, however, the persistently positive viral tests (and respiratory tract shedding) represent ongoing infection with the capacity to generate new and more infectious/virulent mutants, to serve as a source of ongoing spread of virus and even lead to resistance to existing antiviral therapy (Shoham et al. 2023b; Corey et al. 2021; Fourati et al. 2022; Gandhi et al. 2022; Hogan et al. 2022). Passive immunotherapy has proven to be an important component of multimodal antiviral therapy in patients with ongoing, poorly controlled SARS-CoV-2 infection.

3 Passive Immunotherapy for Immunocompromised Patients with COVID-19

Two approaches for passive immunotherapy have been widely used for COVID-19. Laboratory generated monoclonal antibodies (mAbs) directed against the SARS-CoV-2 spike protein and CCP (Bloch et al. 2023). Antiviral mAb products have been effective in immunocompromised people when used as pre-exposure prophylaxis and as treatment early in the course of infection (Al-Obaidi et al. 2022; Yetmar et al. 2021, 2019). However, the emergence of resistant variants that are resistant to the antiviral action of the mAb has emerged as a major limitation of mAb therapy. Due to mutations at the mAb target sites, all of the authorized agents are no longer reliably effective. By contrast, CCP, which is polyclonal and obtained from people who have been recently infected (and often frequently vaccinated) has the advantage of being able to keep up with changing variants. Plasma obtained from people who have been both vaccinated (even with earlier versions of the vaccine) and recovered from COVID-19 contains high levels of neutralization ability, including against the Omicron variant (Sullivan et al. 2022; Leon et al. 2022).

4 Clinical Evidence for CCP Effectiveness in Immunocompromised Patients

Evidence for the effectiveness of CCP in immunocompromised patients was initially derived from case reports and small observational studies. Dozens of case reports and case series provided early evidence of rapid clinical improvement and a mortality benefit in immunosuppressed patients following CCP transfusion (Senefeld et al. 2021). The largest of this type of study was by Thompson et al., who compared the impact of CCP therapy in a retrospective analysis of 966 hematologic malignancy patients with COVID-19 (Thompson et al. 2021). In that study, 143 convalescent plasma recipients were compared with 823 untreated patients. After adjustment for potential confounding factors, CCP treatment was associated with improved 30-day mortality (HR, 0.60; 95% CI, 0.37–0.97). This association remained significant after propensity score matching (HR, 0.52; 95% CI, 0.29–0.92). Mortality among CCP recipients was also lower in patients admitted to intensive care and in recipients of mechanical ventilatory support.

The impact of CCP treatment was assessed in a cohort of 112 patients with hematological malignancies with COVID-19 (Hueso et al. 2022). In that study, propensity score analysis on subgroups of patients treated with CCP (n = 81) or not (n = 120) between May 1, 2020 and April 1, 2021, demonstrated that treatment with CCP was particularly helpful in patients with B cell malignancies. The impact of CCP was also evaluated in a cohort of SOT recipients. In that study, the use of CCP did not lead to a reduction in COVID-19 progression and lethality among kidney transplant recipients (Cristelli et al. 2022). The small sample size, higher severity in the control group, delayed treatment, and use of a low proportion of high-titer CCP may have been important confounders in this analysis.

Several clinical trials and systematic reviews have examined the role of CCP in immunocompromised patients. Denkinger et al. randomized 56 patients with cancer and COVID-19 to standard of care (control arm) or standard of care plus convalescent/vaccinated anti-SARS-CoV-2 plasma (plasma arm) (Denkinger et al. 2023). CCP treatment resulted in a shortened median time to improvement (HR = 2.50; $P = 0.003$) and much lower mortality (HR = 0.28; $P = 0.042$). Lacombe et al. examined the impact of CCP in a trial that included 120 patients who were randomized 1:1 to CCP or usual care (Lacombe et al. 2022). Mortality in immunosuppressed CCP recipients was substantially lower than in recipients of usual care alone. (HR 0.37 [95% CI 0.14–0.97]).

To assess the impact of CCP in immunocompromised patients across multiple trials, Senefeld et al. performed a systematic review and meta-analysis that included over 2,300 patients (Senefeld et al. 2023) (see also Chap. X). Included in this review were 214 patients from three randomized clinical trials, 1,560 patients from five matched cohort studies and another 623 patients from 138 case reports or case series. These studies were analyzed using a standard fixed effects model that compared the observed deaths among patients transfused with CCP with the expected deaths if all patients were assumed to be equally at risk. The results showed an association between CCP use and lower mortality in hospitalized IC patients with COVID-19 (risk ratio, 0.63 [95% CI, 0.50–0.79]). Limitations of this analysis include its lack of individual patient data, differences in the volume and titer of CCP used, and variation in how immunocompromise was defined. Nonetheless, the findings support the conclusion that CCP is safe and beneficial in immunocompromised patients with COVID-19. In another pooled analysis of immunocompromised patients in 3 randomized trials, CCP use lowered mortality compared with standard of care or placebo (RR, 0.71 [CI, 0.51 to 0.98]) (Estcourt et al. 2021, 2022; Bar et al. 2021; Gharbharan et al. 2021).

5 Persistent COVID-19 and CCP

As described earlier, some immunocompromised patients are unable to generate effective antiviral responses and while not succumbing to infection, nonetheless cannot clear it. Such patients typically remain persistently positive for SARS-CoV-2 PCR for several weeks or much longer. Their clinical picture usually includes persistent or relapsing symptoms such as fever, shortness of breath and hypoxemia. They often have radiographic abnormalities on chest X-ray or computed tomography (CT) scan. A distinguishing feature differentiating such patients from others whose abnormalities are solely due to post-viral complications (e.g. bacterial or fungal superinfections, acute respiratory distress syndrome, post-COVID-19 organizing pneumonia, and "long COVID") is the persistence of high levels of SARS-CoV-2 RNA by PCR. Typical patient types for this condition are those with abnormalities in B lymphocyte function or number (Dioverti et al. 2022). These include patients with lymphoid malignancies, hematopoietic stem cell transplant recipients, CAR-T cell recipients, solid organ transplant recipients, people who have received treatment with anti-CD19/20 therapy or other lymphocyte targeted therapies, and certain primary and acquired immunodeficiencies (e.g. common variable immunodeficiency and Good Syndrome).

Data from several observational studies have demonstrated a role for CCP in patients with persistent COVID-19 infection. Hueso et al. reported on a series of 17 consecutive patients with profound B cell lymphopenia and prolonged COVID-19 symptoms, negative SARS-CoV-2 antibodies serology, and positive viral RNA who were treated with 4 units of CCP (Hueso et al. 2020). Within 48 h of transfusion, all but one patient experienced an improvement in clinical symptoms and SARS-CoV-2 RNA levels decreased to below the sensitivity threshold in all 9 evaluated patients. Lang-Meli et al. reported on CCP use in 16 cases of patients with primary antibody deficiency (Lang-Meli et al. 2022). Treatment was associated with a reduction in viral load and improvement of clinical symptoms, even when applied over a week after the onset of infection. One patient required a second course of plasma treatment for persistent infection. Ripol et al. reported on CCP use in 31 immunocompromised patients with persistent COVID-19 infection. Among the 29 patients who received CCP in the inpatient setting, 59% (17 of 29 patients) demonstrated rapid clinical improvement and were discharged within 5 days of vax-plasma transfusion. Among the 7 patients who survived and did not show rapid clinical improvement, the median time to discharge from the date of vax-plasma transfusion was 30 days (range, 7–109 days). Five patients, all of whom were in the intensive care unit at the time of CCP treatment, died despite transfusion (Ripoll et al. 2022). Taken as a whole, these data suggest a role for CCP in such patients. However, CCP alone is not universally effective at resolving persistent viral infection, and in some cases, combination therapies that include direct-acting antiviral agents and repeated cycles of passive immunotherapy have been necessary to control the infection.

6 Guidelines and Recommendations

The US Food and Drug Administration (FDA) provides legal authorization for the use of CCP for the treatment of COVID-19 in patients with immunosuppressive disease or receiving immunosuppressive treatment in either the outpatient or inpatient setting. In addition, several professional societies and clinical guideline panels have provided recommendations regarding CCP use in immunocompromised patients. The Association for the Advancement of Blood & Biotherapies (AABB) suggests CCP transfusion in addition to the usual standard of care for hospitalized patients with COVID-19 and preexisting immunosuppression (Estcourt et al. 2022). The European Conference on Infections in Leukemia (ECIL-9) and the

US-based National Comprehensive Cancer Network (NCCN) recommend CCP for selected patients, notably for SARS-CoV-2 seronegative hematological patients (Cesaro et al. 2022). The Infectious Diseases Society of America (IDSA) suggests against *routine* use of CCP in immunocompromised patients hospitalized with COVID-19 while noting that CCP is a reasonable option for hospitalized immunocompromised patients who do not qualify for other treatments. Additionally, IDSA recommends CCP as an option for early treatment of outpatients, particularly those for whom other therapies such as nirmatrelvir/ritonavir and IV remdesivir are not good options (Bhimraj et al. 2022). The National Institutes of Health (NIH) panel has concluded that there is insufficient evidence for a definitive recommendation for the treatment of COVID-19 in people who are immunocompromised (COVID-19 Treatment Guidelines Panel 2019).

A recent guidance document details the target immunocompromised patient types for CCP use (Bloch et al. 2023). This list includes patients with hematologic malignancies (especially chronic lymphocytic leukemia), solid organ transplant recipients, patients receiving treatment with rituximab or other anti-CD20 therapy, mycophenolate or other antimetabolite therapy, and individuals who have failed to seroconvert following infection and/or vaccination. The transfused product should ideally be high antibody titer CCP obtained from individuals who have had a natural infection within the past 6 months, and have been previously vaccinated. The best dose and duration of therapy are not known, but two units may be preferable to one as two units are likely to provide higher levels of neutralizing antibodies in the bloodstream of recipients. For patients who have not had a brisk clinical and virological response, retreatment may be necessary on a weekly basis of until a response is seen. As indicated above, CCP is typically used as part of combination therapy along with remdesivir in such circumstances.

7 Conclusions and Outlook for the Future

Since the start of the pandemic, CCP has been an important component of therapy for immunocompromised patients with COVID-19. Its deployment has saved countless lives. The role of CCP has evolved during the pandemic. Early on, when other antiviral treatment options were limited or not widely available, CCP served as a crucial front-line therapy. With the advent of effective direct acting antivirals and monoclonal antiviral antibody therapies, CCP took on a role as an alternate therapy for those in whom other effective options were not feasible. With the demise of monoclonal effectiveness, CCP is now again playing a larger role. Additionally, CCP has emerged as an important part of combination therapy for patients with persistent infections.

Additional work is needed to identify more precisely the subset of immunocompromised patients who respond best to CCP, and the ideal dose and duration of therapy. Also needed are reliable assays that can help match stored CCP products against specific viral strains. Lessons learned from the development of CCP should be applied to generating immune-based therapies for other viral infections afflicting immunocompromised patients and as preparation for future viral pandemics.

Sources of Funding for This Work None.

Disclosures Dr. Shoham has received support from Ansun, F2G, and Zeteo (research grants); Pfizer, Adamis, Immunome, Scynexis, and Karius (direct payments).

References

Alexander BT, Hladnik LM, Augustin KM, Casabar E, McKinnon PS, Reichley RM et al (2010) Use of cytomegalovirus intravenous immune globulin for the adjunctive treatment of cytomegalovirus in hematopoietic stem cell transplant recipients. Pharmacotherapy 30(6):554–561. https://doi.org/10.1592/phco.30.6.554

Al-Obaidi MM, Gungor AB, Kurtin SE, Mathias AE, Tanriover B, Zangeneh TT (2022) The prevention of COVID-19 in high-risk patients using Tixagevimab-Cilgavimab (Evusheld): real-world experience at a large academic center. Am J Med. https://doi.org/10.1016/j.amjmed.2022.08.019

Antinori A, Bausch-Jurken M (2023) The burden of COVID-19 in the immunocompromised patient: implications for vaccination and needs for the future. J Infect Dis 228(Suppl 1):S4–S12. https://doi.org/10.1093/infdis/jiad181

Bahremand T, Yao JA, Mill C, Piszczek J, Grant JM, Smolina K (2023) COVID-19 hospitalisations in immunocompromised individuals in the Omicron era: a population-based observational study using surveillance data in British Columbia, Canada. Lancet Reg Health Am 20:100461. https://doi.org/10.1016/j.lana.2023.100461

Bar KJ, Shaw PA, Choi GH, Aqui N, Fesnak A, Yang JB et al (2021) A randomized controlled study of convalescent plasma for individuals hospitalized with COVID-19 pneumonia. J Clin Invest 131(24). https://doi.org/10.1172/JCI155114

Bhimraj A, Morgan RL, Shumaker AH, Baden L, Cheng VCC, Edwards KM et al (2022) Infectious diseases society of America guidelines on the treatment and management of patients with COVID-19. Clin Infect Dis. https://doi.org/10.1093/cid/ciac724

Bloch EM, Focosi D, Shoham S, Senefeld J, Tobian AAR, Baden LR et al (2023) Guidance on the use of convalescent plasma to treat immunocompromised patients with coronavirus disease 2019. Clin Infect Dis 76(11):2018–2024. https://doi.org/10.1093/cid/ciad066

Boyarsky BJ, Werbel WA, Avery RK, Tobian AAR, Massie AB, Segev DL et al (2021) Antibody response to 2-dose SARS-CoV-2 mRNA vaccine series in solid organ transplant recipients. JAMA 325(21):2204–2206. https://doi.org/10.1001/jama.2021.7489

Calderón-Parra J, Múñez-Rubio E, Fernández-Cruz A, García-Sánchez MC, Maderuelo-González E, López-Dosil M et al (2022) Incidence, clinical presentation, relapses and outcome of severe acute respiratory syndrome coronavirus 2 (SARS-CoV-2) infection in patients treated with anti-CD20 monoclonal antibodies. Clin Infect Dis 74(10):1786–1794. https://doi.org/10.1093/cid/ciab700

Cesaro S, Mikulska M, Hirsch HH, Styczynski J, Meylan S, Cordonnier C et al (2023) Update of recommendations for the management of COVID-19 in patients with haematological malignancies, haematopoietic cell transplantation and CAR T therapy, from the 2022 European Conference on Infections in Leukaemia (ECIL 9). Leukemia 37(9):1933–1938. https://doi.org/10.1038/s41375-023-01938-5

Corey L, Beyrer C, Cohen MS, Michael NL, Bedford T, Rolland M (2021) SARS-CoV-2 variants in patients with immunosuppression. N Engl J Med 385(6):562–566. https://doi.org/10.1056/NEJMsb2104756

COVID-19 Treatment Guidelines Panel (2019) Coronavirus disease 2019 (COVID-19) treatment guidelines. National Institutes of Health. Available at https://www.covid19treatmentguidelines.nih.gov/. Accessed 24 Oct 2023

Cristelli MP, Langhi Junior DM, Viana LA, de Andrade LGM, Martins SBS, Dreige YC et al (2022) Efficacy of convalescent plasma to treat mild to moderate COVID-19 in kidney transplant patients: a propensity score matching analysis. Transplantation 106(1):e92–e94. https://doi.org/10.1097/TP.0000000000003962

Denkinger CM, Janssen M, Schäkel U, Gall J, Leo A, Stelmach P et al (2022) Anti-SARS-CoV-2 antibody-containing plasma improves outcome in patients with hematologic or solid cancer and severe COVID-19: a randomized clinical trial. Nat Cancer. https://doi.org/10.1038/s43018-022-00503-w

Denkinger CM, Janssen M, Schäkel U, Gall J, Leo A, Stelmach P et al (2023) Anti-SARS-CoV-2 antibody-containing plasma improves outcome in patients with hematologic or solid cancer and severe COVID-19: a randomized clinical trial. Nat Cancer. 4(1):96–107. https://doi.org/10.1038/s43018-022-00503-w

Dioverti V, Salto-Alejandre S, Haidar G (2022) Immunocompromised patients with protracted COVID-19: a review of "Long Persisters." Curr Transplant Rep 9(4):209–218. https://doi.org/10.1007/s40472-022-00385-y

Dispinseri S, Secchi M, Pirillo MF, Tolazzi M, Borghi M, Brigatti C et al (2021) Neutralizing antibody responses to SARS-CoV-2 in symptomatic COVID-19 is persistent and critical for survival. Nat Commun 12(1):2670. https://doi.org/10.1038/s41467-021-22958-8

Duléry R, Lamure S, Delord M, Di Blasi R, Chauchet A, Hueso T et al (2021) Prolonged in-hospital stay and higher mortality after Covid-19 among patients with non-Hodgkin lymphoma treated with B-cell depleting immunotherapy. Am J Hematol 96(8):934–944. https://doi.org/10.1002/ajh.26209

Estcourt LJ, Turgeon AF, McQuilten ZK, McVerry BJ, Al-Beidh F, Annane D et al (2021) Effect of convalescent plasma on organ support-free days in critically ill patients with COVID-19: a randomized clinical trial. JAMA 326(17):1690–1702. https://doi.org/10.1001/jama.2021.18178

Estcourt LJ, Cohn CS, Pagano MB, Iannizzi C, Kreuzberger N, Skoetz N et al (2022) Clinical practice guidelines from the association for the advancement of blood and biotherapies (AABB): COVID-19 convalescent plasma. Ann Intern Med 175(9):1310–1321. https://doi.org/10.7326/M22-1079

Fourati S, Gautier G, Chovelon M, Soulier A, N'Debi M, Demontant V et al (2022) Persistent SARS-CoV-2 alpha variant infection in immunosuppressed patient, France, February 2022. Emerg Infect Dis 28(7):1512–1515. https://doi.org/10.3201/eid2807.220467

Galmiche S, Luong Nguyen LB, Tartour E, de Lamballerie X, Wittkop L, Loubet P et al (2022) Immunological and clinical efficacy of COVID-19 vaccines in immunocompromised populations: a systematic review. Clin Microbiol Infect 28(2):163–177. https://doi.org/10.1016/j.cmi.2021.09.036

Gandhi S, Klein J, Robertson AJ, Peña-Hernández MA, Lin MJ, Roychoudhury P et al (2022) De novo emergence of a remdesivir resistance mutation during treatment of persistent SARS-CoV-2 infection in an immunocompromised patient: a case report. Nat Commun 13 (1):1547. https://doi.org/10.1038/s41467-022-29104-y

Gharbharan A, Jordans CCE, GeurtsvanKessel C, den Hollander JG, Karim F, Mollema FPN et al (2021) Effects of potent neutralizing antibodies from convalescent plasma in patients hospitalized for severe SARS-CoV-2 infection. Nat Commun 12(1):3189. https://doi.org/10.1038/s41467-021-23469-2

Ghosh S, Champlin RE, Englund J, Giralt SA, Rolston K, Raad I et al (2000) Respiratory syncytial virus upper respiratory tract illnesses in adult blood and marrow transplant recipients: combination therapy with aerosolized ribavirin and intravenous immunoglobulin. Bone Marrow Transplant 25(7):751–755. https://doi.org/10.1038/sj.bmt.1702228

Harpaz R, Dahl RM, Dooling KL (2016) Prevalence of immunosuppression among US adults, 2013. JAMA 316(23):2547–2548. https://doi.org/10.1001/jama.2016.16477

Hogan JI, Duerr R, Dimartino D, Marier C, Hochman SE, Mehta S et al (2022) Remdesivir resistance in transplant recipients with persistent COVID-19. Clin Infect Dis. https://doi.org/10.1093/cid/ciac769

Hueso T, Pouderoux C, Péré H, Beaumont AL, Raillon LA, Ader F et al (2020) Convalescent plasma therapy for B-cell-depleted patients with protracted COVID-19. Blood 136(20):2290–2295. https://doi.org/10.1182/blood.2020008423

Hueso T, Godron AS, Lanoy E, Pacanowski J, Levi LI, Gras E et al (2022) Convalescent plasma improves overall survival in patients with B-cell lymphoid malignancy and COVID-19: a longitudinal cohort and propensity score analysis. Leukemia 36(4):1025–1034. https://doi.org/10.1038/s41375-022-01511-6

Imai M, Iwatsuki-Horimoto K, Hatta M, Loeber S, Halfmann PJ, Nakajima N et al (2020) Syrian hamsters as a small animal model for SARS-CoV-2 infection and countermeasure development. Proc Natl Acad Sci U S A 117(28):16587–16595. https://doi.org/10.1073/pnas.2009799117

Kang SW, Kim JW, Kim JY, Lim SY, Jang CY, Chang E et al (2023) Characteristics and risk factors of prolonged viable virus shedding in immunocompromised patients with COVID-19: a prospective cohort study. J Infect. https://doi.org/10.1016/j.jinf.2023.01.024

Klein SL, Pekosz A, Park HS, Ursin RL, Shapiro JR, Benner SE et al (2020) Sex, age, and hospitalization drive antibody responses in a COVID-19 convalescent plasma donor population. J Clin Invest 130(11):6141–6150. https://doi.org/10.1172/JCI142004

Kwon JH, Tenforde MW, Gaglani M, Talbot HK, Ginde AA, McNeal T et al (2022) MRNA vaccine effectiveness against coronavirus disease 2019 hospitalization among solid organ transplant recipients. J Infect Dis 226(5):797–807. https://doi.org/10.1093/infdis/jiac118

Lacombe K, Thomas H, Porcher R, Mekinian A, Chiarabini T, Georgin-Lavialle S et al (2022) COVID-19 convalescent plasma to treat hospitalised COVID-19 patients with or without underlying immunodeficiency. medRxiv 2022.08.09.22278329

Lang-Meli J, Fuchs J, Mathé P, Ho HE, Kern L, Jaki L et al (2022) Case series: convalescent plasma therapy for patients with COVID-19 and primary antibody deficiency. J Clin Immunol 42(2):253–265. https://doi.org/10.1007/s10875-021-01193-2

Leon J, Merrill AE, Rogers K, Kurt J, Dempewolf S, Ehlers A et al (2022) SARS-CoV-2 antibody changes in patients receiving COVID-19 convalescent plasma from normal and vaccinated donors. Transfus Apher Sci 61(2):103326. https://doi.org/10.1016/j.transci.2021.103326

Ripoll JG, Gorman EK, Juskewitch JE, Razonable RR, Ganesh R, Hurt RT et al (2022) Vaccine-boosted convalescent plasma therapy for patients with immunosuppression and COVID-19. Blood Adv 6(23):5951–5955. https://doi.org/10.1182/bloodadvances.2022008932

Senefeld JW, Klassen SA, Ford SK, Senese KA, Wiggins CC, Bostrom BC et al (2021) Use of convalescent plasma in COVID-19 patients with immunosuppression. Transfusion 61(8):2503–2511. https://doi.org/10.1111/trf.16525

Senefeld JW, Franchini M, Mengoli C, Cruciani M, Zani M, Gorman EK et al (2023) COVID-19 convalescent plasma for the treatment of immunocompromised patients: a systematic review and meta-analysis. JAMA Netw Open 6(1):e2250647. https://doi.org/10.1001/jamanetworkopen.2022.50647

Shoham S, Bloch EM, Casadevall A, Hanley D, Lau B, Gebo K et al (2023a) Transfusing convalescent plasma as post-exposure prophylaxis against severe acute respiratory syndrome coronavirus 2 (SARS-CoV-2) infection: a double-blinded, phase 2 randomized, controlled trial. Clin Infect Dis 76(3):e477–e486. https://doi.org/10.1093/cid/ciac372

Shoham S, Batista C, Ben Amor Y, Ergonul O, Hassanain M, Hotez P et al (2023b) Vaccines and therapeutics for immunocompromised patients with COVID-19. E Clin Med 59:101965. https://doi.org/10.1016/j.eclinm.2023.101965

Singson JRC, Kirley PD, Pham H, Rothrock G, Armistead I, Meek J et al (2022) Factors associated with severe outcomes among immunocompromised adults hospitalized for COVID-19—COVID-NET, 10 states, March 2020–February 2022. MMWR Morb Mortal Wkly Rep 71(27):878–884. https://doi.org/10.15585/mmwr.mm7127a3

Sullivan DJ, Franchini M, Joyner MJ, Casadevall A, Focosi D (2022) Analysis of anti-SARS-CoV-2 Omicron-neutralizing antibody titers in different vaccinated and unvaccinated convalescent plasma sources. Nat Commun 13(1):6478. https://doi.org/10.1038/s41467-022-33864-y

Sun J, Zhuang Z, Zheng J, Li K, Wong RL, Liu D et al (2020) Generation of a broadly useful model for COVID-19 pathogenesis, vaccination, and treatment. Cell 182(3):734–43.e5. https://doi.org/10.1016/j.cell.2020.06.010

Sun J, Zheng Q, Madhira V, Olex AL, Anzalone AJ, Vinson A et al (2022) Association between immune dysfunction and COVID-19 breakthrough infection after SARS-CoV-2 vaccination in the US. JAMA Intern Med 182(2):153–162. https://doi.org/10.1001/jamainternmed.2021.7024

Thompson MA, Henderson JP, Shah PK, Rubinstein SM, Joyner MJ, Choueiri TK et al (2021) Association of convalescent plasma therapy with survival in patients with hematologic cancers and COVID-19. JAMA Oncol. https://doi.org/10.1001/jamaoncol.2021.1799

Werbel WA, Boyarsky BJ, Ou MT, Massie AB, Tobian AAR, Garonzik-Wang JM et al (2021) Safety and immunogenicity of a third dose of SARS-CoV-2 vaccine in solid organ transplant recipients: a case series. Ann Intern Med 174(9):1330–1332. https://doi.org/10.7326/L21-0282

Yek C, Warner S, Wiltz JL, Sun J, Adjei S, Mancera A et al (2022) Risk factors for severe COVID-19 outcomes among persons aged ≥ 18 years who completed a primary COVID-19 vaccination series—465 health care facilities, United States, December 2020–October 2021. MMWR Morb Mortal Wkly Rep 71(1):19–25. https://doi.org/10.15585/mmwr.mm7101a4

Yetmar ZA, Khodadadi RB, Seville MT, Brumble L, O'Horo JC, Ganesh R et al (2022) Outcomes of B-cell-depleted patients with coronavirus disease 2019 treated with antispike monoclonal antibodies. Open Forum Infect Dis 9(7):ofac204. https://doi.org/10.1093/ofid/ofac204

Yetmar ZA, Beam E, O'Horo JC, Ganesh R, Bierle DM, Brumble L et al (2021) Monoclonal antibody therapy for COVID-19 in solid organ transplant recipients. Open Forum Infect Dis 8 (6):ofab255. https://doi.org/10.1093/ofid/ofab255

Shmuel Shoham is a Professor of Clinical Medicine in the Division of Infectious Diseases of the Department of Medicine at Johns Hopkins University School of Medicine. He has over 20 years of experience in management of infections in highly immunocompromised patients and is the author or co-author of over 150 original articles, book chapters and topic reviews. He serves as a reviewer and expert consultant to multiple journals, professional societies, and government agencies and is a member of guideline committees for several national and international organizations. These include the Infectious Diseases Society of America guidelines for COVID-19 and the National Comprehensive Cancer Network guidelines for Prevention and Treatment of Cancer-Related Infections.

Monoclonal Antibody Therapies Against SARS-CoV-2: Promises and Realities

Daniele Focosi

Contents

Abstract Monoclonal antibodies targeting the Spike protein of SARS-CoV-2 have been widely deployed in the ongoing COVID-19 pandemic. I review here the impact of those therapeutics in the early pandemic, ranging from structural classification to outcomes in clinical trials to in vitro and in vivo evidence of basal and treatment-emergent immune escape. Unfortunately, the Omicron variant of concern has completely reset all achievements so far in mAb therapy for COVID-19. Despite the intrinsic limitations of this strategy, future developments such as respiratory delivery of further engineered mAb cocktails could lead to improved outcomes.

D. Focosi (✉)
North-Western Tuscany Blood Bank, Pisa University Hospital, Pisa, Italy
e-mail: daniele.focosi@gmail.com

Current Topics in Microbiology and Immunology (2025) 443: 131–147
https://doi.org/10.1007/82_2024_268

Published Online: 11 August 2024

Abbreviations

CCP	COVID-19 convalescent plasma
EMA	European Medicines Agency
FDA	Food and Drug Administration
mAb	Monoclonal antibody
nAb	Neutralizing antibody
RBD	Receptor-binding domain
RCT	Randomized controlled trial
SARS-CoV-2	Severe acute respiratory syndrome coronavirus 2

1 Introduction

Despite the widespread use of monoclonal antibodies (mAbs) in medicine, relatively few have been developed against viral diseases (see Chap. A for a history of monoclonal antibodies in infectious disease). Notable exceptions include palivizumab for pre-exposure prophylaxis of respiratory syncytial virus (RSV) (Luna et al. 2020), AB1 for post-exposure prophylaxis of rabies virus, and the REGN-EB3 cocktail for the treatment of Ebolavirus disease. The COVID-19 pandemic has been a tremendous boost for developing and deploying antiviral mAb's. Since SARS-CoV-2 Spike protein is currently the best characterized protective antigen, most of mAbs have been directed against the spike protein. As of June 13, 2023, the Coronavirus Antibody Database (CoV-AbDab) contained 12,536 antibodies and nanobodies to SARS-CoV, MERS-CoV, and SARS-CoV-2 (http://opig.stats.ox.ac.uk/webapps/covabdab/). All mAbs thus far approved for clinical use work as infection-neutralizing antibodies (nAb) by binding the Spike protein, and in particular, most of them target epitopes within the ACE2 receptor-binding domain (RBD). In general, when used clinically, those mAbs have proven extremely safe, with no incidence of antibody-dependent enhancement (ADE) of viral infection or antidrug antibodies (ADAs) after single-dose exposures (Syed 2021).

2 mAb Efficacy in Randomized Controlled Trials (RCT)

As with all other SARS-CoV-2 antivirals, the efficacy of mAb was defined in relation to the time of treatment. When deployed in pre- or post-exposure prophylaxis, a reduction in infection rates was the outcome, while for outpatients, efficacy was defined as a reduction in hospitalization rates, and for inpatients, the key metrics were reductions in disease progression or mortality.

As with COVID-19 convalescent plasma (CCP), another form of antibody treatment, clear evidence of the therapeutic efficacy of mAbs was based on studies of seronegative and early-stage inpatients (ACTIV-3/TICO LY-CoV555 Study Group 2020; Isa et al. 2021; Gupta et al. 2021; Fact sheet for healthcare providers: emergency use authorization for bebtelovimab 2022). When results in hospitalized late-stage patients proved disappointing (Lundgren et al. 2021), companies rapidly pivoted to establish efficacy in the outpatient setting, where, in line with the principles of passive serotherapy, they were more effective. This was the case for bamlanivimab (ACTIV-3/TICO LY-CoV555 Study Group 2020; Gottlieb et al. 2021), bamlanivimab + etesivimab casirivimab + imdevimab (Isa et al. 2021; Weinreich et al. 2021; O'Brien et al. 2021; Horby et al. 2021), sotrovimab (Gupta et al. 2021), cilgavimab + tixagevimab (Update on AZD7442 STORM CHASER trial in post-exposure prevention of symptomatic COVID-19 2022; AZD7442 reduced risk of developing severe COVID-19 or death in TACKLE Phase III outpatient treatment trial 2022), regdanvimab (Syed 2021), bebtelovimab (Fact sheet for healthcare providers: emergency use authorization for bebtelovimab 2022), amubarvimab + romlusevimab (Self et al. 2022). Tixagevimab + cilgavimab was the only combination approved for pre-exposure prophylaxis (Update on AZD7442 STORM CHASER trial in post-exposure prevention of symptomatic COVID-19 2022; Levin et al. 2022). Unfortunately, all RCTs of mAb's were conducted before widespread vaccine use. No RCT has been conducted in vaccinated populations, and mAb usage has been deployed in seropositive vaccine recipients without clear evidence supporting their efficacy in this setting.

The explosive nature of the pandemic highlighted some predictable limitations in the development of mAb therapies. From within a broad drug pipeline (Renn et al. 2020), only a few candidates reached initial approval by regulatory authorities in sufficient time to be deployed. The high cost of mAbs—which contrasts with the relatively low expense of the other antibody option, convalescent plasma—has remained prohibitive even for high-income economies in the face of a pandemic. And, even after the few mAbs were finally authorized, many failed to be effective shortly after their introduction because the virus rapidly escaped their narrow specificity with the generation of resistant variants (Focosi and Maggi 2021).

3 Resistance to Anti-Spike mAbs and the Omicron Hurricane

As for any other antiviral therapeutic, resistance to mAbs binding the Spike protein can be either initial (i.e., pre-existing before treatment) or treatment-emergent (i.e., positive selection of immune escape variants by treatment). Caution should be used when drawing conclusions on complex variants of interest (VOIs) and variants of concern (VOCs), because of the multiplicity of variants. The Delta VOC consists of more than 200 sublineages (Focosi et al. 2022a), and the Omicron VOC is more than 1,500. The in vitro sensitivity of individual sublineages to individual clinically authorized anti-Spike mAbs have been extensively detailed for both pre-Omicron

(Focosi et al. 2022b) and Omicron sublineages such as BQ.1.1 (Touret et al. 2022; He et al. 2023), XBB.1.5 (Cao et al. 2022; Imai et al. 2022; Wang et al. 2023a, b; Liu et al. 2021; Qu et al. 2023), XBB.1.16 (Yamasoba et al. 2023), EG.5.1 (Wang et al. 2023c) BA.2.86 (Sheward et al. 2023; Uriu et al. 2023; Yang et al. 2023a; Qu et al. 2023) and the recent "FLip" family (S:L455F + S:F456L) members (Yang et al. 2023a; Qu et al. 2023), but an extensive discussion goes beyond the scope of this chapter. As of this writing (December 2023) sotrovimab has been the most resilient anti-Spike mAb but has finally capitulated to the highly immune-escaping variant JN.1. (Kaku et al. 2023; Yang et al. 2023b).

In November 2021, Omicron BA.1 emerged, and it brought along a much higher number of mutations and deletions than prior VOCs. As a consequence, most clinically approved mAbs suddenly lost their usefulness (Benotmane et al. 2022). Fortunately, all of the small-molecule antivirals—remdesivir, molnupiravir, and nirmatrelvir/ritonavir—have remained effective in vitro against Omicron (Vangeel et al. 2022), but in vivo, while they are quite effective when used early (Jayk Bernal et al. 2022), they are expensive, and sometimes come with safety concerns, such as the mutagenicity of molnupiravir for both human (Zhou et al. 2021) and viral RNA (Sanderson et al. 2023). Given the scarcity of antiviral agents, both the FDA (Food et al. 2021) and the IDSA (IDSA 2022) have recently reevaluated CCP (which is less vulnerable to evasion by variants due to its polyclonal nature (Sullivan et al. 2022, 2023)) for outpatients at risk of progression. For reviews of the effectiveness of CCP, please see Chaps. X and Y.

The US FDA was among the first regulatory authorities to react to the Omicron hurricane, and on Jan 24, 2022 revised its authorizations for 2 mAb treatments—bamlanivimab and etesevimab (administered together) and REGEN-COV (casirivimab and imdevimab)—to limit their use to infections with, or exposures to, a variant that was susceptible in vitro to these treatments (Statement 2022). However, attributing infection to a specific variant requires viral genome sequencing, which is expensive and poorly scalable. Moreover, the long turnaround time for such testing precludes early administration of the mAb. As soon as viral diversification emerged, the resistance of the BA.2 sublineage was also reported (Iketani et al. 2022; Zhou et al. 2022a; Ohashi et al. 2022; Reuters 2022). This led to further restrictions by the FDA on these two mAbs on February 25, 2022 and withdrawal of authorization on April 5, 2022 (ACTIV-3 2022). Sotrovimab deauthorization followed on April 5 (FDA updates Sotrovimab emergency use authorization 2022).

To improve preparedness, bebtelovimab was approved by the FDA for outpatients on Feb 11 on the basis of a phase 2 clinical trial (FDA 2022), but this mAb was also deauthorized on November 30, 2022 (Announces Bebtelovimab and is Not Currently Authorized in Any US Region 2022). Evusheld® was the last mAb to retain its FDA authorization, finally losing that status on January 26, 2023 (FDA announces Evusheld is not currently authorized for emergency use in the U.S. 2023). Despite these several FDA deauthorizations, many US physicians continue to prescribe anti-Spike mAbs (Anderson et al. 2022), based on personal beliefs in discrepancies between in vitro and in vivo activities.

The latter phenomenon has been much amplified across Europe (Focosi and Tuccori 2022), where the EMA has to date not withdrawn a single authorized mAb, and has only sporadically issued alerts about possible resistance (ETF warns that monoclonal antibodies may not be effective against emerging strains of SARS-CoV-2 2023; ETF statement on the loss of activity of anti-spike protein monoclonal antibodies due to emerging SARS-CoV-2 variants of concern 2022). In addition, a plethora of uncontrolled or poorly controlled literature continues to advocate for the efficacy of anti-Spike mAbs based on marginal gains in surrogate endpoints in non-RCTs (Harman et al. 2022; Drysdale et al. 2023) or no change in hospitalization rates of outpatients (Patel et al. 2023). While binding affinity seems a robust surrogate endpoint, advocates of residual in vivo activity argue that the concentrations achieved by mAbs in vivo can overcome several degrees of baseline resistance to neutralization or that alternative effector function still persists (Wu et al. 2022a; Stadler et al. 2022a, b). Neither argument is robust: with IC_{50} above 1,000 (Cao et al. 2022) antigen binding by the mAb is compromised, and effector functions other than neutralization invariably require binding to the virus to activate Fc receptors. Under the current soup of circulating SARS-CoV-2 sublineages (Focosi et al. 2023a), the turnaround time of viral genome sequencing is not compatible with the need for prompt delivery of therapeutics. As a result tools to predict basal efficacy of anti-Spike mAbs based on inferences from regional genomic surveillance have been developed (Focosi 2023).

4 Treatment-Emergent Anti-Spike mAb Resistance

The emergence of anti-Spike mAb resistance during therapy is a common and consistent phenomenon (Focosi et al. 2023b). The discrepancy between the diversity of mutations reported in vitro and the few detected in vivo can be explained by several factors. First, not all mutants selected in vitro are fit enough to compete within the intrahost quasi-species swarm. Second, some mutations existed before treatment. Third, among the viruses characterized as wild-type at baseline, the mutations associated with the highest fold-reduction in neutralization have been the ones most commonly reported. This could reflect more attention by investigators to previously reported mutations, but it is also likely that as soon as one of the highly resistant mutations emerges, the emergence of additional mutations does not provide any further fitness advantage.

Despite the fact that management of the immunocompromised patient with intolerance to small-molecule antivirals has been an important focus of anti-Spike mAb usage, the anti-Spike mAbs authorized so far have not been studied by RCT in the immunocompromised (Focosi et al. 2022c). Immunocompromised patients are at increased risk for anti-Spike mAb treatment-emergent resistance for several reasons—higher basal viral load, higher likelihood of viral persistence because of inadequate immune response, and contraindications to combined therapies with small-molecule antivirals (Casadevall and Focosi 2023). Resistance monitoring is not routinely performed after outpatient anti-Spike mAb treatment, and viral sequencing interpretation requires paying attention to minoritarian sequences (Vellas et al. 2022) Of interest, several cases of anti-Spike mAb-treatment-emergent immune escape were rescued with CCP (Halfmann et al. 2022; Pommeret et al. 2021), which, in contrast to anti-Spike mAbs, is polyclonal and thus more difficult to be defeated by viral evolution.

In many patients the emergence of anti-Spike mAb resistance was associated with only one amino acid substitution in the epitope, establishing the high vulnerability of monotherapy for selecting resistant variants. For some anti-Spike mAbs, the generated footprint is so unique (e.g., S:340X after sotrovimab) that baseline sequencing is not needed to confirm treatment emergence (S:340X being exceedingly rare in GISAID). The transmissibility of such mAb treatment-generated lineages remains a major concern, and it has been formally demonstrated in at least one study (Sabin et al. 2022). Clinical reports of treatment-emergent resistance in response to therapeutic mAbs reviewed here encompass nearly every epitope of the RBD.

Since therapeutic anti-Spike mAbs ultimately trace their ancestry to immune B cells in humans, the emergence of mAb resistance mutations poses the theoretical concern that once these occur in immunosuppressed individuals the mAb-resistant variants could spread to immunocompetent populations, bypass existing immunity, and creating new waves of infection (Casadevall and Focosi 2023). Cocktails of three or more anti-Spike mAbs with non-overlapping epitopes could minimize the chances for immune escape.

5 Perspectives

Modern recombinant mAb technology introduces several modifications to the primary sequence to improve or ablate effector functions and increase circulation half-life (Saunders 2019). These include extending the mAb half-life, ablation of effector functions, expanding antigen specificity, and use of the respiratory delivery route, all of which are elaborated upon below.

5.1 Extending mAb Half-Life

The Fc immunoglobulin region has received considerable attention in efforts to alter the properties of mAbs to improve pharmacokinetic and effector functions of immunoglobulins. The Fc region is responsible for the isotype of the immunoglobulin, the serum half-life of the immunoglobulin, and engaging the cellular Fc receptors to promote phagocytosis, complement activation, and antibody-dependent cell cytotoxicity (ADCC). Fcγ-modified mAbs with the amino acid substitution M252Y/S254T/T256E [YTE], a modification associated with a 2- to fourfold longer half-life, were developed for the prophylaxis of RSV in infants (e.g., nirsevimab (MED18897) (Robbie et al. 2013; Domachowske et al. 2018)). The same technology was used in the development of Evusheld®, the anti-SARS-CoV-2e mAb cocktail approved for pre-exposure prophylaxis. Antibody half-life can also be expanded with the LS mutation (Met428Leu/Asn434Ser), which does not affect ADCC function (Zalevsky et al. 2010; Schäfer et al. 2020), and this modification was used in the anti-Spike mAbs sotrovimab (GSK), and the C-135-LS plus C-144-LS cocktail (BMS) (Schäfer et al. 2020). Adintrevimab/ADG20 (Adagio) and its successor VYD222 (Invivyd) (Kaplon et al. 2022) instead relied on the LA mutation (Met428Leu/Asn434Ala) for half-life extension (Ison et al. 2023; Hao et al. 2023).

5.2 *Ablation of Effector Functions*

The immunoglobulin Fc region can also be modified to reduce effector functions. Such modifications are desirable in clinical situations, where stimulating other aspects of the immune system, such as complement activation or engaging Fc receptors, can trigger side effects. To reduce the risk of both ADE and ADCC, IgG_1 Fc regions can be modified to eliminate binding to FcγRI, FcγRIIa, and FcγRIIIa by changing two amino acids in the CH2 domain (L234A, L235A [LALA]). For IgG_4, FcgR binding can be abrogated by changing Ser228 to Pro in the hinge region and Leu235 to Glu (SPLE or PE mutations) (Newman et al. 2001). Changing P329 to Glu abolishes FcγR and C1q interactions, by disrupting the formation of a proline sandwich motif with the FcγRs, while leaving intact FcRn interactions and Fc stability (Schlothauer et al. 2016). Although FcRn polymorphisms are known to affect IgG half-life (Dalakas and Spaeth 2021), no specific data are available for anti-Spike mAbs.

VIR-7832 is a modification of sotrovimab with the addition of a 3-amino acid GAALIE (G236A, A330L, I332E) to the Fc domain, which enhances binding to FcγRIIa and FcγRIIIa, decreases affinity for FcγRIIb in vitro, and evokes protective CD^{8+} T lymphocytes in vivo (Bournazos et al. 2020; Weitzenfeld et al. 2019). Although ablation of Fc effector function has not been clinically used to date in mAbs to SARS-CoV-2, it is possible that, as we learn more about the pathogenesis of the virus and the mechanisms of antibody-mediated clearance, some of these alterations could find utility in future designs.

5.3 Expanding Antigen Specificity

The experience with SARS-CoV-2 shows that single mAbs are vulnerable to losing effectiveness against a virus as new variants emerge through viral evolution or antibody selection of escape mutants. This represents a major loss both in clinical therapeutic options and in financial investment.

Broadly neutralizing antibodies (bnAbs) target epitopes found in all or most viral variants and are sought after when designing therapies for antigenically variable viruses such as HIV-1 and SARS-CoV-2 (Mishra et al. 2021). At the level of the Betacoronavirus genus, universality in antibody neutralization translates into full coverage of the sarbecovirus subgenus (Shrestha et al. 2021). Candidate pan-sarbecovirus mAbs targeting the Spike protein have been variously referred as cluster VII, class IV, or RBD core cluster II: examples include S2X259 (Tortorici et al. 2021) and DH1047 (Martinez et al. 2022). Other pan-sarbecovirus mAbs belong to the class/cluster I RBM (e.g., S2K146 (Park et al. 2022)), class 3 (e.g., sotrovimab) or bind to the base of the stem-helix at the HR2 boundary in the S2 subunit (e.g., CV3-25 (Hurlburt et al. 2021), 1249A8 (Piepenbrink et al. 2022), and the CC series (Zhou et al. 2022b)). Each of those S2 bnAbs have lower IC_{50} than anti-RBD antibodies, a fact that could promote translation into clinically useful doses, but experiments in animal models suggest protection at lower doses (Pinto et al. 2021; Zhou et al. 2022c), likely thanks to additional effector functions. Notably, each of these pan-sarbecovirus mAbs has retained activity against VOC Omicron sublineages so far (Cameroni et al. 2021).

Combining mAbs creates a polyclonal product. Cocktails have been used against both SARS-CoV-2, Ebola, and rabies virus (CL184, a cocktail of CR57 and CR 4098 mAbs). Apart from protecting the product against viral evolution, cocktails also have the potential to trigger additive or synergistic effects through the action of two or more mAbs, but this comes at a significantly increased cost. Another alternative is to create bispecific antibodies (e.g., 14-H-06 (Ku et al. 2022)) combining 2 Fab regions binding to different epitopes.

5.4 Respiratory Delivery

Immunoglobulins are large protein molecules and only systemic routes have been investigated so far in clinical use. The need to provide mAbs to outpatients generally proved more difficult than for inpatients, since these required the existence of infusion facilities suitable for treating infectious patients, leading to alternative dosing routes such as the subcutaneous (Ronapreve®) and intramuscular (Evusheld®, sotrovimab and adintrevimab) which were eventually approved.

The need for systemic administration is a problem for a therapy targeting a virus that replicates in the nasal epithelium, since the poor affinity of IgGs to the polymeric Ig receptor (pIgR) causes only 1 of 1,000 IgG molecules infused intravenously to reach the respiratory mucosa (Hart et al. 2001; Peebles et al. 1995; Wu et al. 2007). Consequently, the most effective route of mAb delivery against a respiratory pathogen would be directly to the mucosa such as by intranasal or intratracheal administration (Focosi and Maggi 2023). Only 2 clinical trials have been reported so far with respiratory delivery of anti-Spike mAbs. In a phase 1 clinical trial, a single nasal spray of 35B5 conveyed 24-h effective in vitro protection against SARS-CoV-2 VOCs, including the Alpha, Beta, Delta, or Omicron variants in 30 healthy volunteers (Lin et al. 2023). In a single-blind placebo-controlled RCT of post-exposure prophylaxis in healthy adult workers, the anti-Spike mAb SA58 nasal spray reduced laboratory-confirmed symptomatic COVID-19 from 2.8 to 1.04 per 100 person-days (61% efficacy) (Song et al. 2023).

The scFv antibodies are smaller (about 28 kDa vs 150 kDa of mAbs) and less complex (no glycosylation), making them particularly stable and suitable for topical use in outpatient settings. In addition, they are devoid of immunoglobulin Fc and are therefore unlikely to induce antibody-dependent enhancement (ADE) of viral infection. Moreover, because of their rapid production at large scale in non-mammalian expression systems, they can be developed at a relatively lower cost compared with mAbs.

Camelids (camels, dromedaries, alpacas, and llamas) produce antibodies devoid of light chains (Hamers-Casterman et al. 1993). These antibodies—named heavy-chain-only antibodies (HcAbs)—are highly thermostable, and weigh approximately 90 kDa. The variable domain on each heavy chain (VHH or nanobody, 15 kDa) is longer, acting like an extended hand, and is able to reach buried sites unavailable to the short-handed classical antibodies (Muyldermans et al. 2001). However, as they are of camelid origin, they require sophisticated humanization procedures to avoid immunogenic responses, potentially hampering their full development. One possibility is to humanize these camelid nanobodies by fusing them to human Fc regions (Nambulli et al. 2021; Haga et al. 2021; Wu et al. 2022b; Titong et al. 2022; Han et al. 2023; McSweeney et al. 2022; Li et al. 2021).

6 Conclusions

The rapid deployment of multiple mAb therapies during pandemic conditions was a remarkable human accomplishment. However, the experience of the past two years has also shown the limitations of this approach, which might have been anticipated from what we already knew about antibody action, and what soon learned about the antigenic variability of SARS-CoV-2. While a few regulatory authorities promptly issued recommendations to avoid inappropriate usage of anti-Spike mAbs against resistant VOCs as soon as they became locally dominant, much inappropriate usage has continued undisturbed (with the associated wastage of economic resources and unnecessary side effects (Treatments 2022)): this phenomenon—which is totally unacceptable in modern evidence-based medicine—can have unfortunate consequences during a pandemic.

The clinical efficacy of anti-Spike mAbs is limited to patient populations with early and mild disease, as would be expected for a therapy that works primarily as an antiviral agent. Their high cost means that they are unlikely to become significant treatment options for low- and middle-income countries that cannot afford them. Scaling-up of manufacturing is also a bottleneck for high-income economies, which often had troubles in procurement.

Given the persisting and largely unmet disease burden for immunocompromised patients, the mAb pipeline is still at work. AstraZeneca is testing the AZD5156 cocktail (cilgavimab plus AZD3152) in the SUPERNOVA phase I study and AZD3152 monotherapy in the SUPERNOVA phase III RCT (Francica et al. 2023; Webber et al. 2023), and the AZD7158 cocktail (AZD3152 plus AZD3959) is under evaluation at EMA. Invivyd has advanced clinical trials of VYD222 (a reengineered version of Adagio Therapeutics ADG20/adintrevimab), and Brii Biosciences is developing the amubarvimab plus romlusevimab cocktail (Evering et al. 2023).

Several lessons have been learnt, such as the need for combining different (ideally non-overlapping) mAb specificities to minimize immune escape. Recombinant technology has been deployed in order to increase half-life and minimize off-target toxicity. Respiratory delivery seems the next logical step.

Acknowledgements I am grateful to Prof. Arturo Casadevall and Dr. Scott McConnell for their insightful discussion on the topics discussed in this chapter.

Conflict of Interest Declaration None.

Funding None.

References

ACTIV-3/TICO LY-CoV555 Study Group (2020) A neutralizing monoclonal antibody for hospitalized patients with Covid-19. N Engl J Med 384(10):905–914. https://doi.org/10.1056/NEJMoa2033130

ACTIV-3: Therapeutics for Inpatients With COVID-19—Full Text View—ClinicalTrials.gov

Anderson TS, O'Donoghue A, Mechanic O, Dechen T, Stevens J (2022) Administration of anti–SARS-CoV-2 monoclonal antibodies after US food and drug administration deauthorization. JAMA Netw Open 5(8):e2228997-e. https://doi.org/10.1001/jamanetworkopen.2022.28997

AZD7442 reduced risk of developing severe COVID-19 or death in TACKLE Phase III outpatient treatment trial (2022). Accessed online at https://www.astrazeneca.com/content/astraz/media-centre/press-releases/2021/azd7442-phiii-trial-positive-in-covid-outpatients.html on 22 Feb 2022

Benotmane I, Velay A, Thaunat O, Vargas GG, Olagne J, Fafi-Kremer S et al (2022) Pre-exposure prophylaxis with Evusheld™ elicits limited neutralizing activity against the Omicron variant in kidney transplant patients. Kidney Int 22(S0085–2538):00383. https://doi.org/10.1016/j.kint.2022.05.008

Bournazos S, Corti D, Virgin HW, Ravetch JV (2020) Fc-optimized antibodies elicit CD8 immunity to viral respiratory infection. Nature 588(7838):485–490. https://doi.org/10.1038/s41586-020-2838-z

Cameroni E, Bowen JE, Rosen LE, Saliba C, Zepeda SK, Culap K et al (2021) Broadly neutralizing antibodies overcome SARS-CoV-2 Omicron antigenic shift. Nature. https://doi.org/10.1038/s41586-021-04386-2

Cao Y, Jian F, Wang J, Yu Y, Song W, Yisimayi A et al (2022) Imprinted SARS-CoV-2 humoral immunity induces convergent Omicron RBD evolution. Nature. https://doi.org/10.1038/s41586-022-05644-7

Casadevall A, Focosi D (2023) SARS-CoV-2 variants resistant to monoclonal antibodies in immunocompromised patients is a public health concern. J Clin Invest

Dalakas MC, Spaeth PJ (2021) The importance of FcRn in neuro-immunotherapies: from IgG catabolism, FCGRT gene polymorphisms, IVIg dosing and efficiency to specific FcRn inhibitors. Ther Adv Neurol Disord. https://doi.org/10.1177/1756286421997381

Domachowske JB, Khan AA, Esser MT, Jensen K, Takas T, Villafana T et al (2018) Safety, tolerability and pharmacokinetics of MEDI8897, an extended half-life single-dose respiratory syncytial virus prefusion F-targeting monoclonal antibody administered as a single dose to healthy preterm infants. Pediatr Infect Dis J 37(9):886–892. https://doi.org/10.1097/inf.0000000000001916

Drysdale M, Gibbons DC, Singh M, Rolland C, Lavoie L, Skingsley A et al (2024) Real-world effectiveness of sotrovimab for the treatment of SARS-CoV-2 infection during Omicron BA.2 subvariant predominance: a systematic literature review. Infection 52:1–17

ETF statement on the loss of activity of anti-spike protein monoclonal antibodies due to emerging SARS-CoV-2 variants of concern (2022). 9 December 2022 EMA/931457/2022. Accessed online at https://www.ema.europa.eu/en/documents/public-statement/etf-statement-loss-activity-anti-spike-protein-monoclonal-antibodies-due-emerging-sars-cov-2_en.pdf

ETF warns that monoclonal antibodies may not be effective against emerging strains of SARS-CoV-2 (2023). Accessed online at https://www.ema.europa.eu/en/news/etf-warns-monoclonal-antibodies-may-not-be-effective-against-emerging-strains-sars-cov-2 on 24 Jan 2023

Evering TH, Chew KW, Giganti MJ, Moser C, Pinilla M, Wohl DA et al (2023) Safety and efficacy of combination SARS-CoV-2 neutralizing monoclonal antibodies Amubarvimab plus Romlusevimab in nonhospitalized patients With COVID-19. Ann Intern Med 176(5):658–666. https://doi.org/10.7326/m22-3428

Fact sheet for healthcare providers: emergency use authorization for bebtelovimab (2022). Accessed online at https://www.fda.gov/media/156152/download on 21 Feb 2022

FDA (2022) Coronavirus (COVID-19) update: FDA authorizes New monoclonal antibody for treatment of COVID-19 that retains activity against Omicron variant. Accessed online at https://www.fda.gov/news-events/press-announcements/coronavirus-covid-19-update-fda-authorizes-new-monoclonal-antibody-treatment-covid-19-retains on 17 Feb 2022

FDA Announces Bebtelovimab is Not Currently Authorized in Any US Region (2022). Accessed online at https://www.fda.gov/drugs/drug-safety-and-availability/fda-announces-bebtelovimab-not-currently-authorized-any-us-region on 1 Dec 2022

FDA announces Evusheld is not currently authorized for emergency use in the U.S. (2023) Accessed online at https://www.fda.gov/drugs/drug-safety-and-availability/fda-announces-evusheld-not-currently-authorized-emergency-use-us on 3 Feb 2023

FDA Statement (2022) Coronavirus (COVID-19) update: FDA limits use of certain monoclonal antibodies to treat COVID-19 due to the Omicron variant. https://www.fda.gov/news-events/press-announcements/coronavirus-covid-19-update-fda-limits-use-certain-monoclonal-antibodies-treat-covid-19-due-omicron Accessed 8 June 2023

FDA updates Sotrovimab emergency use authorization. March 30, 2022. https://www.fda.gov/drugs/drug-safety-and-availability/fda-updates-sotrovimab-emergency-use-authorization Accessed 9 June 2023

Focosi D, Maggi F (2021) Neutralising antibody escape of SARS-CoV-2 spike protein: risk assessment for antibody-based Covid-19 therapeutics and vaccines. Rev Med Virol. https://doi.org/10.1002/rmv.2231

Focosi D, Maggi F, McConnell S, Casadevall A (2022a) Spike mutations in SARS-CoV-2 AY sublineages of delta variant of concern: implications for the future of the pandemic. Future Microbiol 17:219–221. https://doi.org/10.2217/fmb-2021-0286

Focosi D, Maggi F, Franchini M, McConnell S, Casadevall A (2022b) Analysis of immune escape variants from antibody-based therapeutics against COVID-19: a systematic review. Int J Mol Sci 23(1):29

Focosi D, McConnell S, Casadevall A, Cappello E, Valdiserra G, Tuccori M (2022c) Monoclonal antibody therapies against SARS-CoV-2. Lancet Infect Dis 22(11):00311–00315. https://doi.org/10.1016/S1473-3099(22)00311-5

Focosi D, Quiroga R, McConnell S, Johnson MC, Casadevall A (2023a) Convergent evolution in SARS-CoV-2 Spike creates a variant soup from which new COVID-19 waves emerge. Int J Mol Sci 24(3):2264

Focosi D (2023) A web tool to estimate baaseline anti-Spike monoclonal antibody efficacy based on regional genomic surveillance. Viruses 15(5):1048

Focosi D, Maggi F (2023) Respiratory delivery of passive immunotherapies for SARS-CoV-2 prophylaxis and therapy. Hum Vaccines Immunotherap 19(2)

Focosi D, Tuccori M (2022) Prescription of anti-spike monoclonal antibodies in COVID-19 patients with resistant SARS-CoV-2 variants in Italy. Pathogens (Basel, Switzerland) 11(8). https://doi.org/10.3390/pathogens11080823

Focosi D, McConnell S, Sullivan DJ, Casadevall A (2023b) Analysis of SARS-CoV-2 mutations associated with resistance to therapeutic monoclonal antibodies that emerge after treatment. Drug Resist Updates 100991. https://doi.org/10.1016/j.drup.2023.100991

Francica J, Rosenthal K, Ren K, Flores DJ et al (2023) The SARS-CoV-2 monoclonal antibody AZD3152 potently neutralises historical and currently circulating variants. ECCMID. Copenhagen

Gottlieb RL, Nirula A, Chen P, Boscia J, Heller B, Morris J et al (2021) Effect of bamlanivimab as monotherapy or in combination with etesevimab on viral load in patients with mild to moderate COVID-19: a randomized clinical trial. JAMA 325(7):632–644. https://doi.org/10.1001/jama.2021.0202%JJAMA

Group RC, Horby PW, Mafham M, Peto L, Campbell M, Pessoa-Amorim G et al (2022) Casirivimab and imdevimab in patients admitted to hospital with COVID-19 (RECOVERY): a randomised, controlled, open-label, platform trial. Lancet 399(10325):665–676. https://doi.org/10.1016/S0140-6736(22)00163-5

Gupta A, Gonzalez-Rojas Y, Juarez E, Crespo Casal M, Moya J, Falci DR et al (2021) Early treatment for Covid-19 with SARS-CoV-2 neutralizing antibody sotrovimab. N Engl J Med 385:1941–1950. https://doi.org/10.1056/NEJMoa2107934

Haga K, Takai-Todaka R, Matsumura Y, Song C, Takano T, Tojo T et al (2021) Nasal delivery of single-domain antibody improves symptoms of SARS-CoV-2 infection in an animal model. PLoS Pathog 17(10):e1009542. https://doi.org/10.1371/journal.ppat.1009542

Halfmann PJ, Minor NR, Haddock LA III, Maddox R, Moreno GK, Braun KM et al (2022) Evolution of a globally unique SARS-CoV-2 Spike E484T monoclonal antibody escape mutation in a persistently infected, immunocompromised individual. Virus Evol. https://doi.org/10.1093/ve/veac104

Hamers-Casterman C, Atarhouch T, Muyldermans S, Robinson G, Hamers C, Songa EB et al (1993) Naturally occurring antibodies devoid of light chains. Nature 363(6428):446–448. https://doi.org/10.1038/363446a0

Han Q, Wang S, Wang Z, Zhang C, Wang X, Feng N et al (2023) Nanobodies with cross-neutralizing activity provide prominent therapeutic efficacy in mild and severe COVID-19 rodent models. Virologica Sinica. https://doi.org/10.1016/j.virs.2023.07.003

Hao L, Hsiang T-Y, Dalmat RR, Ireton R, Morton JF, Stokes C et al (2023) Dynamics of SARS-CoV-2 VOC neutralization and novel mAb reveal protection against Omicron. Viruses 15(2):530

Harman K, Nash SG, Webster HH, Groves N, Hardstaff J, Bridgen J et al (2022) Comparison of the risk of hospitalisation among BA.1 and BA.2 COVID-19 cases treated with Sotrovimab in the community in England. Influenza Other Respi Viruses 17(5):e13150. https://doi.org/10.1111/irv.13150

Hart TK, Cook RM, Zia-Amirhosseini P, Minthorn E, Sellers TS, Maleeff BE et al (2001) Preclinical efficacy and safety of mepolizumab (SB-240563), a humanized monoclonal antibody to IL-5, in cynomolgus monkeys. J Allergy Clin Immunol 108(2):250–257. https://doi.org/10.1067/mai.2001.116576

He Q, Wu L, Xu Z, Wang X, Xie Y, Chai Y et al (2023) An updated atlas of antibody evasion by SARS-CoV-2 Omicron sub-variants including BQ.1.1 and XBB. Cell Rep Med 4(4):100991. https://doi.org/10.1016/j.xcrm.2023.100991

Hurlburt NK, Homad LJ, Sinha I, Jennewein MF, MacCamy AJ, Wan Y-H et al (2021) Structural definition of a pan-sarbecovirus neutralizing epitope on the spike S2 subunit. Commun Biol. https://doi.org/10.1038/s42003-022-03262-7

IDSA Guidelines on the Treatment and Management of Patients with COVID-19 (2022). Accessed online at https://www.idsociety.org/practice-guideline/covid-19-guideline-treatment-and-management/ on 9 Feb 2022

Iketani S, Liu L, Guo Y, Liu L, Huang Y, Wang M et al (2022) Antibody evasion properties of SARS-CoV-2 Omicron sublineages. Nature 604(7906):553–556. https://doi.org/10.1038/s41586-022-04594-4

Imai M, Ito M, Kiso M, Yamayoshi S, Uraki R, Fukushi S et al (2022) Efficacy of antiviral agents against Omicron subvariants BQ.1.1 and XBB. N Engl J Med. https://doi.org/10.1056/NEJMc2214302

Isa F, Forleo-Neto E, Meyer J, Zheng W, Rasmussen S, Armas D et al (2021) Repeat subcutaneous administration of REGEN-COV® in adults is well-tolerated and prevents the occurrence of COVID-19. Int J Infect Dis. https://doi.org/10.1016/j.ijid.2022.06.045

Ison MG, Weinstein DF, Dobryanska M, Holmes A, Phelan A-M, Li Y et al (2023) Prevention of COVID-19 following a single intramuscular administration of Adintrevimab: results from a phase 2/3 randomized, double-blind, placebo-controlled trial (EVADE). Open Forum Infect Dis 10(7). https://doi.org/10.1093/ofid/ofad314

Jayk Bernal A, Gomes da Silva MM, Musungaie DB, Kovalchuk E, Gonzalez A, Delos Reyes V et al (2022) Molnupiravir for oral treatment of Covid-19 in nonhospitalized patients. N Engl J Med 386(6):509–520. https://doi.org/10.1056/NEJMoa2116044

Kaku Y, Okumura K, Padilla-Blanco M, Kosugi Y, Uriu K, Alfredo Amolong Hinay J et al (2023) Virological characteristics of the SARS-CoV-2 JN.1 variant. Lancet Infect Dis. https://doi.org/10.1016/S1473-3099(23)00813-7

Kaplon H, Chenoweth A, Crescioli S, Reichert JM (2022) Antibodies to watch in 2022. mAbs 14(1):2014296. https://doi.org/10.1080/19420862.2021.2014296

Ku Z, Xie X, Lin J, Gao P, Sahili AE, Su H et al (2022) Engineering SARS-CoV-2 cocktail antibodies into a bispecific format improves neutralizing potency and breadth. medRxiv. https://doi.org/10.1101/2022.02.01.478504

Levin MJ, Ustianowski A, De Wit S, Launay O, Avila M, Templeton A et al (2022) Intramuscular AZD7442 (tixagevimab-cilgavimab) for prevention of Covid-19. N Engl J Med. https://doi.org/10.1056/NEJMoa2116620

Li C, Zhan W, Yang Z, Tu C, Zhu Y, Song W et al (2021) Broad neutralization of SARS-CoV-2 variants by an inhalable bispecific single-domain antibody. Cell. https://doi.org/10.1016/j.cell.2022.03.009

Lin Y, Yue S, Yang Y, Yang S, Pan Z, Yang X et al (2023) Nasal spray of neutralizing monoclonal antibody 35B5 confers potential prophylaxis against severe acute respiratory syndrome coronavirus 2 variants of concern: a small-scale clinical trial. Clin Infect Dis 76(3):e336–e341. https://doi.org/10.1093/cid/ciac448

Liu Y, Zeng Q, Deng C, Li M, Li L, Liu D et al (2021) Robust induction of B cell and T cell responses by a third dose of inactivated SARS-CoV-2 vaccine. Cell Discov. https://doi.org/10.1038/s41421-022-00373-7

Luna MS, Manzoni P, Paes B, Baraldi E, Cossey V, Kugelman A et al (2020) Expert consensus on palivizumab use for respiratory syncytial virus in developed countries. Paediatr Respir Rev 33:35–44. https://doi.org/10.1016/j.prrv.2018.12.001

Lundgren JD, Grund B, Barkauskas CE, Holland TL, Gottlieb RL, Sandkovsky U et al (2021) A neutralizing monoclonal antibody for hospitalized patients with Covid-19. N Engl J Med 384 (10):905–914. https://doi.org/10.1056/NEJMoa2033130

Martinez DR, Schäfer A, Gobeil S, Li D, De la Cruz G, Parks R et al (2022) A broadly cross-reactive antibody neutralizes and protects against sarbecovirus challenge in mice. Sci Transl Med 14(629):eabj7125. https://doi.org/10.1126/scitranslmed.abj7125

McSweeney MD, Stewart I, Richardson Z, Kang H, Park Y, Kim C et al (2022) Stable nebulization and muco-trapping properties of Regdanvimab/IN-006 support its development as a potent, dose-saving inhaled therapy for COVID-19. Bioeng Transl Med. http://doi.org/10.1002/btm2.10391

Mishra N, Kumar S, Singh S, Bansal T, Jain N, Saluja S et al (2021) Cross-neutralization of SARS-CoV-2 by HIV-1 specific broadly neutralizing antibodies and polyclonal plasma. PLoS Pathog 17(9):e1009958. https://doi.org/10.1371/journal.ppat.1009958

Muyldermans S, Cambillau C, Wyns L (2001) Recognition of antigens by single-domain antibody fragments: the superfluous luxury of paired domains. Trends Biochem Sci 26(4):230–235. https://doi.org/10.1016/s0968-0004(01)01790-x

Nambulli S, Xiang Y, Tilston-Lunel NL, Rennick LJ, Sang Z, Klimstra WB et al (2021) Inhalable nanobody (PiN-21) prevents and treats SARS-CoV-2 infections in Syrian hamsters at ultra-low doses. Sci Adv 7(22). https://doi.org/10.1126/sciadv.abh0319

Newman R, Hariharan K, Reff M, Anderson DR, Braslawsky G, Santoro D et al (2001) Modification of the Fc region of a primatized IgG antibody to human CD4 retains its ability to modulate CD4 receptors but does not deplete CD4+ T cells in chimpanzees. Clin Immunol 98 (2):164–174. https://doi.org/10.1006/clim.2000.4975

O'Brien MP, Forleo-Neto E, Musser BJ, Isa F, Chan K-C, Sarkar N et al (2021) Subcutaneous REGEN-COV antibody combination to prevent Covid-19. N Engl J Med 385:1184–1195. https://doi.org/10.1056/NEJMoa2109682

Ohashi H, Hishiki T, Akazawa D, Kim KS, Woo J, Shionoya K et al (2022) Different efficacies of neutralizing antibodies and antiviral drugs on SARS-CoV-2 Omicron subvariants, BA.1 and BA.2. Antiviral Res. https://doi.org/10.1016/j.antiviral.2022.105372

Park YJ, De Marco A, Starr TN, Liu Z, Pinto D, Walls AC et al (2022) Antibody-mediated broad sarbecovirus neutralization through ACE2 molecular mimicry. Science 375(6579):449–454. https://doi.org/10.1126/science.abm8143

Patel V, Levick B, Boult S, Gibbons DC, Drysdale M, Lloyd EJ et al (2023) Characteristics and outcomes of COVID-19 patients presumed to be treated with sotrovimab in NHS hospitals in England. medRxiv. https://doi.org/10.1101/2023.02.08.23285654

Peebles RS Jr, Liu MC, Lichtenstein LM, Hamilton RG (1995) IgA, IgG and IgM quantification in bronchoalveolar lavage fluids from allergic rhinitics, allergic asthmatics, and normal subjects by monoclonal antibody-based immunoenzymetric assays. J Immunol Methods 179(1):77–86. https://doi.org/10.1016/0022-1759(94)00275-2

Piepenbrink MS, Park J-G, Deshpande A, Loos A, Ye C, Basu M et al (2022) Potent universal-coronavirus therapeutic activity mediated by direct respiratory administration of a Spike S2 domain-specific human neutralizing monoclonal antibody. bioRxiv. https://doi.org/10.1101/2022.03.05.483133

Pinto D, Sauer MM, Czudnochowski N, Low JS, Tortorici MA, Housley MP et al (2021) Broad betacoronavirus neutralization by a stem helix-specific human antibody. Science 373 (6559):1109–1116. https://doi.org/10.1126/science.abj3321

Pommeret F, Colomba J, Bigenwald C, Laparra A, Bockel S, Bayle A et al (2021) Bamlanivimab+ etesevimab therapy induces SARS-CoV-2 immune escape mutations and secondary clinical deterioration in COVID-19 patients with B-cell malignancies. Ann Oncol. https://doi.org/10.1016/j.annonc.2021.07.015

Qu P, Xu K, Faraone JN, Goodarzi N, Zheng Y-M, Carlin C et al (2023) Immune evasion, infectivity, and fusogenicity of SARS-CoV-2 Omicron BA.2.86 and FLip variants. Cell. https://doi.org/10.1016/j.cell.2023.12.026

Qu P, Faraone JN, Evans JP, Zheng YM, Carlin C, Anghelina M et al (2023) Enhanced evasion of neutralizing antibody response by Omicron XBB.1.5, CH.1.1, and CA.3.1 variants. Cell Rep 42(5):112443. https://doi.org/10.1016/j.celrep.2023.112443

Renn A, Fu Y, Hu X, Hall MD, Simeonov A (2020) Fruitful neutralizing antibody pipeline brings hope to defeat SARS-Cov-2. Trends Pharmacol Sci 41(11):815–829. https://doi.org/10.1016/j.tips.2020.07.004

Reuters (2022) GSK-Vir therapy has neutralising activity against Omicron sub-variant, data shows. Accessed online at https://www.reuters.com/business/healthcare-pharmaceuticals/gsk-vir-therapy-works-against-omicron-sub-variant-data-suggests-2022-02-10/ on 15 Feb 2022

Robbie GJ, Criste R, Dall'acqua WF, Jensen K, Patel NK, Losonsky GA et al (2013) A novel investigational Fc-modified humanized monoclonal antibody, motavizumab-YTE, has an extended half-life in healthy adults. Antimicrob Agents Chemother 57(12):6147–6153. https://doi.org/10.1128/aac.01285-13

Sabin AP, Richmond CS, Kenny PA (2022) Emergence and onward transmission of a SARS-CoV-2 E484K variant among household contacts of a bamlanivimab-treated patient. Diagn Microbiol Infect Dis 103(1):115656. https://doi.org/10.1101/2021.10.02.21264415%JmedRxiv

Sanderson T, Hisner R, Donovan-Banfield I, Peacock T, Ruis C (2023) A molnupiravir-associated mutational signature in global SARS-CoV-2 genomes. Nature. https://doi.org/10.1038/s41586-023-06649-6

Saunders KO (2019) Conceptual approaches to modulating antibody effector functions and circulation half-life. Front Immunol 10. https://doi.org/10.3389/fimmu.2019.01296

Schäfer A, Muecksch F, Lorenzi JCC, Leist SR, Cipolla M, Bournazos S et al (2020) Antibody potency, effector function, and combinations in protection and therapy for SARS-CoV-2 infection in vivoIn vivo efficacy of anti–SARS-CoV-2 antibodies. J Exp Med 218(3). https://doi.org/10.1084/jem.20201993

Schlothauer T, Herter S, Koller CF, Grau-Richards S, Steinhart V, Spick C et al (2016) Novel human IgG1 and IgG4 Fc-engineered antibodies with completely abolished immune effector functions. Protein Eng Des Select PEDS 29(10):457–466. https://doi.org/10.1093/protein/gzw040

Self WH, Sandkovsky U, Reilly CS, Vock DM, Gottlieb RL, Mack M et al (2022) Efficacy and safety of two neutralising monoclonal antibody therapies, sotrovimab and BRII-196 plus BRII-198, for adults hospitalised with COVID-19 (TICO): a randomised controlled trial. Lancet Infect Dis 22(5):622–635. https://doi.org/10.1016/S1473-3099(21)00751-9

Sheward DJ, Yang Y, Westerberg M, Öling S, Muschiol S, Sato K et al (2023) Sensitivity of BA.2.86 to prevailing neutralising antibody responses. Lancet Infect Dis. https://doi.org/10.1101/2023.09.02.556033

Shrestha LB, Tedla N, Bull RA (2021) Broadly-neutralizing antibodies against emerging SARS-CoV-2 variants. Front Immunol 12:752003. https://doi.org/10.3389/fimmu.2021.752003

Some States Still Pushing Ineffective Covid Antibody Treatments (2022). Accessed online at https://www.medscape.com/viewarticle/967020. Accessed on 26 April 2022

Song R, Zeng G, Yu J, Meng X, Chen X, Li J et al (2023) Post-exposure prophylaxis with SA58 (anti-SARS-COV-2 monoclonal antibody) nasal spray for the prevention of symptomatic COVID-19 in healthy adult workers: a randomized, single-blind, placebo-controlled clinical study. Emerg Microbes Infect 2212806. https://doi.org/10.1080/22221751.2023.2212806

Stadler E, Chai KL, Schlub TE, Cromer D, Polizzotto MN, Kent SJ et al (2022a) Determinants of passive antibody efficacy in SARS-CoV-2 infection. Lancet Microbe. https://doi.org/10.1016/S2666-5247(23)00194-5

Stadler E, Burgess MT, Schlub TE, Chai KL, McQuilten ZK, Wood EM et al (2022b) Monoclonal antibody levels and protection from COVID-19. Nat Commun. https://doi.org/10.1038/s41467-023-40204-1

Sullivan DJ, Franchini M, Joyner MJ, Casadevall A, Focosi D (2022) Analysis of anti-Omicron neutralizing antibody titers in different convalescent plasma sources. Nat Comm 13(1):6478. https://doi.org/10.1038/s41467-022-33864-y

Sullivan DJ, Franchini M, Senefeld JW, Joyner MJ, Casadevall A, Focosi D (2023) Plasma after both SARS-CoV-2 boosted vaccination and COVID-19 potently neutralizes BQ1.1 and XBB. J General Virol. https://doi.org/10.1101/2022.11.25.517977

Syed YY (2021) Regdanvimab: first approval. Drugs 81(18):2133–2137. https://doi.org/10.1007/s40265-021-01626-7

Titong A, Gallolu Kankanamalage S, Dong J, Huang B, Spadoni N, Wang B et al (2022) First-in-class trispecific VHH-Fc based antibody with potent prophylactic and therapeutic efficacy against SARS-CoV-2 and variants. Sci Rep 12(1):4163. https://doi.org/10.1038/s41598-022-07952-4

Tortorici MA, Czudnochowski N, Starr TN, Marzi R, Walls AC, Zatta F et al (2021) Broad sarbecovirus neutralization by a human monoclonal antibody. Nature 597(7874):103–108. https://doi.org/10.1038/s41586-021-03817-4

Touret F, Giraud E, Bourret J, Donati F, Tran-Rajau J, Chiaravalli J et al (2022) Enhanced neutralization escape to therapeutic monoclonal antibodies by SARS-CoV-2 Omicron sub-lineages. iScience. https://doi.org/10.1016/j.isci.2023.106413

Update on AZD7442 STORM CHASER trial in post-exposure prevention of symptomatic COVID-19 (2022). Accessed online at https://www.astrazeneca.com/media-centre/press-releases/2021/update-on-azd7442-storm-chaser-trial.html on 21 Feb 2022

Uriu K, Ito J, Kosugi Y, Tanaka YL, Mugita Y, Guo Z et al (2023) Transmissibility, infectivity, and immune evasion of the SARS-CoV-2 BA.2.86 variant. Lancet Infect Dis. https://doi.org/10.1016/S1473-3099(23)00575-3

US Food and Drug Administration (FDA) (2021) Clinical memorandum Re: EUA 26382. Product: COVID-19 Convalescent Plasma. Issued on December 27, 2021 at https://www.fda.gov/media/141477/download

Vangeel L, De Jonghe S, Maes P, Slechten B, Raymenants J, André E et al (2022) Remdesivir, Molnupiravir and Nirmatrelvir remain active against SARS-CoV-2 Omicron and other variants of concern. Antiviral Res 198:105252. https://doi.org/10.1016/j.antiviral.2022.105252

Vellas C, Kamar N, Izopet J (2022) Resistance mutations in SARS-CoV-2 Omicron variant after tixagevimab-cilgavimab treatment. J Infect 85(5):e162–e163. https://doi.org/10.1016/j.jinf.2022.07.014

Wang Q, Iketani S, Li Z, Liu L, Guo Y, Huang Y et al (2023b) Alarming antibody evasion properties of rising SARS-CoV-2 BQ and XBB subvariants. Cell 186(2):279–86.e8. https://doi.org/10.1016/j.cell.2022.12.018

Wang X, Jiang S, Jiang S, Li X, Ai J, Lin K et al (2023a) Neutralization of SARS-CoV-2 BQ.1.1 and XBB.1.5 by breakthrough infection sera from previous and current waves in China. bioRxiv. https://doi.org/10.1101/2023.02.07.527406

Wang Q, Guo Y, Zhang RM, Ho J, Mohri H, Valdez R et al. (2023c) Antibody neutralization of emerging SARS-CoV-2: EG.5.1 and XBC.1.6. bioRxiv. https://doi.org/10.1101/2023.08.21.553968

Webber C, Beavon R, Thomas S, Chang LA, Cohen T, Perez J (2023) Trial in progress: a Phase I/III, randomised, modified double-blind, placebo- and active-controlled pre-exposure prophylaxis study of the SARS-CoV-2–neutralising antibody AZD3152 (SUPERNOVA). ECCMID, Copenhagen

Weinreich DM, Sivapalasingam S, Norton T, Ali S, Gao H, Bhore R et al (2021) REGEN-COV antibody combination and outcomes in outpatients with Covid-19. N Engl J Med 385(23):e81. https://doi.org/10.1056/NEJMoa2108163

Weitzenfeld P, Bournazos S, Ravetch JV (2019) Antibodies targeting sialyl Lewis A mediate tumor clearance through distinct effector pathways. J Clin Investig 129(9):3952–3962. https://doi.org/10.1172/jci128437

Wu H, Pfarr DS, Johnson S, Brewah YA, Woods RM, Patel NK et al (2007) Development of Motavizumab, an ultra-potent antibody for the prevention of respiratory syncytial virus infection in the upper and lower respiratory tract. J Mol Biol 368(3):652–665. https://doi.org/10.1016/j.jmb.2007.02.024

Wu MY, Carr EJ, Harvey R, Mears HV, Kjaer S, Townsley H et al (2022a) WHO's therapeutics and COVID-19 living guideline on mAbs needs to be reassessed. Lancet 400(10369):2193–2196. https://doi.org/10.1016/s0140-6736(22)01938-9

Wu X, Wang Y, Cheng L, Ni F, Zhu L, Ma S et al (2022b) Short-term instantaneous prophylaxis and efficient treatment against SARS-CoV-2 in hACE2 mice conferred by an intranasal nanobody (Nb22). Front Immunol 13:865401. https://doi.org/10.3389/fimmu.2022.865401

Yamasoba D, Uriu K, Plianchaisuk A, Kosugi Y, Pan L, Zahradnik J et al (2023) Virological characteristics of the SARS-CoV-2 Omicron XBB.1.16 variant. Lancet Infect Dis 23(6):655–656. https://doi.org/10.1016/S1473-3099(23)00278-5

Yang S, Yu Y, Jian F, Song W, Yisimayi A, Chen X et al (2023a) Antigenicity and infectivity characterization of SARS-CoV-2 BA.2.86. Lancet Infect Dis. https://doi.org/10.1016/S1473-3099(23)00573-X

Yang S, Yu Y, Xu Y, Jian F, Song W, Yisimayi A et al (2023b) Fast evolution of SARS-CoV-2 BA.2.86 to JN.1 under heavy immune pressure. Lancet Infect Dis. https://doi.org/10.1101/2023.11.13.566860

Zalevsky J, Chamberlain AK, Horton HM, Karki S, Leung IWL, Sproule TJ et al (2010) Enhanced antibody half-life improves in vivo activity. Nat Biotechnol 28(2):157–159. https://doi.org/10.1038/nbt.1601

Zhou S, Hill CS, Sarkar S, Tse LV, Woodburn BMD, Schinazi RF et al (2021) β-d-N_4-hydroxycytidine inhibits SARS-CoV-2 through lethal mutagenesis but is also mutagenic to mammalian cells. J Infect Dis 224(3):415–419. https://doi.org/10.1093/infdis/jiab247

Zhou H, Tada T, Dcosta BM, Landau NR (2022) SARS-CoV-2 Omicron BA.2 variant evades neutralization by therapeutic monoclonal antibodies. Viruses. https://doi.org/10.3390/v14061334

Zhou P, Song G, He W-t, Beutler N, Tse LV, Martinez DR et al (2022b) Broadly neutralizing anti-S2 antibodies protect against all three human betacoronaviruses that cause severe disease. Immunity. https://doi.org/10.1016/j.immuni.2023.02.005

Zhou P, Yuan M, Song G, Beutler N, Shaabani N, Huang D et al (2022c) A human antibody reveals a conserved site on beta-coronavirus spike proteins and confers protection against SARS-CoV-2 infection. Sci Transl Med. https://doi.org/10.1126/scitranslmed.abi9215

Daniele Focosi is a hematologist employed as a resident transfusion physician at the largest blood bank in Italy since 2009. He has been a transplant immunologist and immunogeneticist, quality assurance manager, and production manager. He has received awards from the European Federation of Immunogenetics, the European Society of Organ Transplantation, and the Italian Society of Hematology. He has a Ph.D. degree in Clinical and Fundamental Virology, and a master's degree in clinical Trials. He has authored 281 articles indexed in PubMed, for a global *h*-index of 43, on topics ranging from emerging viral infections to new markers of immune competence. He has been licensed as Associate Professor in Microbiology and Clinical Microbiology.

Hyperimmune Globulins in COVID-19

Yasmin Maor and Oren Zimhony

Contents

Abstract The COVID-19 pandemic, resulting from the emergence of the novel coronavirus SARS-CoV-2, posed unprecedented challenges to global health systems as no proven therapy was available. Initially, COVID-19 convalescent plasma (CCP) from recovered COVID-19 patients showed promise as a therapeutic option. However, the efficacy of this approach was closely correlated with the neutralizing antibody titer in the administered plasma and thus effectiveness was not always guaranteed. In response, hyperimmune immunoglobulins (hIG) derived from CCP obtained by apheresis from recovered or vaccinated individuals emerged as a potential alternative. hIG were purified through stringent chromatographic processing from CCP units and displayed varying results in clinical trials, although it

Y. Maor (✉)
Infectious Disease Unit, E. Wolfson Medical Center, Halochamim 62, 58100 Holon, Israel
e-mail: yasminm@wmc.gov.il

Y. Maor
Faculty of Medical and Health Sciences, Tel Aviv University, Tel Aviv, Israel

O. Zimhony
Infectious Diseases Unit, Kaplan Medical Center, Rehovot, Israel
e-mail: Oren_z@clalit.org.il

Faculty of Medicine, Hebrew University and Hadassah, Jerusalem, Israel

Current Topics in Microbiology and Immunology (2025) 443: 149–164
https://doi.org/10.1007/82_2024_277

Published Online: 15 June 2024

seems likely that they improved outcomes compared to placebo or CCP at day 28, particularly in unvaccinated patients. The variability in the effect of hIG likely stems from factors such as the timing of outcome assessment, the administered dose of hIG, the patients' immunological background, and the matching between the variant infecting patients and the neutralization ability of the immunoglobulin batch, which depended on the timing of the CCP collection. Despite logistical challenges and high production costs, hIG showcase advantages over CCP, offering versatility in administration routes and eliminating the need for blood matching, thus facilitating administration in the community, and allowing for variant-specific preparations. hIG appear to be of particular importance in the treatment of immunocompromised patients and patients with persistent COVID-19, although studies in these populations are lacking. Non-human alternatives, such as equine-derived hIG and recombinant hIG, may provide a solution to the logistical challenges of large-scale hIG preparation. Further study is needed to explore these avenues. Establishing the infrastructure for large-scale hIG production independent of plasma donations emerges as a strategic approach for future pandemics, justifying exploration and promotion by health authorities.

Abbreviations

ACE2	Angiotensin-converting enzyme 2
ARDS	Acute respiratory distress syndrome
CI	Confidence interval
CP	Convalescent plasma
CCP	COVID-19 convalescent plasma
CFR	Code of federal regulations
COVID-19	Corona virus disease-2019
EAP	Early Access Program
GMP	Good Manufacturing Practices
hIG	Hyperimmune immunoglobulins
OR	Odds ratio
RBD	Receptor-binding domain
RR	Relative risk
S	Spike protein

1 Introduction

The emergence of the novel coronavirus, SARS-CoV-2, in late 2019 in China led to a global pandemic, causing widespread illness and unprecedented challenges to healthcare systems worldwide (Huang et al. 2020). In the first months of the pandemic, in the absence of effective and established therapeutic options for treating COVID-19, plasma from patients who had recovered from COVID-19, termed convalescent plasma (CCP), emerged as a promising avenue for the early treatment of COVID-19 (Maor et al. 2020; Sullivan et al. 2022; Joyner et al. 2021; Korley et al. 2021). This approach relied on data from previous studies that demonstrated the potential of convalescent plasma (CP) to improve survival in several newly emergent viral diseases including Machupo virus (Bolivian hemorrhagic fever) (Stinebaugh et al. 1966), Junin virus (Argentinian hemorrhagic fever) (Ruggiero et al. 1986), Lassa fever (Frame et al. 1984), and Ebola virus (Mupapa et al. 1999). In the context of corona viruses, CP showed promising results in the treatment of SARS-CoV-1 infection (SARS) (Mair-Jenkins et al. 2015), and the Middle East respiratory syndrome (MERS) (Marano et al. 2016).

As usage of CCP for COVID-19 treatment expanded, it became clear that the level of neutralizing antibody titer present in CCP units affects treatment efficacy. As the antibody titer in the administered plasma increased, treatment outcomes improved. In a study by Maor et al. (2020), patients with moderate and severe COVID-19 were treated with CCP. At the time of administering the plasma, antibody titers in the plasma against the spike protein S1 were unknown. In patients receiving $\leq$4.0-fold ratio of the test/control antibody levels (Euroimmun AG, Luebeck, Germany, product number EI 2606-9601G) 14 days after receiving treatment, 11/30 improved (36.7%) versus 13/19 (68.4%) in patients receiving >4.0-fold ratio [odds ratio (OR) 0.267, 95% confidence interval (CI) 0.079–0.905, $P = 0.03$]. In patients diagnosed >10 days prior to treatment, 4/14 (22.4%) improved in the group receiving $\leq$4.0-fold ratio of the test/control antibody levels antibody versus 6/7 (85.7%) in recipients of the higher titer plasma [OR 0.048, (95% CI, 0.004–0.520), $P = 0.007$]. In a study by Joyner et al. of 3,082 CP recipients with COVID-19 in the EAP (see Chap. 7) in whom antibodies could be measured in remnant plasma samples, death within 30 days after plasma transfusion in patients who had not received mechanical ventilation was related to antibody level in plasma in a dose–response fashion (Joyner et al. 2021). Death occurred in 115 of 515 patients (22.3%) receiving plasma with high antibody titers, 549 of 2,006 patients (27.4%) in the medium-titer group, and 166 of 561 patients (29.6%) in the low-titer group [adjusted relative risk (RR), 0.66; 95% CI, 0.48 to 0.91].

The recognition of the importance of high levels of neutralizing antibodies to gain optimal therapeutic effect led to the exploration of additional methods to obtain high levels of neutralizing antibodies. A popular approach was to synthesize monoclonal antibodies derived originally from B cells of patients or animals infected with SARS-CoV-2. As the pandemic progressed an array of these antibodies usually produced as biclonal antibodies (i.e., pairs of monoclonals) became commercialized. Examples include casirivimab-imdevimab (Weinreich et al. 2021; RECOVERY Collaborative Group 2022), bamlanivimab-etesevimab (Dougan et al. 2022), sotrovimab (Ambrose et al. 2023), and tixagevimab-cilgavimab (Montgomery et al. 2022). All of these preparations were found to improve patient outcomes when administered early in the course of the disease (Weinreich et al. 2021; RECOVERY Collaborative Group 2022; Dougan et al. 2022; Ambrose et al. 2023; Montgomery et al. 2022). Unfortunately, every commercial antibody became obsolete when new SARS-CoV-2 variants emerged that replaced older variants and evaded these antibodies (Guidelines and on the Treatment and Management of Patients with COVID-19, 2024; Jawad et al. 2023). The option we will emphasize here is the production of high-titer antibody preparations through the production of hyperimmune globulins (hIG).

2 Production of hIG

Studies on the production and use of IgG concentrates to prevent and treat infectious diseases flourished in the 1940s and 50s. Pioneering studies were authored by Pauling and Campbell (1942) who described the production of antibodies with specific activity against type III polysaccharide of *Streptococcus pneumonia* from bovine gamma-globulin. IgG therapy derived from human plasma was initially based on the gamma-globulin fraction of plasma that was separated by fractionation processes without chromatographic methods. These were produced from human or animal serum or plasma by fractionating, using water-soluble organic solvents or neutral salts. These products could be administered only by the intramuscular route as when given intravenously they have complement fixation activity termed "anticomplementary activity" caused by aggregated, polymerized IgG molecules (Frommhagen and Fudenberg 1962). This feature limited the full use of these preparations, since compared to the intravenous route, intramuscular administration delays distribution to the bloodstream and other body compartments, and incomplete uptake from muscle can also occur (Janeway and Rosen 1966).

Studies showed that these concentrates of IgG had greater activity than CP or serum (Janeway 1945), and were used on a small scale to treat patients with various infectious diseases, including epidemic typhus (Plotz et al. 1946) and pertussis (Brainerd 1948). These gamma-globulin preparations mainly contained IgG but also had traces of IgA, IgM, and other proteins and were used successfully as prophylaxis for several viral infectious diseases, most notably Hepatitis A, measles, and poliomyelitis (see Chap. 5). Recruitment of hyperimmune donors enabled the formation of *hyperimmune* gamma-globulins and expanded their utility to the prophylaxis of vaccinia, tetanus, pertussis, and for preventing complications of mumps.

The use of ion exchange chromatography led to the capacity to prepare more purified IgG preparations, free of aggregates, that could be administered intravenously (Baumstark et al. 1964).The first approved hIG IgG-based drug other than small-scale treatments provided by blood banks was Anti-D, which has been in continuous use since 1968 to prevent hemolytic disease of the fetus and newborns (Hoppe et al. 1973).

The modern production of hIG involves collecting plasma from recovered and/or immunized individuals (Ali et al. 2021; Burnouf et al. 2022). Blood banks or specialized plasma collection centers play a crucial role in recruiting eligible donors, screening them for eligibility criteria, and collecting plasma through plasmapheresis (Izak et al. 2022). Plasmapheresis involves drawing blood from the donor, using a specialized machine that separates plasma from other blood components, such as red and white blood cells, although simpler methods are available (see Chap. 15). The plasma, enriched with antibodies, is collected, while the remaining blood components are returned to the donor. This process allows for the repeated collection of plasma from a single donor. Donor eligibility is a critical aspect of the collection process.

For COVID management, donors must meet specific criteria, including a documented history of COVID-19, a period of recovery, and the absence of certain medical conditions or medications that might disqualify them (Izak et al. 2022). Later in the pandemic people vaccinated against SARS-CoV-2 could also donate plasma (Burnouf et al. 2022) since antibodies generated by hybrid immunity following vaccination of adults previously infected with SARS-CoV-2 are more cross-reactive and can neutralize circulating Omicron subvariants (Bellusci et al. 2023).

To ensure the safety of the final immunoglobulin product, viral inactivation or reduction steps are employed to eliminate any potential pathogens, including viruses that might be present in the plasma. Thus, industrial purification processes were developed to purify the IgG fraction from plasma donations. High specific activity against the designated target is primarily achieved by screening donors or plasma donations for high antibody titers. Some of the donation programs may involve active donor immunizations. For others, donors could be convalescent individuals with above-threshold titers. All plasma donations in the US are routinely screened for specific viruses according to federal regulations (CFR) Part 610 (https://www.ecfr.gov/current/title-21/part-610) and 640 (https://www.ecfr.gov/current/title-21/part-640).

Following plasma collection and screening, the pooled plasma undergoes a purification process that typically involves several chromatographic as well as viral inactivation steps, resulting in a high purity concentrated IgG product. The final product—IgG—is devoid of IgM and IgA and is packaged into vials in amounts suitable for injection. Most immunoglobulin preparations can be given subcutaneously, by the intramuscular route or intravenously. Restrictions to the route of administration are usually a result of the manufacturer not having studied all routes of administration rather than an established limitation for a particular route of administration. As most treatments are given early in the course of disease, administration by the intramuscular or subcutaneous routes is an advantage enabling easy administration in the community. The main factor for sufficient neutralizing activity in hIG is appropriate selection of donors, as the process of hIG preparation does not use enrichment of the relevant protective antibodies [anti-receptor-binding domain (RBD) in COVID-19] but is merely a method of purification and concentrating the IgG fraction from plasma pools (Novaretti and Dinardo 2011).

In various studies, pools of plasma for preparation of one vial of hIG were collected from tens to hundreds of individuals, thus creating a large diversity of antibodies. This diversity may enhance the potency of hIG compared to CCP. The mean concentrations of IgG in hIG preparations vary significantly among different reports, in the range of 9.60–50 g/l. The IgG percentage was 80–100%.

The antibody titer in a single hIG vial was 3-to-15-fold higher than the level in one plasma unit. Neutralizing activity was 3–10 times higher compared to plasma. The concentration of specific IgG in the final product did not change significantly for at least 14 months (Huang et al. 2020; Ali et al. 2021; Vandeberg et al. 2021; Lachert et al. 2022; Maor et al. 2023).

3 Clinical Data

Several trials have evaluated the efficacy and safety of hIG in treating COVID-19 patients demonstrating mixed results, as was found in studies assessing the use of CCP (Table 1) (see also Chap. 10). In the ITAC study (ITAC (INSIGHT 013) 2022), an international randomized, double-blind, placebo-controlled trial, 593 hospitalized patients with COVID-19 were included from October 2020 to February 2021. These patients were symptomatic for up to 12 days and did not have acute end-organ failure. They were randomly assigned to receive either hIG (400 mg/kg bodyweight, capped at 40 g), or an equivalent volume of saline as placebo, in addition to standard of care. Compared with placebo, the hIG group had similar odds of a favorable outcome on day 7. Infusions were well tolerated, although infusion reactions were more common in the hIG group (18·6% vs. 9·5%).

In another randomized, controlled, multicenter study, conducted from January 2021 to October 2021, 341 patients were randomly assigned to receive 4 g of hIG versus two units of CCP in addition to standard of care that included remdesivir. There was no difference between the groups at day 14, but at day 28, significantly more patients receiving hIG showed improvement compared to those receiving CCP [136 patients (81.9%) and 108 patients (70.6%), respectively; risk difference = 11.3%, 95% CI 1.9%–20.7%]. More unvaccinated patients improved by day 28 in the hIG group (84.1% vs. 66.1%; $P = 0.024$), and survival was better (89.9% vs. 77.4%; $P = 0.066$). No serious adverse events occurred (Maor et al. 2023).

In small studies addressing hIG in patients with acute respiratory distress syndrome (ARDS) or critical illness, there was a trend toward lower mortality at day 28 in the hIG group compared to patients receiving standard of care (Ali et al. 2021; Huygens et al. 2023). In a small study of severely immunocompromised patients, progression of COVID-19 to severe disease was observed in two of 10 patients following administration of hIG, but seven of eight controls treated with intravenous immunoglobulins without COVID-19 antibodies deteriorated. Mortality was observed only in the control group (three of eight patients) (Huygens et al. 2023). This study although small highlights the current relevance of hIG in an era of Omicron and newer variant dominance, immunologic background of vaccination and/or prior Omicron infection, and effective therapeutics such as remdesivir. In these current circumstances, the major benefit of hIG would be for patients who cannot mount an immunologic response to COVID-19 infection due to inherited or acquired immunological deficiency.

Table 1 Studies assessing hIG in COVID-19 patients

Authors	Study design	Number of patients	hIG for RBD: preparation route and dose	Patient eligibility	Period	Comparator and endpoint	Main outcome
Lopardo et al. (2021) eClinical Medicine (Lopardo et al. 2021)	Double-blind, randomized study	245 118—intervention 123—placebo	Equine polyclonal antibodies; Two IV doses 4 mg/kg 48h apart	Moderate-to-severe COVID-19	8/21 to 10/21	Placebo; Improvement of 2 points in the WHO 8-scale clinical score	Higher rate of clinical improvement particularly in patients with severe disease and no COVID-19 antibodies at baseline
Ali et al. (2021) eClinical Medicine (Ali et al. 2021)	Single center, phase I/II, randomized controlled, single-blinded study	50 40—intervention 10—placebo	hIG; Four IV dosage arms (0.15, 0.20, 0.25, 0.30 g/kg)	Severe or critical COVID-19 with ARDS	6/20 to 2/21	Standard of care; 28-day mortality, patient's clinical status on an ordinal scale	Lower mortality, clinical improvement with hIG
ITAC (INSIGHT 013) Study group. (2022). Lancet (ITAC (INSIGHT 013 2022)	International randomized, double-blind study	593 310—intervention 292—placebo	hIV; IV 400 mg/kg capped at 40 g	Symptomatic COVID-19 for up to 12 days and did not have acute end-organ failure, receiving standard of care including remdesivir	10/20 to 2/21	Saline as placebo; Day 7 seven-category ordinal scale	Similar odds of a favorable outcome at day seven between the two groups
Maor et al. (2023) Clin Infect Dis (Maor et al. 2023)	Randomized controlled, open-label, phase II,	341 166—intervention 153—CCP	hIG (Kamada); 4g IV	Up to 4 days from hospitalization at screening and up to 10 days from COVID-19	1/21 to 10/21	Two CCP units; Improvement of 2 points in	No difference between the groups at day 14 Significantly more patients and particularly

(continued)

Table 1 (continued)

Authors	Study design	Number of patients	hIG for RBD: preparation route and dose	Patient eligibility	Period	Comparator and endpoint	Main outcome
	non-inferiority multicenter study			symptom onset, and oxygen saturation at room air <95% at screening		the WHO 8-scale clinical score on day 14	unvaccinated patients improved by day 28 in the hIGgroup compared to placebo, and survival was significantly better
Alemany et al. (2023). eClinicalMedicine (Alemany et al. 2023)	Multicenter, randomized controlled, double-blind study	461 152—1g hIG 153—2g hIG 156—placebo	hIG 10 mL subcutaneous with 1 g or 2 g	Asymptomatic, unvaccinated adults with confirmed SARS-CoV-2 infection within 5 days of diagnosis	04/21 to 12/21	Saline as placebo; Development of symptoms by day 14	No difference between groups by day 14
Huygens et al. (2023). J. Infect. Dis (Huygens et al. 2023)	Randomized controlled study	18 hIG—10 IVIG—8	150 mL:100 mg/mL	Severely immunocompromised hospitalized, COVID-19 patients	04/21 to 07/21	IVIG; Severe COVID-19, evaluated up to day 28 after treatment	Severe COVID-19 developed in 2 of 10 patients receiving hIG and 7 of 8 receiving IVIG

Another potential use for hIG might be to halt deterioration following infection with COVID-19 in the community, but the efficacy of this strategy still needs to be demonstrated. In a large multicenter, randomized, double-blind, placebo-controlled study, conducted between April and December 2021, involving 461 asymptomatic, unvaccinated adults with confirmed SARS-CoV-2 infection within 5 days, the administration of 1–2 g hIG did not reduce the number of patients who developed symptomatic COVID-19 infection by day 14. Of note, in this study, hIG were administered subcutaneously. No severe adverse events were reported (Alemany et al. 2023).

There are significant differences in these studies regarding the target population, which ranged from asymptomatic individuals to hospitalized patients with mild-to-critical disease. The dose of hIG varied between the studies as did the variants addressed, the route of administration, and the background SARS-CoV-2 immunity of the subjects enrolled. Additionally, the time point of assessing outcomes differed between studies.

Despite these limitations, it appears that hIG can improve the outcome of patients, particularly hospitalized individuals, and that the appropriate assessment point is 28 days rather than earlier time points. The higher and clear impact of hIG in unvaccinated patients strongly suggests that hIG confers an antiviral effect. The superiority of hIG over CCP beyond 14 days may be related to an extended antiviral effect of hIG preparations. Another explanation may be the slower clearance time of the virus in the presence of immunoglobulin G, thus delaying the immunoglobulins' effect (Focosi et al. 2023; Klingler et al. 2020).

Although the main effect of hIG is viral neutralization, additional effects have been suggested such as inflammatory attenuation similar to that seen in the administration of immunoglobulins to modify the inflammatory effect of Streptococcus Group A in toxic shock syndrome (Parks et al. 2018; Sriskandan et al. 2006). Further studies are needed to better understand the effects of hIG on viral clearance in the affected lungs and possibly the inflammatory response that accounts for response to hIG in COVID-19 patients. It seems that hIG are more effective in patients with reduced immunity to COVID-19, whether due to lack of immunization to SARS-CoV-2, or reduced immune function caused by immune-compromising diseases or drugs, but studies addressing this population regarding the effectiveness of hIG are still lacking.

4 Challenges and Considerations

Despite the promising results observed in some clinical trials, challenges persist for the widespread implementation of hIG for COVID-19 treatment. The logistics involved in large-scale plasma collection that needs constant updating due to the ongoing evolution of SARS-CoV-2 variants are challenging. Lots of hIG are manufactured from CCP collected from many donors, usually pools of 100–1000 L of plasma. At the beginning of the pandemic most plasma donors were not exposed to SARS-CoV-2 and therefore could not donate, leading to shortage in available CCP for hIG production. As the pandemic increased most CCP donations were prioritized for emergency use in hospitals. Now as most people have been infected by SARS-CoV-2 and/or vaccinated against COVID-19, blood banks are experiencing lack of plasma donors due to fatigue relating to the prolonged COVID-19 pandemic (Al-Riyami et al. 2022; Monteiro et al. 2023). Preparing large quantities of hIG in accordance with quality assurance, and Good Manufacturing Practices (GMP) and regulation, may take up to 6 months from plasma donation. Meanwhile, new SARS-CoV-2 variants may emerge, potentially reducing the efficacy of even the most recently produced hIG product. In this regard, despite the fact the CCP is less efficacious than hIG, CCP can be more readily adapted to new variants through plasma donations from patients who more recently contracted COVID-19 (Maor et al. 2023). Another challenge with hIG preparations is the high cost of production. Consequently, this treatment is less accessible for low-income countries.

Despite these reservations, hIG carry advantages over CCP. They have a higher IgG titer compared to CCP and can be administered intramuscularly, or subcutaneously with no need for blood typing, making them easily administered in the community. A library of variant-specific hIG can be prepared and stored, as plasma and globulin preparations relying on recent donation are less effective against previous variants (Bellusci et al. 2023; Awasthi et al. 2019). This may particularly serve immune compromised patients with persistent COVID-19 who still carry prior variants that may no longer circulate. Several reports have demonstrated the utility of combination treatment of patients with persistent COVID-19 with antivirals combined with antibody therapy (Brosh-Nissimov et al. 2024; Maruki et al. 2024; Meijer et al. 2024), though the added contribution of antibodies beyond antivirals in those cases, is uncertain.

As the pandemic progressed, the majority of people either recovered from COVID-19, or were immunized against SARS-CoV-2, or both. Currently standard intravenous immunoglobulin preparations collected without information regarding previous COVID-19 history of the donors and their immunization status have reasonable levels of anti-SARS-CoV-2 antibodies and may serve as a therapeutic options for preventing or treating COVID-19 with a threefold higher neutralizing antibody level compared to CCP (Karbiener et al. 2021).

5 Non-human Alternatives to Produce hIG

An alternative to producing hIG, overcoming the need for many plasma donors, is to generate them from animals, as was done extensively in the early twentieth century (see Chap. 5). Equine polyclonal antibodies containing highly purified F (ab')2 fragments have been demonstrated to be safe and well tolerated (Zylberman et al. 2020; Gupta et al. 2022). During the COVID-19 pandemic, a study was conducted in which horses were immunized with the receptor-binding domain of the viral Spike glycoprotein. The equine product showed around 50 times the potency in in vitro sero-neutralization assays than the average of CCP samples. In a double-blind, randomized, placebo-controlled trial performed in 245 patients during August to October 2020, the equine polyclonal antibodies were moderately effective in improving outcomes of hospitalized adult patients with moderate and severe COVID-19 pneumonia at day 28, and displayed a good safety profile with no serious adverse reactions (Gupta et al. 2022; Lopardo et al. 2021). Similar successful studies demonstrating the feasibility of producing hyperimmune anti-SARS-CoV-2 antibodies were also performed in rabbits (Pakdemirli et al. 2021), sheep (Findlay-Wilson et al. 2023), and bovine milk (Duman and Karav 2023).

Another alternative is to produce recombinant hIG using microfluid technology and molecular genomic strategies for capturing diverse mammalian antibody repertoires. These methods generate thousands of diverse mixtures of recombinant antibodies, enriched for specificity and activity against therapeutic targets such as specific SARS-CoV-2 variants. Each hIG product comprises thousands to tens of thousands of antibodies derived from convalescent or vaccinated human donors or immunized mice. Live SARS-CoV-2 neutralization assays suggested that the neutralizing titer of recombinant hIG was between 44- and 1,767-fold higher than corresponding CCP (Keating et al. 2021; Mizrahi et al. 2022; Wayham et al. 2023).

Another approach is immunoadhesin-based immunotherapy that targets the angiotensin-converting enzyme 2 (ACE2) receptor. This strategy employs antibody-like molecules that consist of an engineered binding domain fused to an Fc portion on an antibody, thus mediating Fc-effector functions. A computational approach was taken to design a superior ACE2 binder, with orders of magnitude increases in neutralization capacity compared with human ACE2 adhesins against the RBD. This immunoadhesin was active in neutralization of the Delta and Omicron variants, but has not yet been tested in clinical studies (Cohen-Dvashi et al. 2022).

6 Conclusions

Trials evaluating the efficacy of hIG in treating COVID-19 patients demonstrate mixed results. Likely, some of the differences between study outcome relate to the timing of assessment and treatment, dose of hIG administered, and the degree of matching between the variant infecting patients and the neutralizing capacity of the batch of immunoglobulin relating to the timing of the CCP collection.

Yet, it seems that hIG were effective in improving 28-day outcomes in immunized and non-immunized patients, reduced mortality in unvaccinated patients, and were superior to CCP. In all studies hIG were not associated with severe adverse events. Producing hIG is expensive and complicated, but they may provide the most accessible antibody-based therapeutic options for patients. The need for hIG treatment seems to be particularly important for the prevention and treatment of persistent COVID-19 in severely immunocompromised patients. Producing hIG in animals can be an efficient and cost-effective method to quickly produce hIG, and further study is needed to assess the specific indications for this treatment. When considering future pandemics, having the infrastructure to produce large quantities of hIG independent on plasma donations, prior to the availability of vaccines and other treatment options, may be an effective and quick way to begin controlling the pandemic. Health authorities should explore and promote the availability of this technology.

Acknowledgements None.

Author Contributions Y.M. drafted the chapter and both Y.M. and O.Z edited it.

Funding None.

Conflict of Interest Declaration Yasmin Maor received writing fees from MHS; honoraria for lectures from Pfizer, MSD, Gilead, and Medison; honoraria for advisory board meetings from MSD; and travel grants from Pfizer. Oren Zimhony-None.

References

Alemany A, Millat-Martinez P, Corbacho-Monné M, Suñer C, Galvan-Casas C, Carrera C et al (2023) Subcutaneous anti-COVID-19 hyperimmune immunoglobulin for prevention of disease in asymptomatic individuals with SARS-CoV-2 infection: a double-blind, placebo-controlled, randomised clinical trial. EClinicalMedicine 10(57):101898

Ali S, Uddin SM, Shalim E, Sayeed MA, Anjum F, Saleem F et al (2021) Hyperimmune anti-COVID-19 IVIG (C-IVIG) treatment in severe and critical COVID-19 patients: a phase I/II randomized control trial. EClinicalMedicine 4(36):100926

Ali S, Uddin SM, Ali A, Anjum F, Ali R, Shalim E et al (2021) Production of hyperimmune anti-SARS-CoV-2 intravenous immunoglobulin from pooled COVID-19 convalescent plasma. Immunotherapy 13(5):397–407

Al-Riyami AZ, Burnouf T, Wood EM, Devine DV, Oreh A, Apelseth TO et al (2022) International society of blood transfusion survey of experiences of blood banks and transfusion services during the COVID-19 pandemic. Vox Sang 117(6):822–830

Ambrose N, Amin A, Anderson B, Barrera-Oro J, Bertagnolli M, Campion F et al (2023) Neutralizing monoclonal antibody use and COVID-19 infection outcomes. JAMA Netw Open 6(4):e239694

Awasthi M, Golding H, Khurana S (2023) Severe acute respiratory syndrome coronavirus 2 hyperimmune intravenous human immunoglobulins neutralizes omicron subvariants BA.1, BA.2, BA.2.12.1, BA.3, and BA.4/BA.5 for treatment of coronavirus disease 2019. Clin Infect Dis 76(3):e503–e506

Baumstark JS, Laffin RJ, Bardawil WA (1964) A preparative method for the separation of 7s gamma globulin from human serum. Arch Biochem Biophys 108:514–522

Bellusci L, Golding H, Khurana S (2023) Comparison of SARS-CoV-2 hyperimmune immunoglobulins following infection plus vaccination vs infection. JAMA Netw Open 6(8): e2327307

Bellusci L, Golding H, Khurana S (2023) Therapeutic potential of convalescent plasma and hyperimmune immunoglobulins against SARS-CoV-2 BQ.1, BQ.1.1, and XBB variants. J Clin Invest 133(8)
Brainerd H (1948) The treatment of pertussis and pneumonia complicating pertussis; the role of hyperimmune gamma globulin and sulfadiazine. J Pediatr 32(1):30–34
Brosh-Nissimov T, Ma'aravi N, Leshin-Carmel D, Edel Y, Ben Barouch S, Segman Y et al (2024) Combination treatment of persistent COVID-19 in immunocompromised patients with remdesivir, nirmaltrevir/ritonavir and tixegavimab/cilgavimab. J Microbiol Immunol Infect 57(1):189–194
Burnouf T, Gathof B, Bloch EM, Bazin R, de Angelis V, Patidar GK et al (2022) Production and Quality assurance of human polyclonal hyperimmune immunoglobulins against SARS-CoV-2. Transfus Med Rev 36(3):125–132
Cohen-Dvashi H, Weinstein J, Katz M, Eilon-Ashkenazy M, Mor Y, Shimon A et al (2022) Anti-SARS-CoV-2 immunoadhesin remains effective against Omicron and other emerging variants of concern. iScience 25(10):105193
Dougan M, Azizad M, Mocherla B, Gottlieb RL, Chen P, Hebert C et al (2022) A randomized, placebo-controlled clinical trial of Bamlanivimab and Etesevimab together in high-risk ambulatory patients with COVID-19 and validation of the prognostic value of persistently high viral load. Clin Infect Dis 75(1):e440–e449
Duman H, Karav S (2023) Bovine colostrum and its potential contributions for treatment and prevention of COVID-19. Front Immunol 16(14):1214514
Findlay-Wilson S, Easterbrook L, Smith S, Pope N, Aldridge M, Humphries G et al (2023) Refinement of an ovine-based immunoglobulin therapy against SARS-CoV-2, with comparison of whole IgG versus F(ab')2 fragments. Sci Rep 13(1):13912
Focosi D, Franchini M, Nicastri E, Sullivan DJ, Casadevall A (2023) Convalescent plasma versus hyperimmune immunoglobulins. Clin Infect Dis 77(9):1356–1357
Frame JD, Verbrugge GP, Gill RG, Pinneo L (1984) The use of Lassa fever convalescent plasma in Nigeria. Trans R Soc Trop Med Hyg 78(3):319–324
Frommhagen LH, Fudenberg H (1962) The role of aggregated gamma-globulins in the anticomplementary activity of human and animal sera. J Immunol 89:336–343
Gupta D, Ahmed F, Tandel D, Parthasarathy H, Vedagiri D, Sah V et al (2022) Equine immunoglobulin fragment F(ab')2 displays high neutralizing capability against multiple SARS-CoV-2 variants. Clin Immunol 237:108981
Hoppe HH, Mester T, Hennig W, Krebs HJ (1973) Prevention of Rh-immunization. Modified production of IgG anti-Rh for intravenous application by ion exchange chromatography (IEC). Vox Sang 25(4):308–316
Huang C, Wang Y, Li X, Ren L, Zhao J, Hu Y et al (2020) Clinical features of patients infected with 2019 novel coronavirus in Wuhan, China. Lancet 395(10223):497–506
Huygens S, Hofsink Q, Nijhof IS, Goorhuis A, Kater AP, Te Boekhorst PAW et al (2023) Hyperimmune globulin for severely immunocompromised patients hospitalized with coronavirus disease 2019: a randomized, controlled trial. J Infect Dis 227(2):206–210
IDSA Guidelines on the Treatment and Management of Patients with COVID-19 [Internet]. [cited 2024 Feb 3]. Available from: https://www.idsociety.org/practice-guideline/covid-19-guideline-treatment-and-management/
ITAC (INSIGHT 013) Study Group (2022) Hyperimmune immunoglobulin for hospitalised patients with COVID-19 (ITAC): a double-blind, placebo-controlled, phase 3, randomised trial. Lancet 399(10324):530–540
Izak M, Gendelman V, Bransburg-Zabary S, Stoyanov E, Gat R, Cohen D et al (2022) Qualifying coronavirus disease 2019 convalescent plasma donors in Israel. Vox Sang 117(2):185–192
Janeway CA (1945) Use of concentrated human serum gamma-globulin in the prevention and attenuation of measles. Bull N Y Acad Med 21(4):202–222
Janeway CA, Rosen FS (1966) The gamma globulins. IV. Therapeutic uses of gamma globulin. N Engl J Med 275(15):826–831

Jawad B, Adhikari P, Podgornik R, Ching W-Y (2023) Impact of BA.1, BA.2, and BA.4/BA.5 Omicron mutations on therapeutic monoclonal antibodies. Comput Biol Med 167:107576

Joyner MJ, Carter RE, Senefeld JW, Klassen SA, Mills JR, Johnson PW et al (2021) Convalescent plasma antibody levels and the risk of death from Covid-19. N Engl J Med 384(11):1015–1027

Karbiener M, Farcet MR, Schwaiger J, Powers N, Lenart J, Stewart JM et al (2021) Highly potent SARS-CoV-2 neutralization by intravenous immunoglobulins manufactured from post-COVID-19 and COVID-19-vaccinated plasma donations. J Infect Dis 224(10):1707–1711

Keating SM, Mizrahi RA, Adams MS, Asensio MA, Benzie E, Carter KP et al (2021) Generation of recombinant hyperimmune globulins from diverse B-cell repertoires. Nat Biotechnol 39 (8):989–999

Klingler J, Weiss S, Itri V, Liu X, Oguntuyo KY, Stevens C et al (2022) Role of IgM and IgA antibodies in the neutralization of SARS-CoV-2. medRxiv

Korley FK, Durkalski-Mauldin V, Yeatts SD, Schulman K, Davenport RD, Dumont LJ et al (2021) Early convalescent plasma for high-risk outpatients with Covid-19. N Engl J Med 385 (21):1951–1960

Lachert E, Lasocka J, Bielawski A, Sulkowska E, Guz K, Pyrc K et al (2022) Human intramuscular hyperimmune gamma globulin (hIHGG) anti-SARS-CoV-2-characteristics of intermediates and final product. Viruses 14(6)

Lopardo G, Belloso WH, Nannini E, Colonna M, Sanguineti S, Zylberman V et al (2021) RBD-specific polyclonal F(ab)2 fragments of equine antibodies in patients with moderate to severe COVID-19 disease: a randomized, multicenter, double-blind, placebo-controlled, adaptive phase 2/3 clinical trial. EClinicalMedicine 11(34):100843

Mair-Jenkins J, Saavedra-Campos M, Baillie JK, Cleary P, Khaw F-M, Lim WS et al (2015) The effectiveness of convalescent plasma and hyperimmune immunoglobulin for the treatment of severe acute respiratory infections of viral etiology: a systematic review and exploratory meta-analysis. J Infect Dis 211(1):80–90

Meijer SE, Paran Y, Belkin A, Ben-Ami R, Maor Y, Nesher L, Hussein K, Rahav G, Brosh-Nissimov T (2024) Persistent COVID-19 in immunocompromised patients—Israeli Society of Infectious Diseases consensus statement on diagnosis and management. Clin Microbiol Infect S1198-743X(24)00204-0

Maor Y, Cohen D, Paran N, Israely T, Ezra V, Axelrod O et al (2020) Compassionate use of convalescent plasma for treatment of moderate and severe pneumonia in COVID-19 patients and association with IgG antibody levels in donated plasma. EClinicalMedicine 9(26):100525

Maor Y, Shinar E, Izak M, Rahav G, Brosh-Nissimov T, Kessler A et al (2023) A randomized controlled study assessing convalescent immunoglobulins vs convalescent plasma for hospitalized patients with coronavirus 2019. Clin Infect Dis 77(7):964–971

Marano G, Vaglio S, Pupella S, Facco G, Catalano L, Liumbruno GM et al (2016) Convalescent plasma: new evidence for an old therapeutic tool? Blood Transfus 14(2):152–157

Maruki T, Nomoto H, Iwamoto N, Yamamoto K, Kurokawa M, Iwatsuki-Horimoto K et al (2024) Successful management of persistent COVID-19 using combination antiviral therapy (nirmatrelvir/ritonavir and remdesivir) and intravenous immunoglobulin transfusion in an immunocompromised host who had received CD20 depleting therapy for follicular lymphoma. J Infect Chemother

Mizrahi RA, Lin WY, Gras A, Niedecken AR, Wagner EK, Keating SM et al (2022) GMP manufacturing and IND-enabling studies of a recombinant hyperimmune globulin targeting SARS-CoV-2. Pathogens 11(7)

Monteiro TH, Ferreira Í de J da R, Junior ACFP, Chocair HS, Ferreira JD (2023) Barriers and motivations for blood donation: an integrative review. Hematol Transfus Cell Therapy

Montgomery H, Hobbs FDR, Padilla F, Arbetter D, Templeton A, Seegobin S et al (2022) Efficacy and safety of intramuscular administration of tixagevimab-cilgavimab for early outpatient treatment of COVID-19 (TACKLE): a phase 3, randomised, double-blind, placebo-controlled trial. Lancet Respir Med 10(10):985–996

Mupapa K, Massamba M, Kibadi K, Kuvula K, Bwaka A, Kipasa M et al (1999) Treatment of Ebola hemorrhagic fever with blood transfusions from convalescent patients. International scientific and technical committee. J Infect Dis 179(Suppl 1):S18–S23
Novaretti MCZ, Dinardo CL (2011) Immunoglobulin: production, mechanisms of action and formulations. Rev Bras Hematol Hemoter 33(5):377–382
Pakdemirli A, Çalışkan E, Hacıoğlu S, Danyer E, Kardoğan Ö, Kurt ZE et al (2021) History repeats itself: horse originated hyperimmune sera production against SARS CoV-2. Turk J Med Sci. 51(5):2263–2273
Parks T, Wilson C, Curtis N, Norrby-Teglund A, Sriskandan S (2018) Polyspecific intravenous immunoglobulin in clindamycin-treated patients with streptococcal toxic shock syndrome: a systematic review and meta-analysis. Clin Infect Dis 67(9):1434–1436
Pauling L, Campbell DH (1942) The manufacture of antibodies in vitro. J Exp Med 76(2):211–220
Plotz H, Bennett BL, Tabet F (1946) Effect of concentrated hyperimmune rabbit serum on circulating agent in louse borne typhus. Proc Soc Exp Biol Med 63(1):176–178
RECOVERY Collaborative Group. Casirivimab and imdevimab in patients admitted to hospital with COVID-19 (RECOVERY): a randomised, controlled, open-label, platform trial. Lancet 399(10325):665–676
Ruggiero HA, Pérez Isquierdo F, Milani HA, Barri A, Val A, Maglio F et al (1986) Treatment of Argentine hemorrhagic fever with convalescent's plasma. 4433 cases. Presse Med 15 (45):2239–2242
Sriskandan S, Ferguson M, Elliot V, Faulkner L, Cohen J (2006) Human intravenous immunoglobulin for experimental streptococcal toxic shock: bacterial clearance and modulation of inflammation. J Antimicrob Chemother 58(1):117–124
Stinebaugh BJ, Schloeder FX, Johnson KM, Mackenzie RB, Entwisle G, De Alba E (1966) Bolivian hemorrhagic fever. Am J Med 40(2):217–230
Sullivan DJ, Gebo KA, Shoham S, Bloch EM, Lau B, Shenoy AG et al (2022) Early outpatient treatment for Covid-19 with convalescent plasma. N Engl J Med 386(18):1700–1711
Vandeberg P, Cruz M, Diez JM, Merritt WK, Santos B, Trukawinski S et al (2021) Production of anti-SARS-CoV-2 hyperimmune globulin from convalescent plasma. Transfusion 61(6):1705–1709
Wayham NP, Niedecken AR, Simons JF, Chiang YY, Medina-Cucurella AV, Mizrahi RA et al (2023) A potent recombinant polyclonal antibody therapeutic for protection against new severe acute respiratory syndrome coronavirus 2 variants of concern. J Infect Dis 228(5):555–563
Weinreich DM, Sivapalasingam S, Norton T, Ali S, Gao H, Bhore R et al (2021) REGEN-COV antibody combination and outcomes in outpatients with Covid-19. N Engl J Med 385(23):e81
Zylberman V, Sanguineti S, Pontoriero AV, Higa SV, Cerutti ML, Morrone Seijo SM et al (2020) Development of a hyperimmune equine serum therapy for COVID-19 in Argentina. Medicina (B Aires) 80(Suppl 3):1–6

Yasmin Maor received her M.D. degree from Tel Aviv University and completed her residency in Internal Medicine and Infectious Diseases at the Sheba Medical Center, Ramat-Gan, Israel. For the past 10 years she had served as Head of the Infectious Disease Unit at The E. Wolfson Medical Center, Holon, Israel. Her research focuses on nosocomial infections, *Staphylococcus aureus* and virulence mechanisms of *Group B Streptococcus*. She has published over 70 scientific articles. During the COVID-19 pandemic she was one of the leaders of COVID-19 research in Israel focusing on antibody treatment and other novel therapies and diagnostics for COVID-19. She led the Israeli National Study for convalescent plasma in COVID-19 patients and the study comparing convalescent plasma to convalescent hyper-immunoglobulins. She also serves as an Advisor to the Israeli Ministry of Health regarding vaccines, infectious disease, and COVID-19.

Oren Zimhony received his M.D. degree from Ben Gurion University of the Negev in 1987. He completed his internship and Internal Medicine residency in Kaplan Medical Center, the faculty of Medicine, Hebrew University. From 1995 to 1997, he was a Fellow in Infectious diseases and from 1997 to 2000 a Research Fellow in the laboratory of Professor William Jacobs, both at the Albert Einstein College of Medicine (AECOM). His paper on the anti-tuberculous mechanism of action of pyrazinamide and its effects on fatty acid synthetase was published in Nature Medicine. This study is complemented by biochemical and structural studies of key enzymes for fatty acid synthesis at the Department of Chemical and Structural Biology at the Weizmann Institute of Science. His clinical research focuses on resistant Enterobacterales. During the COVID-19 pandemic, he was a leading researcher focusing on antibody therapy, and serves as an Advisor to the Israeli Ministry of Health on epidemic management focusing on therapeutics.

The Logistics of Using Convalescent Plasma

The Role of the Patient Advocate During a Pandemic: The Case of Convalescent Plasma

Chaim Lebovits

Abstract The onset of the COVID-19 pandemic confronted medicine with several difficulties, including a lack of specific therapeutic options, the absence of out-of-hospital testing facilities to diagnose the condition, and the sudden extraordinary need for intensive medical care that overwhelmed most hospitals. Early in the pandemic, many physicians recognized that using antibodies harvested from recovering patients was a treatment that had a proven track record for many diseases and that might be used to manage the disease at least as a stopgap until more specific medicines for COVID-19 were developed. But using convalescent plasma raised many additional complications, most especially the logistics that needed to be put in place to collect and distribute such plasma. Unlike drugs ordered from a pharmacy, plasma and other blood products are procured by a complex process that depends intensely on interaction with the public, the provider of all blood products that are directly provided to patients. Blood components such as convalescent plasma, intended to be used immediately without major processing, are entirely supplied by donations from the public. This form of treatment can therefore benefit from patient advocates, especially if they are experienced in solving problems of logistics and in the process of matching supply to demand that is more commonly encountered in the business world than in medicine. In this chapter, one patient advocate, Chaim Lebovits, describes the process of mobilizing the population, interacting with blood banks and hospitals, and successfully channeling thousands of units of plasma from volunteers recovering from COVID-19 to patients in hospitals. Starting in New York City in early 2020 and initially working with communities with which Mr. Lebovits was familiar, the efforts steadily spread across many parts of the US. The model described here, which uses patient advocates to serve as a link between patients, blood banks, and hospitals in the service of gathering and distributing high-titer convalescent plasma to patients is likely to be relevant to the next pandemic.

C. Lebovits (✉)
20 Wiener Drive Unit 301, Monsey, NY 10952, USA
e-mail: chaiml@cslshoes.com

Current Topics in Microbiology and Immunology (2025) 443: 167–172
https://doi.org/10.1007/82_2024_271

Published Online: 15 June 2024

Medicine is a complex art that is based on established scientific principles that change and evolve as the underlying science progresses. Physicians need to use all tools at their disposal to prevent, cure, and manage the disease, always recognizing that each patient differs in the level of complexity and that the same condition treated with the same treatment will not necessarily have the same outcome in every patient. The healthcare system combines the use of data, diagnostic tools, and physical resources to enable practitioners to practice their art and serve as many patients as humanly possible.

But when a new disease emerges, the system is challenged by lack of experience, of relevant data and diagnostic tools and, if the disease is of pandemic proportions such as COVID-19, may also experience insufficient physical resources such as hospital beds and staffing. Moreover, the logistical systems required to get treatments to patients in a timely fashion may not be easily available.

This is where an outsider, a non-medical professional, a person operating outside of the constraints of the medical care system, who is used to getting things done even without proper systems and data sets in place, and who has credentials and trust in the community, can really help. Since the mid-1990s, I have been active as a patient advocate to help patients in my community with complex medical conditions get access to higher and more coordinated levels of care.

Although there was no firm data on what COVID-19 convalescent plasma (CCP) might do for COVID-19 patients, many physicians knew, from the vast experience with convalescent serum accumulated between 1890 and 1940 (see Chaps. 1 and 6), that the recovering patient was a resource for other patients with active disease because their recovery was virtually always accompanied by the production of antibodies that could neutralize or otherwise interfere with viral or bacterial reproduction, and by harvesting these antibodies, doctors could help patients. But having said this, how does one proceed? Several questions emerge, such as how do we know who qualifies as a donor, i.e. who has antibodies to COVID-19? And once we know that, we are likely to face challenges both in getting blood banks to collect CCP and getting physicians to transfuse it.

This particular set of problems, which involve the interactions of many systems—the licensing authorities such as the FDA, blood banks, hospitals, medical organizations, and, most importantly, the community of convalescent plasma donors—can take advantage of the patient advocate who plays a key support role for the healthcare system and healthcare professionals. Patient advocates may not know medicine, and certainly don't practice medicine, but in spite of these limitations they can play a key role in linking the patient to the physician or team that has the most expertise in their particular condition, making sure the patients' expectations are managed, helping to ensure that patients understand the medical processes and translating medical terminology into plain English. Patient advocates can also encourage and help patients to provide the healthcare system with all information pertinent to their condition, such as records and imaging.

My involvement with CCP began when I was contacted on March 19, 2020, by a friend who is also an eminent infectious disease specialist, Dr. Shmuel Shoham of Johns Hopkins University Medical School. He wondered whether I might be able to help another distinguished infectious disease specialist, Dr. Liise-anne Pirofski of the Albert Einstein College of Medicine, obtain convalescent plasma to treat her many patients ill with COVID-19 at Montefiore Hospital in the Bronx, where she practiced. I have had the privilege of working with Dr. Shoham on many occasions to help guide some of the physicians at our local community hospitals in how best to handle complex patients with infectious disease issues.

My answer was to say I was sure we could do it, perhaps not fully realizing the challenges that lay ahead.

I came to learn that getting convalescent plasma to patients was a process full of challenges. To list just a few of them:

1. The need to identify and recruit potential donors.
2. The complexity of addressing the many requirements of Institutional Review Boards (IRBs) for consent.
3. Adhering to the changing positions of the Food and Drug Administration (FDA) in approving the use of CCP.
4. Working with blood banks to make sure that collection capacity is sufficient.
5. Aligning blood bank protocols with the requirements and conditions set by local hospitals.
6. Communicating with hospital networks to discuss potential treatment options and weigh challenges and determine what treatments are effective.
7. Networking with hospital networks and blood banks to make sure that physicians who want to use CCP have access to it.

A major challenge was the FDA requirement to have proof that the CCP donor actually had COVID-19. In March 2020, there were very few COVID-19 tests available so we knew that this would be a huge challenge for us. In the particular circumstance of Montefiore Hospital in New York City, we encountered an additional constraint. The hospital's IRB required (before the Expanded Access Program was approved by the FDA in August 2020) that the consenting and testing of donors all be done at Montefiore Hospital, which was quite a challenge when people were being advised to isolate and not travel! I explained to the hospital that if they wanted donors, they would have to allow testing nearer to the donor's location, since traveling for an hour each way, often multiple times, for an individual to find out whether they were qualified to donate represented an insurmountable obstacle to obtaining convalescent plasma.

The blood banks also needed to be on board with the testing algorithm. Eventually, in conversations with the FDA, we were able to make the point that the presence of antibodies to COVID-19 was proof of prior infection and that requiring PCR-proven presence of the virus could be waived. Montefiore graciously sent mobile testing units to locations that we chose, so that consenting and testing could be done locally in communities that had experienced the earliest outbreaks.

Further efforts included conversations that included Dr. Shoham, Dr. Michael Joyner of the Mayo Clinic, and me with Dr. David Reich, President of Mount Sinai Hospital of New York City, to expand our donor pool. Mount Sinai and Montefiore were together able to test 400 potential donors daily by April 2020, just the second month of the pandemic. All our recruited donors had experienced symptomatic disease and contributed an average of three plasma units per donation.

During my several years of advocating for patients, I had the privilege of working with many physicians in leadership positions at Mount Sinai Hospital, including Dr. Reich, as well as with Dr. Nicole Bouvier, who was one of the first physicians to treat COVID-19 patients with CCP, and Dr. Jeffery Bander, Chief of Cardiology at Mount Sinai West and Medical Director for Network Development. My experience with these physicians taught me that they knew how to get things done efficiently.

But a major challenge was the process of antibody testing at a time when there was no approved assay, and many competing blood tests, including one developed by Mount Sinai.

The next bottleneck we encountered was that there was only one blood bank in the New York City area that collected plasma using plasmapheresis equipment, which allows donors to donate more frequently than whole blood donors because the process returns red blood cells to the patient. Thus the typical plasmapheresis donation provides three units of plasma rather than the single unit emerging from a whole blood donation. But that single blood bank (The New York Blood Center) had only limited slots available, and their collection locations were far from where our largest number of potential donors were located. To make matters worse, staff availability was limited. Our solution was to work with the philanthropic community to pledge donations to pay for overtime, so that the blood bank employees collecting plasma could provide the extra hours needed to expand capacity.

We also discovered that different hospital networks used different blood banks as their primary blood suppliers. This required us to contact many additional blood banks within a two-hour drive of the Greater New York area to increase collection capacity and to make sure that any hospital trying to obtain CCP should be able to get it on the same day as the request was made. We reached out to American Red Cross, who, by mid-April 2020, had collected only a handful of units. Unable to fill the orders they were receiving, they agreed to work with us to expand their reach.

The American Red Cross and Vitalant were thus persuaded to add further capacity to their blood banks in New Jersey and Pennsylvania, and we also partnered with Miller Keystone in Pennsylvania and Delmarva in Delaware to add even more capacity. While we worked with these additional regional blood banks, the New York City blood banks also worked to increase their capacity. Since we had more donors than collection capacity, we prioritized donors based on titer levels, so that the units transfused should have the highest titers and the best chance of being therapeutic.

Our initial experience with the orthodox Jewish community in the New York area showed that antibody titer levels were highest in people whose work exposed them to people with COVID-19. Thus, we began to prioritize outreach to hospital employees—from physicians to EMS Personnel to janitorial workers—as they had the most exposure to COVID-19. An added advantage was that such individuals, who work in caring professions and are unlikely to be deterred by needle sticks, are more likely to donate plasma. For similar reasons, we later made special efforts to reach out to firefighters and other first responders.

We also had to decide whether to try to set up direct donation (to a specific patient, so that the patient's family could help with the recruiting of potential donors), a process generally resisted by hospitals and blood banks, or donations to a general pool. Notwithstanding that many donors would prefer to direct their donations to specific individuals, we encouraged general donations to ensure that a sufficient number of units would be available to treat every COVID-19 patient whose physician felt that CCP could be beneficial to their recovery.

We learned that the severity of symptoms was not always correlated with antibody titer level, but that, as noted above, individuals with frequent viral exposure, such as emergency medical technicians, often had high antibody levels, and this was true even if they had been asymptomatic. Our encouragement of antibody testing in such healthcare workers had the added benefit of helping ambulance systems to know which of their responders were likely to have immunity to COVID-19 and could be prioritized to respond to calls for assistance that were likely COVID-19-related.

Having increased our collection capacity, we now turned to the need to increase our recruiting efforts. But that effort met another roadblock, the simple lack of capacity to test potential donors for antibody titers. We reached out to the leadership at Mayo Clinic Laboratories, directed by Dr. William G. Morice, because of the Mayo Clinic leadership of the Expanded Access Program (see Chapter 8) to see if they could help us increase our titer testing capacity, and within hours they agreed to test as many as 10,000 specimens a day.

We set up almost daily phone calls between physicians at some of the highest ranking hospitals in the country with their counterparts in community hospitals, so that we could solve challenges and share experiences of different therapeutics available at that point. These connections reflected both the relationships I had established in my patient advocacy work and the efforts of many physicians at large academic centers, such as Dr. Shoham at Johns Hopkins and Dr. Joyner at the Mayo Clinic to reach out to many local community hospitals.

We strove to offer CCP to patients as soon as they tested positive, and because the FDA until December 2021 insisted that all CCP treatment be on an inpatient basis, we began to see if we could provide CCP to admitted patients as soon as the decision had been made to admit, and were still in the Emergency Room, following the well-established principle that antibody treatment works best the earlier in the progression of the disease it is given.

The experience that we had gained in Greater New York area led us to work with American's Blood Centers (BCA) and blood banks around the country to expand our initiatives across the country. We began by setting up convalescent plasma donation drives in different states wherever there were large Orthodox Jewish communities. We made linkages with local religious communities in Atlanta, Detroit, Baltimore, and Miami. Eventually, Bill Block, President of the Blood Centers of America, asked me to help his organization members collect CCP.

Another strategy we used was to place HIPAA forms at the bigger COVID-19 testing sites nationally, allowing laboratories to get consent from patients to share their contact information with us, so we could follow up with every COVID-19 positive patient 45 days after testing positive (roughly 30 days after recovering) to see if they were interested in donating CCP.

Of the many things we learned, and that is absolutely essential in preparing for a future pandemic, is the importance of being able to test antibodies and to perform blood typing in a mobile setting, allowing an expansion of the locations in which convalescent plasma can be transfused.

If I were asked what I would do to prepare for a future pandemic, I would emphasize the need to

- Identify potential treatments and preventive measures.
- Identify the most efficient way to administer these interventions.
- Identify the most efficient way to educate both potential patients and physicians about treatments and what is known of their efficacy and risk.
- Assemble informal groups from different communities so that they can discuss freely what their experiences are with proposed programs and the challenges they have experienced in their communities.
- Be sure to link people working on logistics with those working on regulatory matters. These two groups are likely to have opposite viewpoints on almost every relevant topic, and yet it is essential that operational aspects and regulatory compliance operate hand in glove.

To summarize: The COVID-19 experience with convalescent plasma revealed the value of incorporating logistics and community organization experience into the overall public health and clinical strategy for disease control. These skill sets are sometimes found outside of the medical community, but can be brought together with the expertise of healthcare providers to efficiently control disease in the confusion and difficult circumstances of a new pandemic.

Declarations No conflicts to declare.

Chaim Lebovits is the CEO of The CSL Group Inc., a shoe import and wholesale company. He makes use of the logistical skills developed in his manufacturing career and his connections in the medical community to help patients with complex medical conditions get access to medical care, as well as to help medical professionals and medical systems streamline the needs of certain patients in the community.

Blood Banking Capacity in Low-and Middle-Income Countries: Covid-19 Convalescent Plasma in Context

Evan M. Bloch

Contents

Abstract Blood transfusion capacity in low- and middle-income countries (LMICs), encompassing both the safety and adequacy of the blood supply, is limited. The challenges facing blood banks in LMICs include regulatory oversight, blood donor selection, collection procedures, laboratory testing, and

E. M. Bloch (✉)
Department of Pathology, Johns Hopkins University School of Medicine, Baltimore, MD, USA
e-mail: Ebloch2@jhmi.edu

Transfusion Medicine, Johns Hopkins University School of Medicine, Baltimore, MD, USA

Department of Pathology, Johns Hopkins Bloomberg School of Public Health (Joint Appt. International Health), 600 N. Wolfe Street/Carnegie 446 D1, Baltimore, MD 21287, USA

Current Topics in Microbiology and Immunology (2025) 443: 173–184
https://doi.org/10.1007/82_2024_266

Published Online: 22 May 2024

post-transfusion surveillance. A high proportion of LMICs are unable to fully meet clinical demands for blood products, and many do not meet even the minimum threshold of collection (10 units per 1000 population). Suboptimal clinical transfusion practices, in large part due to a lack of training in transfusion medicine, contribute to blood wastage. During the COVID-19 pandemic, high- and LMICs alike experienced blood shortages, in large part due to quarantine and containment measures that impeded donor mobility. COVID-19 convalescent plasma (CCP) was particularly appealing for the treatment of patients with COVID-19 in LMICs, as it is a relatively inexpensive intervention and makes use of the existing blood collection infrastructure. Nonetheless, the challenges of using CCP in LMICs need to be contextualized among broad concerns surrounding blood safety and availability. Specifically, reliance on first time, family replacement and paid donors, coupled with deficient infectious disease testing and quality oversight, increase the risk of transfusion transmitted infections from CCP in LMICs. Furthermore, many LMICs are unable to meet general transfusion needs; therefore, CCP collection also risked exacerbation of pervasive blood shortages.

Abbreviations

CCP	COVID-19 convalescent plasma
COVID-19	Coronavirus Disease 2019
HBV	Hepatitis B virus
HCV	Hepatitis C virus
HICs	High-income countries
HIV	Human immunodeficiency virus
LMICs	Low- and middle-income countries
NAT	Nucleic acid testing
PCR	Polymerase chain reaction
PEPFAR	President's Emergency Plan for AIDS Relief
POC	Point of care
PR	Pathogen reduction
SARS-CoV-2	Severe acute respiratory syndrome coronavirus 2
TMA	Transcription mediated amplification
TTIs	Transfusion transmitted infections
VNRBDs	Voluntary non-remunerated blood donors
WHO	World Health Organization
US	United States

1 Introduction

Blood transfusion is integral to the clinical management of diverse patient populations. In 2013, the World Health Organization (WHO) included blood and blood components in its list of essential medicines (WHO 2013). Despite wide recognition of the importance of blood safety and availability, disparities in blood banking capacity are found between high- and low-middle-income countries (LMICs). COVID-19 convalescent plasma (CCP) was particularly appealing for LMICs given the lack of alternative treatments for COVID-19, its relatively low cost compared to alternatives, and anticipated barriers to the procurement, distribution, and uptake of SARS-CoV-2 vaccines. Even after SARS-CoV-2 vaccines became available, LMICs often experienced low vaccine coverage (Wells and Galvani 2022). Similarly, other treatments for COVID-19 such as monoclonal antibodies and antiviral agents, remained cost-prohibitive—and consequently unavailable—in LMICs. By contrast, CCP could—at least in theory—be collected locally using a blood collection infrastructure that already existed prior to the pandemic. Nonetheless, the use of CCP in LMICs needs to be understood in the context of blood transfusion safety and availability in LMICs.

Blood transfusion in high-income countries (HICs) is remarkably safe. Spurred by the tragedy of transfusion-transmitted human immunodeficiency virus (HIV), attainment of the current high level of blood transfusion safety follows decades of overhaul of all aspects of the blood transfusion continuum spanning regulatory oversight, donor selection, testing, clinical practice, and post-transfusion surveillance. Further, a robust infrastructure for blood collection and component diversification has largely assured sufficient blood supply in HICs to meet clinical demands (Roberts et al. 2019). This is in stark contrast to LMICs where systemic challenges have long strained the ability to meet the clinical need for transfusion (Roberts et al. 2019; Weimer et al. 2019). An estimated 119 countries, representing almost two-thirds of the world, lack sufficient blood supply to meet clinical needs (Roberts et al. 2019). This deficiency bears directly on the role of all forms of convalescent plasma use in LMICs given that its use further detracts from an already inadequate supply, which is coupled with a higher infectious risk of blood products than in HICs. The scale of CCP collection in LMICs during the pandemic was far lower than in HICs. An understanding of the transfusion practices in LMICs is important if we would like to advance the use of convalescent plasma.

2 Situational Analysis, National Policy and Governance Pertaining to Blood Transfusion Services

In 2021 WHO report on global blood access and availability, only three-quarters of respondent countries had national blood policies, 59% had a multiyear national strategic plan for blood safety and two-thirds (66%) had enacted legislation around blood banking (WHO 2022). Despite progress in blood transfusion safety over the past two decades, these WHO estimates suggest that many countries still lack formal oversight of blood transfusion services.

3 Blood Donor Recruitment

Recruitment of a sufficient number of safe blood donors remains a major challenge in LMICs; this affected the procurement of CCP, directly. Stringent COVID-19 lockdown and quarantine policies in many LMICs during the early phase of the pandemic exacerbated pre-existing donor-donation shortages, thus limiting the ability to prioritize CCP (Barnes et al. 2022).

Given pervasive shortages of blood, both hospitals and blood collection services in LMICs frequently rely on paid replacement donors (i.e., typically comprising friends or family of the intended recipient) to replenish the blood supply. In Africa, family replacement donation accounts for approximately a third of all donations (WHO 2022). Both paid and replacement donors represent higher infectious risks to recipients than voluntary non-remunerated blood donors (VNRBDs) (Grady and Chalmers 1964; Busch et al. 2019). The transition to VNRBDs has been central to blood safety in HICs. This follows research in the 1960s that demonstrated a four-fold increase in post-transfusion hepatitis in patients who received blood from paid donors as compared to those who received blood from VNRBDs (Grady and Chalmers 1964). Individuals who are compensated for blood transfusion are disincentivized from admitting any high-risk behavior or medical history that might otherwise result in their deferral or removal from blood donation. Similarly, replacement donors, are deemed to be higher risk than VNRBDs (Bloch et al. 2020a). Given the acuity of need for a friend or family member to donate, individuals are unlikely to self-defer even if they are known to harbor risk for infection (or in some cases already have knowledge of infection).

Another challenge in LMICs is donor retention with a disproportionate collection from first time donors, who are at higher risk for harboring transfusion-transmissible infections (TTIs). By contrast, repeat blood donors (RBDs) are at lower risk due to being repeatedly subjected to questioning for high-risk behaviors, limited physical examination, and laboratory screening for TTIs. Given the pervasive blood deficit in LMICs, donor retention is arguably more pressing than the transition to exclusive VNRBDs (Allain 2010, 2011; Allain and Sibinga 2016). Collectively, reliance on first time, paid, and replacement donors confers a higher risk of TTIs, which needs to be taken into account when considering the decision to transfuse CCP or other blood products.

4 Blood Collection

Blood in LMICs is mostly collected as whole blood after which it may or may not be used to prepare blood components such as platelets or plasma. During the pandemic, apheresis technology was the major mode of collection of CCP in HICs. Apheresis technology enables the highly efficient collection of a desired component (in this case plasma) with the return of the unwanted cellular fractions (e.g., red blood cells) back to the donor. In so doing, a single apheresis collection can yield as much as 600–800 mL of plasma (2–3 units of CCP) (Bloch et al. 2020b). By contrast, whole blood collection yields less than a single unit of plasma, which may then be pooled with other whole blood donations to produce an equivalent unit of CCP. More significantly, whole blood collection confers a minimum deferral of 8 weeks following donation, while an apheresis plasma donation can be undertaken as frequently as twice a week. Apheresis is widely available in HICs, where it is increasingly used to collect blood components (notably platelets and plasma). By contrast, given the high cost and technical expertise needed for apheresis, its use is comparatively limited in LMICs, whereby some countries lack any capacity in this regard (Patidar et al. 2022). The recent development of less costly, gravity-driven, membrane-based methods for separating plasma from red cells, which may have greater application in LMICs, is described in Chap. 15 (P6).

5 Blood Donor Testing

5.1 Transfusion Transmissible Infections

Transfusion-associated infectious risk is vastly different between high and LMICs. As way of example, the estimated risk of the major transfusion transmitted viruses in the US and other HICs is well below 1 in a million transfusions, thus approximating the risks of a lightning strike or an airplane fatality (Busch et al. 2019; Steele et al. 2021; Carson et al. 2012). The well-documented higher risks of TTIs in LMICs (Ahmad et al. 2021; Morar et al. 2016; Lefrere et al. 2011; Jayaraman et al. 2010) are due to several factors. Foremost, is the higher incidence and prevalence of the major TTIs in LMICs, Weimer et al. (2019) which include both the major transfusion viruses (i.e., HIV, HCV, and HBV) as well as other pathogens that are not commonly encountered in HICs but are widely endemic in LMICs such as *Plasmodium* (malaria) and *T. cruzi* (Chagas disease). Second, as noted above, the donor pool in LMICs comprises a high proportion of first time, replacement, and paid donors. Third, testing is often suboptimal. Finally, quality oversight and post-transfusion surveillance are mostly absent in LMICs; this accounts for the paucity of data pertaining to transmission, thus limiting the ability to improve on extant processes (Busch et al. 2019).

5.2 *Infectious Testing*

Laboratory-based blood donor testing is suboptimal in LMICs. This reflects the broader challenges related to pathology and laboratory services in LMICs (Wilson et al. 2018). The deficiencies run the gamut from the assays in use for donor screening, incomplete testing algorithms (i.e., low uptake of repeat and/or confirmatory testing in the event of an initial reactive test result), the use of testing approaches (e.g., point of care [POC]/rapid tests) that have not been validated for blood donor screening and lack of quality oversight (e.g., proficiency testing). Further, with rare exception, donor testing in LMICs is serology based, despite being undertaken in countries with the highest incidences of the major TTIs, thus conferring risk of transmission in the pre-seroconversion window period. Nucleic acid testing (NAT) using PCR or transcription mediated amplification is well established in HICs, where they have been in use for over 20 years. NAT enables early detection of HIV, HBV, and HCV (i.e., prior to development of antibodies). Nonetheless, high cost and technical complexity are major barriers to its wider implementation (Bloch et al. 2020c). Given the resource constraints in LMICs, selected low-cost approaches have been attempted for donor screening. POC ("rapid") testing is one such approach, despite low sensitivity and specificity (Laperche 2013; Bloch et al. 2014; Prugger et al. 2016). Since CCP is a plasma derivative, suboptimal infectious testing increases the risk conveyed to recipients of CCP in LMICs. Likewise, the lack of capacity to test for SARS-CoV-2 antibodies means that high-titer CCP, which is critical to effectiveness (see Chap. 6, P8) cannot be identified and prioritized for transfusion.

5.3 *Pathogen Reduction*

Pathogen reduction (PR), which encompasses a variety of approaches (e.g., photochemical inactivation, solvent detergent treatment, nanofiltration) to mitigate infectious risk, is increasingly in use in HICs. PR has the advantage of addressing multiple classes of pathogens with a single intervention, thus minimizing the risk of established, emerging, and re-emerging pathogens. There are already approved PR technologies for platelets and plasma that are in use in HICs (notably in the US Europe and parts of the Middle East). PR has been raised as a possible contributing factor to negative findings in clinical trials of CCP (Focosi et al. 2022; Focosi and Casadevall 2022). Several PR technologies have been evaluated for their impact on the integrity and neutralizing capacity of immunoglobulins. The few published studies suggest that there is little in-vitro effect with methylene blue (Larrea et al. 2022; Raster et al. 2021), and variable—albeit small—effects of several other methods, notably Riboflavin and UV light (Kostin et al. 2021; Focosi and Franchini 2021). PR is promising for LMICs given the limitations in LMICs surrounding donor selection and testing. In a pivotal trial in Ghana, the incidence of transfusion transmitted malaria was significantly lower in recipients of pathogen reduced whole blood than recipients of standard (untreated) whole blood (4% vs 22%, respectively) (Allain et al. 2016). Despite the advantages of PR, the high cost and technological complexity (notably equipment) are formidable barriers to its wider adoption (Ware et al. 2018).

6 Transfusion Practices in LMICs

Even in HICs, there is inconsistent training in transfusion medicine, a lack of national, standardized curricula, and low levels of knowledge despite transfusion being one of the most performed medical procedures (Panzer et al. 2013; Haspel et al. 2015). Not surprisingly, gaps in training have also been reported across LMICs (Rambiritch et al. 2021). Suboptimal clinical transfusion practices in LMICs, in large part due to the lack of training in transfusion medicine, contribute to blood wastage, straining extant blood inventories. There are opportunities for improvement, and diverse training platforms have been developed across a range of settings (Rambiritch et al. 2021; Louw 2014; Tagny et al. 2013). Unfortunately, uptake has been limited given a host of competing priorities and limited support. The low level of knowledge pertaining to transfusion medicine in general affects the use of CCP, impacting decisions to collect, distribute, and—ultimately—transfuse CCP, appropriately.

7 Blood Access and Availability

There is a sharp divide between access and availability of blood between high and LMICs. Globally, 40% of the approximately 118 million annual donations originate in HICs, which represent only 16% of the world's population (WHO 2023). A historic metric was recommended by the WHO to collect a minimum of 10 units per 1000 population (Tapko et al. 2006). As of 2023, the WHO reported a blood donation rate per 1000 population in HICs, upper-middle-income countries, lower-middle-income, and low-income countries as 31.5, 16.4, 6.6 donations, and 5.0 donations, respectively (WHO 2023). One study estimated that a quarter (26%) of maternal hemorrhage related deaths in sub-Saharan Africa were ascribed to lack of blood (Bates et al. 2008). A study of global blood needs and availability, concludes that "every country in central, eastern, and western sub-Saharan Africa, Oceania, and south Asia had insufficient blood to meet their needs" (Roberts et al. 2019). The profound transfusion deficit in LMICs is a major barrier to the widespread adoption of CCP, given the need to prioritize the collection and processing of conventional blood products to support diverse patient populations (e.g., surgery, oncology, obstetrics, and trauma).

8 Post-Transfusion Surveillance

Hemovigilance is an umbrella term that refers to quality oversight of all facets of the blood transfusion continuum from donor selection through post-transfusion surveillance. The latter is critically lacking in LMICs. Consequently, the true burden of transfusion-associated adverse events is unknown. The reported incidence of TTIs deserves scrutiny given the limitations surrounding data capture and collation in LMICs (Morar et al. 2016). Systems are still lacking to conduct post-transfusion surveillance.

9 Economic Landscape: Funding Blood Transfusion Safety in LMICs

Blood transfusion services in LMICs are impacted by systemic economic challenges. A high proportion of LMICs receive external aid, through a variety of mechanisms (Ifland et al. 2018). In the mid-2000s, there was a transient increase in support for blood safety whereby, 14 countries in Africa and the Caribbean received US funding (collectively $437 million) through the President's Emergency Plan for AIDS Relief (PEPFAR) over a 10-year period to strengthen their national transfusion services (Chevalier et al. 2016). Although PEPFAR was focused on HIV/AIDS mitigation, the advancement in blood safety and availability conferred wide ranging benefits. That funding support peaked around 2013 and subsequently waned (Ifland et al. 2018). Even prior to the discontinuation of funding, the model of external aid was questioned (Ala et al. 2012). As of 2020, 24 of the 39 respondents (61.5%) in the WHO survey of countries in the African region were still receiving external funding (WHO 2022). Germane to CCP, the core mission to provide safe and adequate blood supply is not being met, as a consequence of the financial insufficiency within blood systems in LMICs. It is unlikely that efforts to make CCP available would be funded in the current economic landscape.

10 Conclusion: The Myriad of Challenges in Transfusion Services in LMICs Impact the Adoption of Convalescent Plasma

Blood transfusion safety and availability are very different in high and LMICs The drive to mobilize CCP early in the COVID-19 pandemic highlighted the related obstacles that are routinely encountered in LMICs. Indeed, many of the publications pertaining to CCP in LMICs, focused on the challenges of its deployment, rather than its efficacy (Bloch et al. 2021a, b). Convalescent plasma can be harvested locally even in remote or resource-limited settings, drawing on an extant blood collection infrastructure. The overarching concern is the net risk—versus benefit—of its use, amidst the higher risk of TTIs, allied with the diversion of blood products away from services that are critically dependent on the blood supply, which already suffer from shortages. Further, blood transfusion services in LMICs remain underfunded, and insufficiently equipped to scale CCP (or future plasma derivatives) to meet population demands in a time of pandemic. Innovative, low-cost solutions are needed for low-resourced settings, such as are proposed in Chap. 15 (P10). Personnel capacity is another barrier to the wider adoption of CCP, given limited training in evidence-based transfusion practice, even for conventional blood products, let alone plasma derivatives. Each of the described challenges is soluble but will require greater investment in the greater blood transfusion infrastructure. The use of CCP and convalescent plasma in LMICs may be revisited during future outbreaks and pandemics, but the decision to transfuse needs to be weighed carefully against the described risks and broader considerations.

Declarations and Disclosures
The author was supported in part by the National Heart Lung and Blood Institute (1K23HL15182) and is a co-investigator on a US government funded clinical trial evaluating Mirasol Pathogen Reduction Technology. The author reports personal fees and non-financial support from Grifols, Abbott, UptoDate, Tegus, and Health Advances outside of the submitted work.

The author is a member of the U.S. Food and Drug Administration (FDA) Blood Products Advisory Committee. Any views or opinions expressed in this chapter are Dr. Bloch's and are based on his own scientific expertise and professional judgment; they do not necessarily represent the views of the Blood Products Advisory Committee or the formal position of the FDA and also do not bind or otherwise obligate or commit either the Advisory Committee or the FDA to the views expressed.

References

Ahmad Y, Heroes AS, Hume HA, Farouk M, Owusu-Ofori A, Gehrie EA, Goel R, Ness PM, Tobian AAR, Bloch EM (2021) Bacterial contamination of blood products in Africa. Transfusion 61:767–780

Ala F, Allain JP, Bates I, Boukef K, Boulton F, Brandful J, Dax EM, El Ekiaby M, Farrugia A, Gorlin J, Hassall O, Lee H, Loua A, Maitland K, Mbanya D, Mukhtar Z, Murphy W, Opare-Sem O, Owusu-Ofori S, Reesink H, Roberts D, Torres O, Totoe G, Ullum H, Wendel S (2012) External financial aid to blood transfusion services in sub-Saharan Africa: a need for reflection. PLoS Med 9:e1001309

Allain J-P (2010) Volunteer safer than replacement donor blood: a myth revealed by evidence. ISBT Sci Ser 5:169–175

Allain JP (2011) Moving on from voluntary non-remunerated donors: who is the best blood donor? Br J Haematol 154:763–769

Allain JP, Sibinga CT (2016) Family donors are critical and legitimate in developing countries. Asian J Transfus Sci 10:5–11

Allain JP, Owusu-Ofori AK, Assennato SM, Marschner S, Goodrich RP, Owusu-Ofori S (2016) Effect of Plasmodium inactivation in whole blood on the incidence of blood transfusion-transmitted malaria in endemic regions: the African Investigation of the Mirasol System (AIMS) randomised controlled trial. Lancet 387:1753–1761

Barnes LS, Al-Riyami AZ, Ipe TS, Bloch EM, Sibinga CS, Eichbaum QG (2022) COVID-19 and the impact on blood availability and transfusion practices in low- and middle-income countries. Transfusion 62:336–345

Bates I, Chapotera GK, McKew S, van den Broek N (2008) Maternal mortality in sub-Saharan Africa: the contribution of ineffective blood transfusion services. BJOG 115:1331–1339

Bloch EM, Kipiani E, Shadaker S, Alkhazashvili M, Gvinjilia L, Kuchuloria T, Chitadze N, Keating SM, Gamkrelidze A, Turdziladze A, Getia V, Nasrullah M, Averhoff F, Izoria M, Skaggs B (2020a) Blood transfusion safety in the country of Georgia: collateral benefit from a national hepatitis C elimination program. Transfusion 60:1243–1252

Bloch EM, Shoham S, Casadevall A, Sachais BS, Shaz B, Winters JL, van Buskirk C, Grossman BJ, Joyner M, Henderson JP, Pekosz A, Lau B, Wesolowski A, Katz L, Shan H, Auwaerter PG, Thomas D, Sullivan DJ, Paneth N, Gehrie E, Spitalnik S, Hod EA, Pollack L, Nicholson WT, Pirofski LA, Bailey JA, Tobian AA (2020b) Deployment of convalescent plasma for the prevention and treatment of COVID-19. J Clin Invest 130:2757–2765

Bloch EM, Gehrie EA, Ness PM, Sugarman J, Tobian A (2020c) Blood transfusion safety in low-resourced countries: aspiring to a higher standard. Ann Intern Med 173:482–483

Bloch EM, Goel R, Wendel S, Burnouf T, Al-Riyami AZ, Ang AL, DeAngelis V, Dumont LJ, Land K, Lee CK, Oreh A, Patidar G, Spitalnik SL, Vermeulen M, Hindawi S, Van den Berg K, Tiberghien P, Vrielink H, Young P, Devine D, So-Osman C (2021b) Guidance for the procurement of COVID-19 convalescent plasma: differences between high- and low-middle-income countries. Vox Sang 116:18–35

Bloch EM, Shah A, Kaidarova Z, Laperche S, Lefrere JJ, van Hasselt J, Zacharias P, Murphy EL, Anglophone Africa Transfusion Research G (2014) A pilot external quality assurance study of transfusion screening for HIV, HCV and HBsAG in 12 African countries. Vox Sang 107:333–342

Bloch EM, Goel R, Montemayor C, Cohn C, Tobian AAR (2021a) Promoting access to COVID-19 convalescent plasma in low- and middle-income countries. Transfus Apheresis Sci Off J World Apheresis Assoc Off J Eur Soc Haemapheresis 60:102957

Busch MP, Bloch EM, Kleinman S (2019) Prevention of transfusion-transmitted infections. Blood 133:1854–1864

Carson JL, Grossman BJ, Kleinman S, Tinmouth AT, Marques MB, Fung MK, Holcomb JB, Illoh O, Kaplan LJ, Katz LM, Rao SV, Roback JD, Shander A, Tobian AA, Weinstein R, Swinton McLaughlin LG, Djulbegovic B (2012) Clinical transfusion medicine committee of the A. Red blood cell transfusion: a clinical practice guideline from the AABB*. Ann Intern Med 157:49–58

Chevalier MS, Kuehnert M, Basavaraju SV, Bjork A, Pitman JP (2016) Progress toward strengthening national blood transfusion services—14 Countries, 2011–2014. MMWR Morb Mortal Wkly Rep 65:115–119

Focosi D, Casadevall A (2022) Convalescent plasma in outpatients with COVID-19. Lancet Respir Med 10:226–228

Focosi D, Franchini M (2021) Impact of pathogen-reduction technologies on COVID-19 convalescent plasma potency. Transfus Clin Biol 28:132–134

Focosi D, Franchini M, Pirofski LA, Burnouf T, Paneth N, Joyner MJ, Casadevall A (2022) COVID-19 convalescent plasma and clinical trials: understanding conflicting outcomes. Clin Microbiol Rev 35:e0020021

Grady GF, Chalmers TC (1964) Risk of Post-Transfusion Viral Hepatitis. N Engl J Med 271:337–342

Haspel RL, Lin Y, Mallick R, Tinmouth A, Cid J, Eichler H, Lozano M, van de Watering L, Fisher PB, Ali A, Parks E, Investigators B-T (2015) Internal medicine resident knowledge of transfusion medicine: results from the BEST-TEST international education needs assessment. Transfusion 55:1355–1361

Ifland L, Bloch EM, Pitman JP (2018) Funding blood safety in the 21st century. Transfusion 58:105–112

Jayaraman S, Chalabi Z, Perel P, Guerriero C, Roberts I (2010) The risk of transfusion-transmitted infections in sub-Saharan Africa. Transfusion 50:433–442

Kostin AI, Lundgren MN, Bulanov AY, Ladygina EA, Chirkova KS, Gintsburg AL, Logunov DY, Dolzhikova IV, Shcheblyakov DV, Borovkova NV, Godkov MA, Bazhenov AI, Shustov VV, Bogdanova AS, Kamalova AR, Ganchin VV, Dombrovskiy EA, Volkov SE, Drozdova NE, Petrikov SS (2021) Impact of pathogen reduction methods on immunological properties of the COVID-19 convalescent plasma. Vox Sang 116:665–672

Laperche S (2013) Francophone African group for research in blood T. Multinational assessment of blood-borne virus testing and transfusion safety on the African continent. Transfusion 53:816–826

Larrea L, Castro E, Navarro L, Vera B, Francés-Gómez C, Sánchez-Sendra B, Giménez Á, Castelló E, Collado M, Vayá MJ, Mirabet V, Callao V, Ortiz-de-Salazar MI, Roig R, Geller R, Arbona C (2022) Preservation of anti-SARS-CoV-2 neutralising antibodies in convalescent plasma after pathogen reduction with methylene blue and visible light. Blood Transfus 20:206–212

Lefrere JJ, Dahourouh H, Dokekias AE, Kouao MD, Diarra A, Diop S, Tapko JB, Murphy EL, Laperche S, Pillonel J (2011) Estimate of the residual risk of transfusion-transmitted human immunodeficiency virus infection in sub-Saharan Africa: a multinational collaborative study. Transfusion 51:486–492

Louw VJ (2014) Transfusion education and practice in South Africa. Transfus Apher Sci 51:6–7

Morar MM, Pitman JP, McFarland W, Bloch EM (2016) The contribution of unsafe blood transfusion to human immunodeficiency virus incidence in sub-Saharan Africa: reexamination of the 5% to 10% convention. Transfusion 56:3121–3132

Panzer S, Engelbrecht S, Cole-Sinclair MF, Wood EM, Wendel S, Biagini S, Zhu Z, Lefrère JJ, Andreu G, Zunino T, Cabaud JJ, Rouger P, Garraud O, Janetzko K, Müller-Steinhardt M, van der Burg P, Brand A, Agarwal P, Triyono T, Gharehbaghian A, Manny N, Zelig O, Takeshita A, Yonemura Y, Fujihara H, Nollet KE, Ohto H, Han KS, Nadarajan VS, Berlin G, Sandler SG, Strauss RG, Reesink HW (2013) Education in transfusion medicine for medical students and doctors. Vox Sang 104:250–272

Patidar GK, Thachil J, Dhiman Y, Oreh A, Vrielink H, van den Berg K, Grubovic Rastvorceva RM, So-Osman C, Al-Riyami AZ (2022) Management of blood transfusion services in low-resource countries. Vox Sang 117:1375–1383

Prugger C, Laperche S, Murphy EL, Bloch EM, Kaidarova Z, Tafflet M, Lefrere JJ, Jouven X (2016) Screening for transfusion transmissible infections using rapid diagnostic tests in Africa: a potential hazard to blood safety? Vox Sang 110:196–198

Rambiritch V, Vermeulen M, Bell H, Knox P, Nedelcu E, Al-Riyami AZ, Callum J, van den Berg K, Forum obotESotAGT (2021) Transfusion medicine and blood banking education and training for blood establishment laboratory staff: a review of selected countries in Africa. Transfusion 61:1955–1965

Raster J, Zimmermann K, Wesche J, Aurich K, Greinacher A, Selleng K (2021) Effect of methylene blue pathogen inactivation on the integrity of immunoglobulin M and G. Transfus Med Hemother 48:148–153

Roberts N, James S, Delaney M, Fitzmaurice C (2019) The global need and availability of blood products: a modelling study. Lancet Haematol 6:e606–e615

Steele WR, Dodd RY, Notari EP, Haynes J, Anderson SA, Williams AE, Reik R, Kessler D, Custer B, Stramer SL (2021) HIV, HCV, and HBV incidence and residual risk in US blood donors before and after implementation of the 12-month deferral policy for men who have sex with men. Transfusion 61:839–850

Tagny CT, Murphy EL, Lefrère J-J, Network ObotFABTR (2013) The francophone Africa blood transfusion research network: a five-year report (2007–2012). Transfus Med 23: 442–424

Tapko J, Mainuka P, Diarra-Nama AJ (2006) Status of blood safety in the WHO African region: report of the 2006 survey. http://www.afro.who.int/en/divisions-a-programmes/dsd/health-technologies-a-laboratories.html. WHO Regional Office for Africa, Brazzaville, Republic of Congo

Ware AD, Jacquot C, Tobian AAR, Gehrie EA, Ness PM, Bloch EM (2018) Pathogen reduction and blood transfusion safety in Africa: strengths, limitations and challenges of implementation in low-resource settings. Vox Sang 113:3–12

Weimer A, Tagny CT, Tapko JB, Gouws C, Tobian AAR, Ness PM, Bloch EM (2019) Blood transfusion safety in sub-Saharan Africa: a literature review of changes and challenges in the 21st century. Transfusion 59:412–427

Wells CR, Galvani AP (2022) The global impact of disproportionate vaccination coverage on COVID-19 mortality. Lancet Infect Dis 22:1254–1255

WHO (2013) Guidelines on management of blood and blood components as essential medicines, Annex 3, TRS No 1004 [monograph on the internet]. Available from: https://www.who.int/publications/m/item/blood-and-blood-components-as-essential-medicines-annex-3-trs-no-1004

WHO (2022) WHO African region status report on blood availability, safety and quality Brazzaville, Republic of Congo: WHO Regional Office for Africa

WHO (2023) Blood safety and availability [monograph on the internet]. 2023. Available from: https://www.who.int/news-room/fact-sheets/detail/blood-safety-and-availability

Wilson ML, Fleming KA, Kuti MA, Looi LM, Lago N, Ru K (2018) Access to pathology and laboratory medicine services: a crucial gap. The Lancet 391:1927–1938

Evan M. Bloch is an Associate Professor of Pathology at Johns Hopkins University School of Medicine, where he serves as the Associate Director of Transfusion Medicine and Medical Director of the Apheresis Service. He also holds a joint appointment in International Health at the Johns Hopkins Bloomberg School of Public Health. Originally from South Africa, Dr. Bloch's medical training and subsequent clinical practice spurred an interest in blood transfusion safety in resource-constrained settings. To this end, he has been actively involved in research, education and operational outreach related to advancement of global blood transfusion safety. Dr. Bloch has a specific interest in the interplay between blood transfusion and infectious diseases, particularly in low and low-middle-income countries where transfusion transmitted infection remains a metric of suboptimal policies and practices.

HemoClear: A Practical and Cost-Effective Alternative to Conventional Convalescent Plasma Retrieval Methods

Arno Nierich, Rosita Bihariesingh, and Rakesh Bansie

Contents

Abstract Convalescent plasma has increasingly been used to treat various viral infections and confer post-exposure prophylactic protection during the last decade and has demonstrated favorable clinical outcomes in patients infected with Severe Acute Respiratory Syndrome Coronavirus 2 (SARS-CoV-2) during the recent

A. Nierich (✉)
Department of Anesthesiology, Academic Hospital Paramaribo, Paramaribo, Suriname

Chief Medical Officer Hemoclear, Zwolle, The Netherlands
e-mail: arno.nierich@hemoclear.com

R. Bihariesingh
Department of Anesthesiology & Intensive Care, Academic Hospital Paramaribo, Paramaribo, Suriname
e-mail: rbihariesingh@azp.sr

R. Bansie
Department of Anesthesiology & Internal Medicine, Academic Hospital Paramaribo, Paramaribo, Suriname
e-mail: rbansie@azp.sr

Current Topics in Microbiology and Immunology (2025) 443: 185–201
https://doi.org/10.1007/82_2024_276

Published Online: 11 August 2024

COVID-19 pandemic. The pandemic has highlighted the need for cost-effective, accessible, and easy-to-use alternatives to conventional blood plasmapheresis techniques, allowing hospitals to become more self-sufficient in harvesting and transfusing donor plasma into recipients in a single setting. To this end, the use of a membrane-based bedside plasmapheresis device (HemoClear) was evaluated in an open-label, non-randomized prospective trial in Suriname in 2021, demonstrating its practicality and efficacy in a low-to middle-income country. This paper will review the use of this method and its potential to expedite the process of obtaining convalescent plasma, especially during pandemics and in resource-constrained settings.

Glossary of Abbreviations

COVID-19	Coronavirus disease 2019
CP	Convalescent plasma
ICU	Intensive Care
IgA	Immunoglobulin A
IgG	Immunoglobulin G
IgM	Immunoglobulin M
MC	Medium Care
RBC	Red Blood Cells
SARS-CoV-2	Severe Acute Respiratory Syndrome Coronavirus 2
TPE	Therapeutic Plasma Exchange

1 Introduction

Therapeutic plasma exchange is a procedure in which plasma is removed from a patient and purified using extracorporeal mechanisms (Hafer et al. 2016). Historically, therapeutic plasma exchange was widely utilized during the 1918 influenza pandemic, with documented clinical improvement in treated patients (Luke et al. 2006). Originally, the basis for plasma exchange was the removal of injurious substances from a patient's plasma with replacement of volume with donor plasma or appropriate fluids; however, convalescent plasma donation, in which plasma from recovered donors containing protective antibodies is transfused into an unwell recipient was later developed to treat infections and confer post-exposure prophylaxis, while preserving the donors red cells, thus permitting more frequent donations (Bloch et al. 2020; Garraud et al. 2016; Focosi et al. 2020a).

The therapeutic use of convalescent plasma relies upon transfusing antibodies generated by a previously infected patient into a patient with an active infection, directly neutralizing pathogens (Bloch et al. 2020; Selvi 2020). Such passive immunity can be achieved using various products including whole blood, human or animal plasma, pooled immunoglobulins, and monoclonal antibodies as described in other chapters in this book, but human plasma is the most immediately available product, and is thus an efficient and cost-effective way to provide protection from viral infections (Selvi 2020).

As noted in Chap. 5 (P6), passive immunization has been used since the late 1800s to treat serious bacterial infections (Garraud et al. 2016; Selvi 2020). The initial focus was on using antitoxins to neutralize toxins produced by bacteria, followed later by antisera used to neutralize the bacteria themselves, but these latter treatments were largely superseded by the advent of antibiotic therapy (Behring and Kitasato 1890; Finland 1982). The pre-COVID historical role of convalescent plasma in viral infections is more modest, but not without success, as reviewed in Chap. 5 (P6) (Casadevall and Scharff 1995; WHO Blood Regulators Network 2014; Delamou et al. 2016; Griensven et al. 2016; Edwards et al. 2016). Human immunoglobulin from recovered donors continues to be used to produce immunoglobulins to prevent tetanus, hepatitis B, and rabies, as reviewed in Chap. 6 (P8) (Garraud et al. 2016).

This article will review the use of a novel plasmapheresis method in the context of pandemics and the potential to expedite the process of obtaining convalescent plasma, especially during pandemics and in resource-constrained settings.

2 Convalescent Plasma Use in COVID-19

As reviewed in several other chapters in this book, convalescent plasma emerged as a prospective therapeutic avenue during the recent global COVID-19 pandemic (Bloch et al. 2020; Focosi et al. 2020b; Tiberghien et al. 2020; Li et al. 2020). But the pandemic laid bare the inadequacy of existing healthcare frameworks in responding to extensive emergency health scenarios and demonstrated that with a complex biological product obtaining swift and efficacious treatment was a challenge (Focosi et al. 2020a).

Randomized trials of convalescent plasma showed mixed findings, often because trials with negative findings used plasma with low antibody levels and/or treated late in the course of illness. But other trials, especially those that used plasma early and in sufficient strength, showed lower mortality and reduced risk of hospitalization (Sullivan et al. 2022; Tworek et al. 2021; Ipe et al. 2022; Casadevall et al. 2021; Salazar et al. 2019; Klassen et al. 2021a; Bihariesingh et al. 2021; Allahyari et al. 2021; Libster et al. 2021). Most trials, and much of clinical practice, use one to two units of 200 ml of plasma infused at a rate of 100 to 200 ml/hour, though these dosages are often individualized based on the recipient's body weight and the antibody titer of the donated plasma (Focosi et al. 2020a). Resistance to using convalescent plasma has often stemmed from the complexity of the process of obtaining the plasma using plasmapheresis, a topic we will address in this chapter.

2.1 Donor Selection and Optimal Donation Window

The selection of suitable convalescent plasma donors is based on optimal blood antibody titers, with higher titers associated with a greater reduction in COVID-19-related mortality in recipients (Focosi et al. 2020a; Tiberghien et al. 2020; Ipe et al. 2022; Klassen et al. 2021a; Joyner et al. 2020, 2021). As reviewed in Chap. 9 (P7) (Kunze et al.), patients geographically located in the same area are likely subject to infection with similarly mutated viruses, and multi-center studies indicate benefit in selecting donors who live near the recipient and presumably have been exposed to local viral strains (Focosi et al. 2020a; Selvi 2020).

In individuals infected with SARS-CoV-2, IgM and IgA antibodies can be detected as early as 5 days after infection and IgG antibodies at around 14 days, with seroconversion occurring at a mean of 8 to 21 days (Bloch et al. 2020; Garraud et al. 2016; Delamou et al. 2016). Neutralizing antibody levels have been found to peak at approximately 1 month after infection and decline thereafter. Lee et al. reported peak antibody levels at 31 to 35 days following symptom onset in infected individuals and Xu et al. demonstrated peak concentrations at week four, followed by a slow decline in titers, suggesting that this may be the optimal point for post-infection convalescent plasma donation to increase the likelihood of a high antibody yield (Lee et al. 2021; Natarajan et al. 2021; Xu et al. 2021). The persistence of relatively high titers for several months following symptom onset suggests that beyond peak antibody levels, donations could continue to yield adequate neutralizing antibody titers for a longer duration should adequate numbers of donors in the peak window not be identified (Tiberghien et al. 2020).

The specific antibody composition of convalescent plasma differs over time after the primary infection of the donor and this may also play a role in its efficacy of treatment. IgG dynamics appear to follow a pattern of a peak after 4 weeks, with a plateau and decline after 8 weeks with persistence at lower levels. By contrast, IgM levels show a peak 2 to 5 weeks after initial symptom onset and a decline between 3 to 5 weeks. Virus-specific IgM becomes undetectable in almost all cases by around 6 weeks after disease onset (Post et al. 2020).

The availability of IgM in convalescent plasma seems to be important, as in ICU-admitted COVID-19 patients with severe progression, a delayed specific IgM antibody response is observed (Shen et al. 2020). Early harvesting of antibodies after the initial infection of the donor allows a mixture of both IgM and IgG if the processing is done before 8 weeks. However, current practices in acquiring convalescent plasma often include collections undertaken after 12 weeks, resulting in plasma containing only IgG antibodies. The HemoClear method allows early in-hospital harvesting of convalescent plasma which contains both IgM and IgG. Among COVID-19 patients with severe progression, a delayed specific IgM antibody response has been linked to worse outcome (Shen et al. 2020). Thus the broader antibody profile of HemoClear plasma may have played a role in the efficacy found in severely ill patients in a study in Suriname (Bihariesingh-Sanchit et al. 2023).

Mortality reduction in COVID-19-infected patients has been found to be statistically significant when patients are transfused with convalescent plasma within 2 to 3 days of hospitalization, at a stage where pathology is mainly driven by ongoing viral replication (Tiberghien et al. 2020; Salazar et al. 2019; Klassen et al. 2021a; Libster et al. 2021). This is thought to be related to the concentration of neutralizing antibodies which may prevent both viral entry and replication (Tiberghien et al. 2020).

3 Current Plasmapheresis Methods

Current plasmapheresis methodologies predominantly encompass two main techniques: centrifugal based and membrane based (Redant et al. 2021; Bacal et al. 2021; Delgado and Romero 2020). Centrifugal-based techniques utilize centrifugal force to separate whole blood into components of varying densities, allowing low-density plasma to be separated for donation (Redant et al. 2021; Bacal et al. 2021; Delgado and Romero 2020). Despite being the most widely used technique, centrifugal techniques have several limitations, including high operating costs, complex specimen-handling techniques, non-portability of equipment, and labor intensity (Redant et al. 2021; Bacal et al. 2021). Membrane-based techniques utilize membranes manufactured with specific pore sizes allowing the passage of specific blood components while retaining others. Both systems perform comparable, adequate, and safe therapeutic plasma exchange (TPE) procedures. However, centrifugal-based plasmapheresis is more efficient in terms of plasma removal and takes significantly less time in comparison to membrane-based plasmapheresis (Hafer et al. 2016; Bacal et al. 2021; Kes et al. 2016). Although established and scaled-up, challenges related to these techniques' selectivity and long-term impact on blood cells and their constituents remain. The centrifugal forces or high flow rates through a membrane during extraction may stress blood cells leading to cell damage with hemolysis and platelet activation and may harm the donor plasma (Bacal et al. 2021).

4 Barriers to Convalescent Plasma Donation

Challenges in selecting appropriate convalescent plasma donors include the heterogeneity of neutralizing antibody titers in recovered patients. Screening the general population for high titer levels through general blood donations is possible, but is the least financially viable, until a high fraction of the population achieves high levels of antibodies due to infection and/or immunization. Recruiting donors recently discharged from hospital following recovery from COVID-19 infection presents a higher likelihood of identifying donors with high titers and is, therefore, the more cost-effective, especially in the beginning of a pandemic. However, patients requiring hospitalization are often frail, elderly, or have multiple comorbidities, so are often not eligible to act as convalescent plasma donors (Focosi et al. 2020a). A third option is to contact individuals confirmed as COVID-19-infected via testing who are not ill enough for hospitalization; however, studies have shown that patients with less severe infections tend to have lower antibody titers (Focosi et al. 2020a).

A Ugandan study underscored non-clinical barriers to procuring donor plasma for convalescent plasma treatment, with 12.8% of eligible donors electing not to participate due to various factors such as time constraints, fear, and concerns regarding stigmatization (Muttamba et al. 2021). The findings of this study are particularly pertinent as securing sufficient volumes of donor convalescent plasma to address viral disease outbreaks hinges on the efficacy of the blood donation system, a facility often inaccessible or poorly accessible in low- and middle-income countries (Bloch et al. 2021). Further, the high cost of centrifugal plasmapheresis equipment and the level of technical expertise required to operate this equipment is a significant barrier to the successful implementation of such management strategies and is not present in low- to middle-income countries in a sufficient capacity (Bloch et al. 2021; Gifford et al. 2018). In addition, a rapid increase in the volume of required blood and plasma donation, processing, storage, and distribution as a result of emergencies such as pandemics places an insurmountable burden on existing infrastructure.

The above barriers underscore the necessity for more cost-effective alternatives to conventional plasmapheresis techniques which demand lower operational expertise and can be deployed in low- to middle-income countries in preparation for future viral epidemics or pandemics. In view of this, an alternative membrane-based bedside plasmapheresis device (HemoClear) was rigorously evaluated in an open-label, non-randomized prospective trial in Suriname in 2021, demonstrating practicality and efficacy (Bihariesingh et al. 2021).

5 HemoClear

HemoClear is a gravity-driven, membrane-based, microfiltration blood product separation device, initially piloted in 2018 for washing shed blood in cardiac surgery (Hoetink et al. 2020). Its success in bedside apheresis, cost-effectiveness, and practicality led to its implementation for the collection of other blood products, including plasma for convalescent plasma treatment.

The HemoClear device separates whole blood into concentrated blood cells and plasma by passing blood through a multilayered cross-flow membrane module, with a relatively large membrane pore size of 2.2 Mu (Bihariesingh et al. 2021; Osemwengie et al. 2022; Amenge et al. 2022). The construction of the device and its blood separation process are depicted in Figs. 1 and 2. The closed system

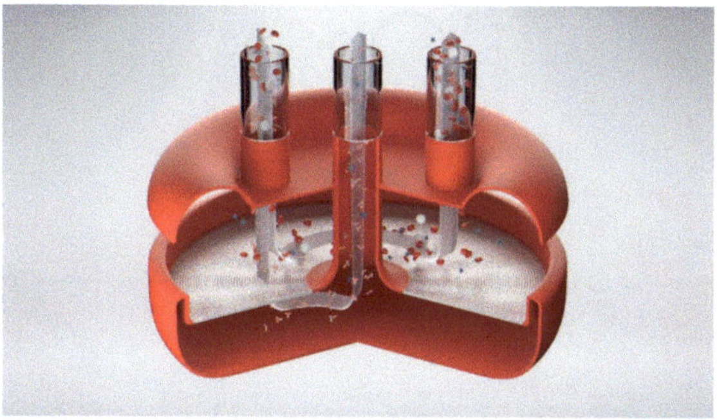

Fig. 1 Schematic illustration of the HemoClear cross-flow membrane filtration device. Arrow left in: blood is separated in a cellular and non-cellular part; central arrow out: plasma with a small amount of platelets; right arrow out: platelet-leucocyte-rich red blood cell part

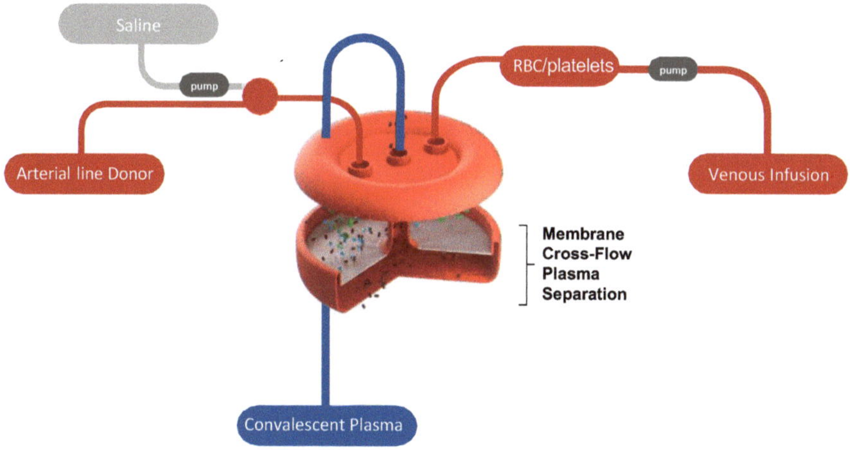

Fig. 2 Schematic diagram of the production process of convalescent plasma. The driving force of the mean arterial pressure of the patient by the arterial line as an inlet is a continuous option beside the one-meter gravity force of a blood volume above the filter as an intermittent method. Abbreviation: RBC, red blood cells

filtration process prevents possible contamination (Osemwengie et al. 2022). The device is portable, disposable, and contains no electrical components, with the microfiltration process taking place by one meter of gravity force alone, either produced by the donor mean arterial pressure or by placing the donor blood one meter above the filter on an infusion rod (Osemwengie et al. 2022). This renders it ideal for use in resource-constrained environments such as low- to middle-income countries or poor, rural, or military settings where an uninterrupted power supply is not guaranteed (Bihariesingh-Sanchit et al. 2023). The ability to generate convalescent plasma at the bedside removes the need for external blood centers, thus

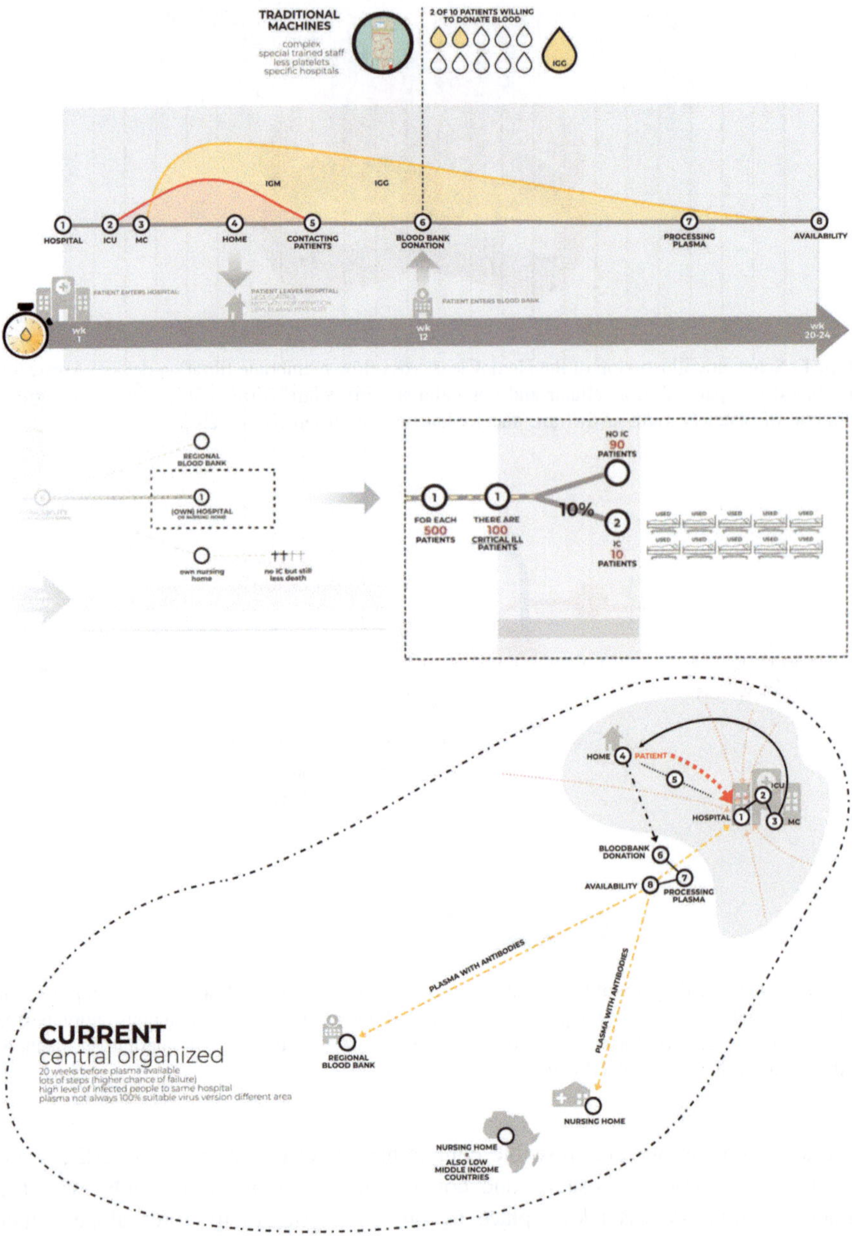

Fig. 3 Current supply chain workflow of conventional plasma donation methods

avoiding the logistic, transport, processing, and storage constraints accompanying their use. The current convalescent plasma donation and the HemoClear donation process and their differences are illustrated in Figs. 3 and 4.

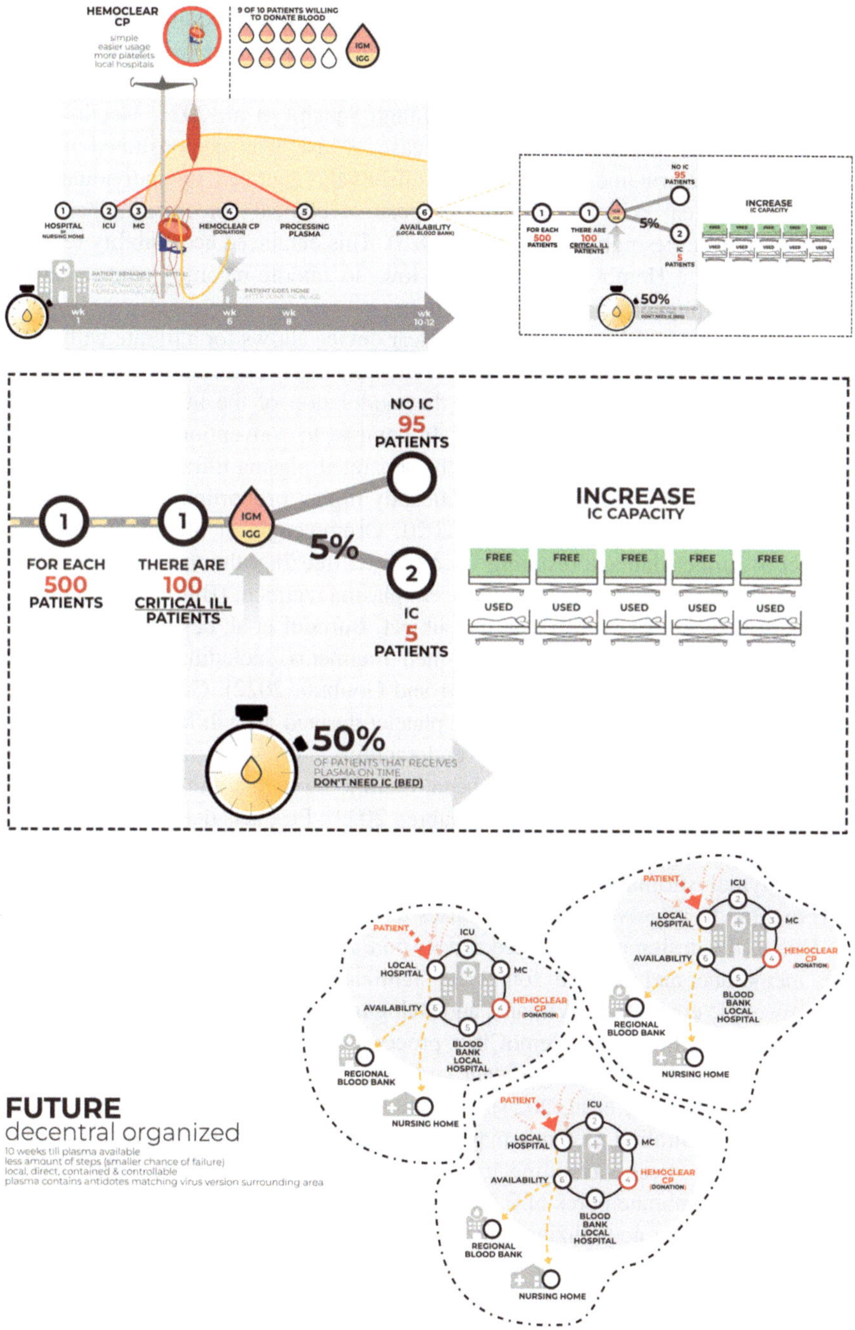

Fig. 4 Future supply chain workflow of HemoClear plasma donation method

The ease of use of the HemoClear device means that it requires little technical laboratory expertise or training, and the rapidity of the process in comparison to conventional techniques means that many more units of whole blood can be separated over a shorter period (Bihariesingh-Sanchit et al. 2023; Hoetink et al. 2020; Osemwengie et al. 2022). This ease of use was demonstrated in trials undertaken in Suriname, where it was effectively operated by individuals with minimal technical skill and resulted in positive clinical outcomes (Bihariesingh et al. 2021; Bihariesingh-Sanchit et al. 2023). This enhanced accessibility reaffirms the utility of the HemoClear method in low- to middle-income countries and in emergency settings such as viral pandemics.

The technology utilized in the HemoClear device allows for a filtrate with a high particulate content, which includes platelets and high molecular weight proteins, which may play an important role in the modulation of the immune response to COVID-19 (Osemwengie et al. 2022). In contrast to conventional techniques that do not filter platelets into the plasma, the separated plasma filtrate generated using the HemoClear device contains a significantly higher proportion of intact platelets (Gifford et al. 2018; Hoetink et al. 2020; Osemwengie et al. 2022). Although subject to further damage resulting from plasma freezing, the immunologic effects of platelets may persist in the convalescent plasma recipient (Burnouf and Goubran 2022). In their extensive work on the subject, Burnouf et al. described a multitude of benefits of transfused platelets and their fragments, including a notable role in tissue repair and regeneration (Burnouf and Goubran 2022). Growth factors identified in human blood products include platelet-derived growth factor, transforming growth factor and vascular endothelium growth factor found in platelet alpha-granules, which have been found to modulate immune responses and stimulate cell growth (Burnouf and Goubran 2022). Products derived from platelets have been utilized ex vivo as a growth medium to effectively propagate human mesenchymal stromal cells, and their role in immune cell activation is being investigated to improve patient responses to immunotherapy (Barro et al. 2022).

Due to the design of the device and its pore sizes, proteins pass readily through the membrane and can be found in significant concentrations in the filtrate. (Osemwengie et al. 2022; Williams and Balogun 2014). A recent in-vitro study that compared centrifugal and HemoClear processes on their ability to separate antibodies showed no significant differences in total concentrations (Bihariesingh-Sanchit et al. 2024). The antibodies generated by the conventional centrifugal or filtration methods may act differently, due to the processing method in which shear stresses resulting from centrifugal forces and passage through relatively tiny membrane pores of 0.4 Mu may play a role. The clinical impact of this difference on virus neutralizing effects requires further research.

Another difference is the ease of implementation of HemoClear in a potentially contaminated area such as an isolation ward within a hospital or ICU. There is no need for transporting a conventional plasmapheresis team into the contaminated area with all the logistical burdens related to this procedure. A lack of plasmapheresis staffing by local blood banks further prevents large-scale procedures (Williams and Balogun 2014). In contrast, the HemoClear method lacks these shortcomings and allows scale-up when needed.

6 Costs of Convalescent Plasma

One of the barriers of implementing CP therapy is costs. The conventional methods vary greatly in cost between healthcare systems and limited data are available. A cost calculation from a low- and middle-income country, such as Bali, showed a price of $135 per unit of plasma (Noerwidayati and Pribadi 2023). In high-income countries, such as the United Kingdom, USA, and Canada, the costs range from $514 to as high as $1,880 per unit (Aleccia 2021; Jorda et al. 2022; Health Technology Wales 2020; Tse et al. 2023). The HemoClear method costs in Suriname were well below $300 per unit of CP which makes it an attractive alternative based on cost alone (Bihariesingh-Sanchit et al. 2023).

7 Discussion

Convalescent plasma transfusion has been demonstrated to be an effective and cost-effective means of conferring passive immunity against a broad range of viruses in infected individuals and was thus implemented as a method of treatment and immunomodulation in patients infected with SARS-CoV-2 in the recent global pandemic, with promising efficacy and safety results demonstrated in multiple trials (Selvi 2020; Sullivan et al. 2022; Tworek et al. 2021; Ipe et al. 2022; Casadevall et al. 2021; Salazar et al. 2019; Klassen et al. 2021a, b; Allahyari et al. 2021; Joyner et al. 2020, 2021; Bihariesingh-Sanchit et al. 2023).

Studies have associated convalescent plasma donations containing high antibody titers with improved mortality outcomes for SARS-CoV-2-infected recipients (Focosi et al. 2020a; Tiberghien et al. 2020; Ipe et al. 2022; Klassen et al. 2021a; Joyner et al. 2021; Lee and Hsueh 2020; Lee et al. 2021). Peak neutralizing antibody levels have been demonstrated at approximately 4 weeks following symptom onset, indicating that this is the optimal point at which plasma for convalescent transfusion should be obtained from donors (Xu et al. 2021; Lee et al. 2021). In emergency settings such as the recent global pandemic, this recommendation warrants rapid mobilization of resources to enable the acquisition of plasma donations before the decline of antibody levels in donors, necessitating large-scale pandemic-readiness and ease of access to equipment that will facilitate this process.

The HemoClear method utilizes membrane-based apheresis techniques to separate blood components, allowing donated plasma to be utilized for convalescent therapy. The HemoClear device is a low-cost alternative to traditional centrifugal or membrane-based blood separation techniques, and its portability and accessibility favor its use in lower resourced countries (Bihariesingh-Sanchit et al. 2023; Osemwengie et al. 2022). Convalescent plasma generated at the bedside removes the need for external blood centers with their associated expenses and constraints, allowing rapid mobilization of convalescent plasma therapy in situations such as viral pandemics, where large volumes of plasma need to be obtained promptly.

Trials in Suriname have revealed the value of the HemoClear device in its ability to be operated at the bedside by non-skilled operators such as nurses and thus remove some of the previously identified barriers to convalescent plasma use in low- to middle-income countries or in pandemic healthcare settings with limited dedicated and skilled labor to perform conventional plasmapheresis (Bihariesingh et al. 2021; Bihariesingh-Sanchit et al. 2023). The use of the HemoClear device reduces reliance on external blood-processing centers, allowing hospitals to become more self-sufficient in their ability to harvest plasma for donation and provide convalescent plasma transfusions to recipients in a single setting.

Convalescent plasma obtained from donors using the HemoClear method has been demonstrated to contain a higher concentration of intact, functional platelets than conventional methods (Gifford et al. 2018; Hoetink et al. 2020; Osemwengie et al. 2022). The regenerative and immunomodulatory effects of transfused platelets and their derived growth factors continue to be investigated and may play a role in the improved response of convalescent plasma recipients to viral infections such as COVID-19 (Osemwengie et al. 2022; Burnouf and Goubran 2022; Barro et al. 2021; Burnouf 2018). If this is the case, then the plasma filtrate obtained using the HemoClear device may prove superior to that derived using other blood-processing techniques.

The HemoClear method, employed as a local supply chain for convalescent plasma donation and production, as illustrated in Fig. 5, could result in greater global applicability and availability of convalescent plasma during future new virus pandemics.

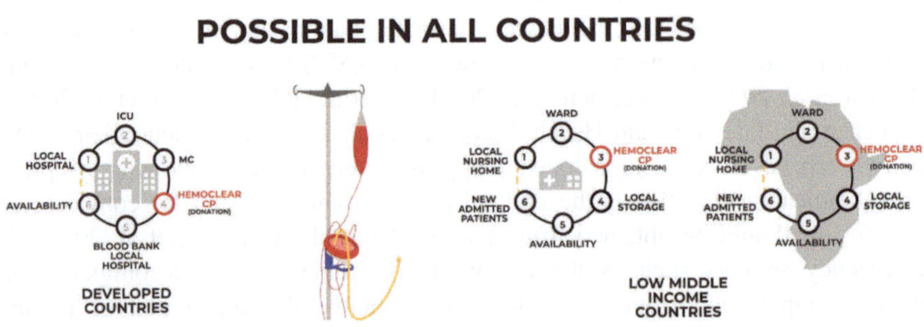

Fig. 5 Impact of a local supply chain workflow of HemoClear plasma donation method on global applicability and availability of convalescent plasma

8 Conclusion

Convalescent plasma therapy has shown promising results in the reduction of disease progression and mortality rates in COVID-19-infected individuals, and the possibility of its use in future viral disease outbreaks. The recent pandemic has highlighted the need for cost-effective, accessible, and easy-to-use alternatives to conventional centrifugation techniques.

The HemoClear device is cost-effective, portable, requires little technical expertise, and can be used in multiple settings even without access to a stable electricity supply, rendering it ideal for use in low- to middle-income countries by staff without in-depth knowledge of blood-processing techniques. The speed of whole blood separation by this method can aid in rapid turnaround, allowing donations to be made and transfused to recipients within the optimal donation window. Many more donations can be processed using the HemoClear method when compared with conventional methods in the same timeframe.

The use of the HemoClear device can promote self-sufficiency in hospitals, using recovered patients as donors for newly infected patients without the expense, time consumption, and processing needs of conventional blood bank plasma separation techniques. These characteristics make HemoClear a viable and effective alternative to expensive conventional blood plasmapheresis machines, allowing lesser-resourced countries to be prepared for similar viral outbreaks without the need for extensive existing infrastructure.

Acknowledgements Illustrations are created by Gert Cruysen by translating content to visuals (www.gertcruysen.nl).

Conflicts of Interest
Dr. Nierich is the inventor of the HemoClear filter and holds stock ownership in HemoClear BV, Dr. Stolteweg 70, 8025 AZ Zwolle, Netherlands.

The other authors have no conflicts to declare.

References

Aleccia J (2021) The hype has faded, but don't count out convalescent plasma in Covid battle. KHN News 1–6. Available from: https://khn.org/news/article/convalescent-plasma-covid-battle-super-donors/

Allahyari A, Seddigh-Shamsi M, Mahmoudi M, Amel Jamehdar S, Amini M, Mozdourian M et al (2021) Efficacy and safety of convalescent plasma therapy in severe COVID-19 patients with acute respiratory distress syndrome. Int Immunopharmacol 93:107239. Available from: https://linkinghub.elsevier.com/retrieve/pii/S1567576920337061

Amenge J, Scherphof S, Osemwengie D, Nierich A, Lagerberg JW (2022) Comparison of washing efficiency and recovery of blood cells between centrifugation, coarse filtration and microfiltration techniques to prepare autologous blood for transfusion. J Blood Med 13:549–558.

Available from: https://www.dovepress.com/comparison-of-washing-efficiency-and-recovery-of-blood-cells-between-c-peer-reviewed-fulltext-article-JBM

Bacal CJO, Maina JW, Nandurkar HH, Khaleel M, Guijt R, Chang Y et al (2021) Blood apheresis technologies—a critical review on challenges towards efficient blood separation and treatment. Mater Adv 2(22):7210–7236. Available from: http://xlink.rsc.org/?DOI=D1MA00859E

Barro L, Burnouf PA, Chou ML, Nebie O, Wu YW, Chen MS et al (2021) Human platelet lysates for human cell propagation. Platelets 32(2):152–162. https://doi.org/10.1080/09537104.2020.1849602

Behring H, Kitasato H (1890) Ueber das Zustandekommen der Diphtherie-Immunität und der Tetanus-Immunität bei Thieren. DMW - Dtsch Medizinische Wochenschrift 16(49):1113–1114. https://doi.org/10.1055/s-0029-1207589

Bihariesingh R, Bansie R, Fröberg J, Ramdhani N, Mangroo R, Bustamente D et al (2021) Mortality reduction in ICU-admitted COVID-19 patients in suriname after treatment with convalescent plasma acquired via gravity filtration. Anesth Clin Res 2(2):1–12. Available from: https://www.sciencerepository.org/mortality-reduction-in_ACR-2021-2-103

Bihariesingh-Sanchit R, Bansie R, Ramdhani N, Mangroo R, Bustamente D, Diaz E et al (2023) Reduced mortality among COVID-19 ICU patients after treatment with hemoclear convalescent plasma in Suriname. Casadevall A, editor. MBio 14(2). https://doi.org/10.1128/mbio.03379-22

Bihariesingh-Sanchit R, Bansie R, Sam STA, Lieveld C, Simons A, Kromosoeto J et al (2024) Gravity-driven crossflow membrane filtration and mechanical centrifugation yield plasma with equivalent total and COVID-19-specific immunoglobulin content. World J Adv Res Rev 21 (02):1107–1115. Available from: https://wjarr.com/sites/default/files/WJARR-2024-0527.pdf

Bloch EM, Shoham S, Casadevall A, Sachais BS, Shaz B, Winters JL et al (2020) Deployment of convalescent plasma for the prevention and treatment of COVID-19. J Clin Invest 130 (6):2757–2765. http://www.jci.org/articles/view/138745

Bloch EM, Goel R, Montemayor C, Cohn C, Tobian AAR (2021) Promoting access to COVID-19 convalescent plasma in low- and middle-income countries. Transfus Apher Sci 60(1):102957. Available from: https://linkinghub.elsevier.com/retrieve/pii/S1473050220302731

Burnouf T (2018) Multifaceted regenerative lives of 'expired' platelets. ISBT Sci Ser 13(3):323–330

Burnouf T, Goubran HA (2022) Regenerative effect of expired platelet concentrates in human therapy: an update. Transfus Apher Sci 61(1):103363. https://doi.org/10.1016/j.transci.2022.103363

Casadevall A, Dragotakes Q, Johnson PW, Senefeld JW, Klassen SA, Wright RS et al (2021) Convalescent plasma use in the usa was inversely correlated with covid-19 mortality. Elife 10:1–14

Casadevall A, Scharff MD (1995) Return to the past: the case for antibody-based therapies in infectious diseases. Clin Infect Dis 21:150–161. Available from: https://academic.oup.com/cid/article/21/1/150/402600

Delamou A, Haba NY, Mari-Saez A, Gallian P, Ronse M, Jacobs J et al (2016) Organizing the donation of convalescent plasma for a therapeutic clinical trial on ebola virus disease: the experience in guinea. Am J Trop Med Hyg 95(3):647–653. https://doi.org/10.4269/ajtmh.15-0890

Delgado DR, Romero E (2020) A comparative analysis between nanomembrane-based therapeutic plasmapheresis and other plasmapheresis methods and their potential use during the COVID-19 pandemic. Altern Integr Med 9(3):1–5

Edwards T, Semple MG, De Weggheleire A, Claeys Y, De Crop M, Menten J et al (2016) Design and analysis considerations in the Ebola_Tx trial evaluating convalescent plasma in the treatment of Ebola virus disease in Guinea during the 2014–2015 outbreak. Clin Trials J Soc Clin Trials 13(1):13–21. https://doi.org/10.1177/1740774515621056

Finland M (1982) Gaps in therapy for infectious diseases: a historical perspective. J Infect Dis 145 (3):401–407. Available from: http://www.jstor.org/stable/30109497

Focosi D, Anderson AO, Tang JW, Tuccori M (2020) Convalescent plasma therapy for COVID-19: state of the art. Clin Microbiol Rev 33(4). https://doi.org/10.1128/CMR.00072-20

Focosi D, Anderson AO, Tang JW, Tuccori M (2020) Convalescent plasma therapy for COVID-19: state of the art. Clin Microbiol Rev 33(4). Available from: https://cmr.asm.org/content/33/4/e00072-20

Garraud O, Heshmati F, Pozzetto B, Lefrere F, Girot R, Saillol A et al (2016) Plasma therapy against infectious pathogens, as of yesterday, today and tomorrow. Transfus Clin Biol 23 (1):39–44. https://doi.org/10.1016/j.tracli.2015.12.003

Gifford SC, Strachan BC, Xia H, Vörös E, Torabian K, Tomasino TA et al (2018) A portable system for processing donated whole blood into high quality components without centrifugation. Bönig H, editor. PLoS One 13(1):e0190827. https://doi.org/10.1371/journal.pone.0190827

van Griensven J, De Weiggheleire A, Delamou A, Smith PG, Edwards T, Vandekerckhove P et al (2016) The use of ebola convalescent plasma to treat ebola virus disease in resource-constrained settings: a perspective from the field. Clin Infect Dis 62(1):69–74. https://doi.org/10.1093/cid/civ680

Hafer C, Golla P, Gericke M, Eden G, Beutel G, Schmidt JJ et al (2016) Membrane versus centrifuge-based therapeutic plasma exchange: a randomized prospective crossover study. Int Urol Nephrol 48(1):133–138

Health Technology Wales (2020) Convalescent plasma: a costing review

Hoetink A, Scherphof SF, Mooi FJ, Westers P, van Dijk J, van de Leur SJ et al (2020) An in vitro pilot study comparing the novel HemoClear gravity-driven microfiltration cell salvage system with the conventional centrifugal XTRA^{TM} autotransfusion device. Anesthesiol Res Pract 2020:1–10. Available from: https://www.hindawi.com/journals/arp/2020/9584186/

Ipe TS, Ugwumba B, Spencer HJ, Le T, Ridenour T, Armitage J et al (2022) Treatment of COVID-19 patients with two units of convalescent plasma in a resource-constrained state. Lab Med 1–6. https://doi.org/10.1093/labmed/lmac055/6623577

Jorda A, Kussmann M, Kolenchery N, Siller-Matula JM, Zeitlinger M, Jilma B et al (2022) Convalescent plasma treatment in patients with Covid-19: a systematic review and meta-analysis. Front Immunol 13:1–9. Available from: https://doi.org/10.3389/fimmu.2022.817829/full

Joyner MJ, Bruno KA, Klassen SA, Kunze KL, Johnson PW, Lesser ER et al (2020) Safety update: COVID-19 convalescent plasma in 20,000 hospitalized patients. Mayo Clin Proc 95 (9):1888–1897. Available from: https://linkinghub.elsevier.com/retrieve/pii/S0025619620306510

Joyner MJ, Carter RE, Senefeld JW, Klassen SA, Mills JR, Johnson PW et al (2021) Convalescent plasma antibody levels and the risk of death from Covid-19. N Engl J Med 384(11):1015–1027. https://doi.org/10.1056/NEJMoa2031893

Kes P, Janssens ME, Bašić-Jukić N, Kljak M (2016) A randomized crossover study comparing membrane and centrifugal therapeutic plasma exchange procedures. Transfusion 56(12):3065–3072

Klassen SA, Senefeld JW, Johnson PW, Carter RE, Wiggins CC, Shoham S et al (2021a) The effect of convalescent plasma therapy on mortality among patients with COVID-19: systematic review and meta-analysis. Mayo Clin Proc 96(5):1262–1275. https://doi.org/10.1016/j.mayocp.2021.02.008

Klassen SA, Senefeld JW, Senese KA, Johnson PW, Wiggins CC, Baker SE et al (2021b) Convalescent plasma therapy for COVID-19: a graphical mosaic of the worldwide evidence. Front Med 8:2–10. https://doi.org/10.3389/fmed.2021.684151/full

Lee WT, Girardin RC, Dupuis AP, Kulas KE, Payne AF, Wong SJ et al (2021) Neutralizing antibody responses in COVID-19 convalescent sera. J Infect Dis 223(1):47–55

Lee PI, Hsueh PR (2020) Emerging threats from zoonotic coronaviruses-from SARS and MERS to 2019-nCoV. J Microbiol Immunol Infect

Li L, Zhang W, Hu Y, Tong X, Zheng S, Yang J et al (2020) Effect of convalescent plasma therapy on time to clinical improvement in patients with severe and life-threatening COVID-19. JAMA 324(5):460. Available from: https://jamanetwork.com/journals/jama/fullarticle/2766943

Libster R, Pérez Marc G, Wappner D, Coviello S, Bianchi A, Braem V et al (2021) Early high-titer plasma therapy to prevent severe Covid-19 in older adults. N Engl J Med 384(7):610–618. https://doi.org/10.1056/NEJMoa2033700

Luke TC, Kilbane EM, Jackson JL, Hoffman SL (2006) Meta-analysis: convalescent blood products for spanish influenza pneumonia: a future H5N1 treatment? Ann Intern Med 145 (8):599. https://doi.org/10.7326/0003-4819-145-8-200610170-00139

Muttamba W, Lusiba J, Namakula LO, Byakika-Kibwika P, Ssali F, Ddungu H et al (2021) Feasibility of collecting and processing of COVID-19 convalescent plasma for treatment of COVID-19 in Uganda. Hosuru Subramanya S, editor. PLoS One 16(6):e0252306. https://doi.org/10.1371/journal.pone.0252306

Natarajan H, Crowley AR, Butler SE, Xu S, Weiner JA, Bloch EM et al (2021) Markers of polyfunctional SARS-CoV-2 antibodies in convalescent plasma. Diamond MS, editor. MBio 12(2):1–14. https://doi.org/10.1128/mBio.00765-21

Noerwidayati E, Pribadi F (2023) An analysis of convalescent plasma unit cost calculation using time-driven activity-based costing (TDABC) method. Bali Med J. 12(1):331–339

Osemwengie D, Lagerberg JW, Vlaar R, Gouwerok E, Go M, Nierich AP et al (2022) Recovery of platelet-rich red blood cells and acquisition of convalescent plasma with a novel gravity-driven blood separation device. Transfus Med 32(1):53–63. https://doi.org/10.1111/tme.12830

Post N, Eddy D, Huntley C, van Schalkwyk MCI, Shrotri M, Leeman D et al (2020) Antibody response to SARS-CoV-2 infection in humans: a systematic review. PLoS One 15(12):1–27. https://doi.org/10.1371/journal.pone.0244126

Redant S, De Bels D, Ismaili K, Honoré PM (2021) Membrane-based therapeutic plasma exchange in intensive care. Blood Purif 50(3):290–297

Salazar E, Christensen PA, Graviss EA, Nguyen DT, Castillo B, Chen J et al (2019) Significantly decreased mortality in a large cohort of coronavirus disease 2019 (COVID-19) patients transfused early with convalescent plasma containing high-titer anti–severe acute respiratory syndrome coronavirus 2 (SARS-CoV-2) spike protein IgG. Am J Pathol 191(1):90–107. Available from: https://linkinghub.elsevier.com/retrieve/pii/S0002944020304892

Selvi V (2020) Convalescent plasma: a challenging tool to treat COVID-19 patients—a lesson from the past and new perspectives. Biomed Res Int 2020:1–8. Available from: https://www.hindawi.com/journals/bmri/2020/2606058/

Shen L, Wang C, Zhao J, Tang X, Shen Y, Lu M et al (2020) Delayed specific IgM antibody responses observed among COVID-19 patients with severe progression. Emerg Microbes Infect 9(1):1096–1101. https://doi.org/10.1080/22221751.2020.1766382

Sullivan DJ, Gebo KA, Shoham S, Bloch EM, Lau B, Shenoy AG et al (2022) Early outpatient treatment for Covid-19 with convalescent plasma. N Engl J Med

Tiberghien P, de Lamballerie X, Morel P, Gallian P, Lacombe K, Yazdanpanah Y (2020) Collecting and evaluating convalescent plasma for COVID-19 treatment: why and how? Vox Sang 115(6):488–494. https://doi.org/10.1111/vox.12926

Tse P, Jamula E, Cook R, Fergusson D, Cushing M, Arnold D (2023) Cost-effectiveness of convalescent plasma for COVID-19 : a cost-effectiveness analysis of the CONCOR-1 randomized trial

Tworek A, Jaroń K, Uszyńska-Kałuża B, Rydzewski A, Gil R, Deptała A et al (2021) Convalescent plasma treatment is associated with lower mortality and better outcomes in high-risk COVID-19 patients—propensity-score matched case-control study. Int J Infect Dis 105:209–215

WHO Blood Regulators Network (2014) Position paper on collection and use of convalescent plasma or serum as an element in middle east respiratory syndrome coronavirus response [Internet]. WHO. Available from: http://www.who.int/blood-products/brn/BRN_PositionPaperConvPlasmaMERSCoV_March2014.pdf

Williams ME, Balogun RA (2014) Principles of separation: indications and therapeutic targets for plasma exchange. Clin J Am Soc Nephrol 9(1):181–190

Xu X, Nie S, Wang Y, Long Q, Zhu H, Zhang X et al (2021) Dynamics of neutralizing antibody responses to SARS-CoV-2 in patients with COVID-19: an observational study. Signal Transduct Target Ther 6(1):197. https://doi.org/10.1038/s41392-021-00611-6

Dr. Arno Nierich started his career as Anesthesiology and Intensivist after his training at the free University of Brussels, Belgium. In 1996, he was appointed as consultant at the Utrecht University Medical Center, The Netherlands. During these years, he finished his Ph.D. on the topic of anesthetic management during Off-Pump Cardiac surgery by the so-called Octopus method. From 2000 to 2021, he served as Senior Consultant in Cardiac Anesthesia & Intensive Care at Isala Clinics in Zwolle, the Netherlands. He obtained a Master's degree in Business Innovation and became Head of Medical Innovation Isala. He is the author of > 60 scientific papers, including publications in Circulation and JAMA. He has shifted his clinical career to the global world as an innovator and inventor of MedTech devices. He is recently appointed as visiting Professor of Anesthesiology & Intensive Care at the Academic Hospital of Paramaribo, Suriname.

Dr. Rosita Bihariesingh is a Cardio-anesthesiologist and Intensivist in the Academic Hospital Paramaribo, Suriname, a small Dutch-speaking country on the northern coast of South America with a population of 600,000. Since 2006, she has been part of a small dedicated team responsible for all pre- and postoperative care for cardiac and pulmonary surgery patients. Since 2022 she is also the chief trainer for anesthesiology residents in Suriname. Working in a third world/ developing country demands dedication and inventiveness because of scarce resources. Blood management, plasmapheresis, and convalescent plasma in the COVID period are among her fields of interest. She recently received the Jan Taco te Gussinklo Award for her work on obtaining convalescent plasma using the HemoClear filter for treatment of COVID-19 patients.

Dr. Rakesh Bansie is an Internist/Clinical Immunologist in the Academic Hospital of Paramaribo. In the wide array of systemic diseases, he considers the multi-disciplinary approach as one of the best methods for achieving optimal care for patients even in resource-limited settings.

Passive Antibody and the Next Pandemic

Generating the Evidence Base for Convalescent Plasma Use for a New Infectious Disease

Hyunah Yoon and Liise-anne Pirofski

Contents

Abstract Severe acute respiratory syndrome coronavirus 2 (SARS-CoV-2) swept across the world in the waning months of 2019 and emerged as the cause of the coronavirus disease 19 (COVID-19) pandemic in early 2020. The use of convalescent plasma (CP) for prior respiratory pandemics provided a strong biological rationale for the rapid deployment of COVID-19 convalescent plasma (CCP) in early

---

Both authors contributed equally and share co-corresponding authorship.

Dr. Pirofski and Dr. Yoon were co-principal investigators at Montefiore Medical Center of an NIH-funded multicenter randomized clinical trial (CONTAIN) and observational studies of COVID-19 Convalescent Plasma therapy in hospitalized patients with COVID-19. Their work on COVID-19 convalescent plasma has contributed to the accumulating evidence of how to use convalescent plasma in a pandemic.

H. Yoon · L. Pirofski (✉)
Division of Infectious Diseases, Albert Einstein College of Medicine and Montefiore Medical Center, Bronx, NY, USA
e-mail: l.pirofski@einsteinmed.edu

H. Yoon
e-mail: hyoon@montefiore.org

L. Pirofski
Department of Microbiology and Immunology, Albert Einstein College of Medicine and Montefiore Medical Center, Bronx, NY, USA

Current Topics in Microbiology and Immunology (2025) 443: 205–217
https://doi.org/10.1007/82_2024_275

Published Online: 9 August 2024

2020 when no validated treatments or prior immunity existed. CCP is an antiviral agent, with its activity against SARS-CoV-2 stemming from specific antibodies elicited by the virus. Early efforts to investigate the efficacy of CCP in randomized clinical trials (RCTs) that targeted hospitalized patients with COVID-19 did not demonstrate the overall efficacy of CCP despite signals of benefit in certain subgroups, such as those treated earlier in disease. In contrast, studies adhering to the principles of antibody therapy in their study design, choice of patient population, and product qualification, i.e., those that administered high levels of specific antibody during the viral phase of disease in immunocompromised or very early in immunocompetent individuals, demonstrated benefits. In this chapter, we leverage the knowledge gained from clinical studies of CCP for COVID-19 to propose a framework for future studies of CP for a new infectious disease. This framework includes obtaining high-quality CP and designing clinical studies that adhere to the principles of antibody therapy to generate a robust evidence base for using CP.

Glossary

ACTIV	Accelerating COVID-19 Therapeutic Interventions and Vaccines
ACTT	Adaptive COVID-19 Treatment Trial
CCP	COVID-19 Convalescent Plasma
CP	Convalescent Plasma
COVID-19	Coronavirus Disease 2019
DSMB	Data Safety Monitoring Board
IRB	Institutional Review Board
NHLBI	National Heart, Lung, and Blood Institute
NIAID	National Institute of Allergy and Infectious Diseases
NIH	National Institutes of Health
RCT	Randomized Controlled Trial
RECOVERY	Randomized Evaluation of COVID-19 Therapy
REMAP-CAP	Randomized, Embedded, Multifactorial Adaptive Platform Trial for Community-Acquired Pneumonia
US	United States

1 The Building Blocks of a CP Research Framework

1.1 Diagnostics

Studying antiviral agents for a new infectious disease will depend upon the availability of accurate diagnostic tests. The initial development and rollout of SARS-CoV-2 diagnostic platforms in the US encountered numerous roadblocks (World Health 2020; McAdam 2020). Eventually, multiple antigen detection platforms became available (Peeling et al. 2022; Weissleder et al. 2020), but quantitative measures of SARS-CoV-2 burden were not developed for clinical use, nor were quantitative and/or functional assessments of SARS-CoV-2 antibody immunity. Since CCP is most beneficial in the early (viral) phase of COVID-19, it is logical to posit that its use in studies could have been optimized with objective measures of viral load and antibody immunity, as was done in some monoclonal antibody trials (Weinreich et al. 2020; Dougan et al. 2021).

A new infectious disease can be caused by a known or an unknown agent. For a known agent, diagnostic platforms to identify the causative agent as well as serological assays to measure specific antibody are likely to be available. However, when an unknown agent is responsible for an outbreak, neither diagnostic nor serological tools will be immediately available. Although this was the case when COVID-19 struck in November–December 2019 in China, rapid analysis of respiratory samples made it possible to identify a viral agent, SARS-CoV-2 as the cause of COVID-19 within weeks of the onset of the disease (Zhu et al. 2020; Wu et al. 2020). While the approach to studying CP will differ for a known and unknown agent, CP can theoretically be deployed in either case since those who have recovered will have produced a specific antibody response.

1.2 Obtaining CP

Collection and characterization of CP: Community engagement and efforts to encourage recovered persons to donate plasma are essential to the development of an adequate supply of CP as described in Chap. 13. Many potential CCP donors came forth during the first wave of the COVID-19 pandemic (Yoon et al. 2021). Although precise guidance on the timing of CP collection must be devised for the relevant agent, the CP should contain a mature antibody response. While early isotypes, such as IgM and IgA exhibited the ability to neutralize SARS-CoV-2 (Gasser et al. 2021; Sterlin et al. 2021), affinity-matured IgG is most likely to exhibit neutralizing activity and as well as non-neutralizing functional activities that may enhance viral clearance and dampen inflammation (Casadevall et al. 2020). Affinity-matured antibodies are most likely to be present in the plasma of recovered individuals in weeks to a month after recovery, but there was ample evidence that SARS-CoV-2 antibody levels wane, particularly in non-immune individuals (Self et al. 2020). In the event of an outbreak with a known agent for which a vaccine is available, plasma from vaccinated individuals and/or vaccinated individuals who recovered from the disease may have the highest amount of antibody and the most antibody function (Ripoll et al. 2022; Ordaya et al. 2022; Vickers et al. 2021).

Vetting CP for use: A standardized approach to identifying donors, obtaining CP, and rapidly quantifying antibody content and functional activity should be developed as soon as possible in an infectious disease emergency as described in Chaps. 6 and 12. This will require a period of optimization for each component of this workflow, e.g., identifying the best donors and developing the best assay platforms. The support of an overarching infrastructure and network is needed to ensure that the process of vetting CP is efficient, streamlined, and completed in as short a timeframe as possible.

CP used for a known agent should adhere to the quantitative principle of antibody therapy (Cecil 1939) and be vetted for binding and functional activity. Based on extensive evidence from CCP use, CP should be "high titer" (Joyner et al. 2021; Focosi and Franchini 2021; Libster et al. 2021). Higher titer CCP had the best correlation with antiviral neutralization (Salazar et al. 2020). As of early 2024, available CCP data suggest that the best way to identify CP with the highest level of neutralizing activity may be to select plasma with specific binding measurements in the upper 20th percentile of donated units based on a standardized platform (Sullivan and Casadevall 2023). However, non-neutralizing activities of CP are also likely to mediate beneficial effects (Herman et al. 2022; Natarajan et al. 2021; Bandopadhyay et al. 2021) and this is an area that should be investigated for CP collected in the future. While the need for local CP must be determined for each causative agent (Kunze et al. 2021), it is likely that CP with a high titer and high functional activity will be active against both local and non-local agents (Sullivan and Casadevall 2023; Li et al. 2022).

Compared to small molecule antiviral agents, which must be studied against a known agent to determine their antiviral activity, in theory CP could be used to treat an unknown agent. This is because CP contains specific antibodies to the causative agent of a disease provided the agent induces an antibody response. The development of specific antibodies is a hallmark of the immune response of an infectious agent for those who can generate antibodies. Therefore, CP from individuals who recover from even an unknown infectious disease will contain specific antibodies with the potential to mitigate the causative agent upon transfer to affected persons (Casadevall and Pirofski 2020).

2 Study and Trial Design

The Rational Approach to Study Design: The fundamental biological rationale for employing CP remains consistent for both known and potential infectious agents. Therefore, the historical and collective body of evidence from studies that adhered to the principles of antibody therapy should form the basis for the rational design of CP studies for a new infectious disease. In this context, assessing the presence of clinical equipoise—genuine uncertainty—that justifies the design of an RCT should be grounded in the accumulated body of evidence. It is most sensible to assess the equipoise concerning antibody therapy as a function of the immunological status of individuals and the level of immunity within the population.

Initial CCP trials predominantly included hospitalized, non-immunocompromised patients with hypoxia, who it was later learned were beyond the viral phase during which an antiviral agent would be most effective (Simonovich et al. 2020; RECOVERY Collaborative Group 2021; Ortigoza et al. 2022; Investigators REMAP-CAP Investigators 2021; Bégin et al. 2021; Agarwal et al. 2020; Self et al. 2022). For this reason, clinical equipoise remains, as the studies were neither designed nor powered to effectively test the hypothesized mechanism. Demonstrating this point, smaller RCTs that strategically minimized disease spectrum heterogeneity showed that high titer CCP was effective when administered early in the disease to targeted groups such as the elderly (Libster et al. 2021), ambulatory patients (Libster et al. 2021; Sullivan et al. 2022) and in hospitalized patients with early acute respiratory distress syndrome (Misset et al. 2023). While these findings are encouraging, enough uncertainty may remain for CP use in non-immunocompromised hospitalized and/or hypoxic patients to support the conduct of a well-designed, placebo-controlled RCT of high titer CP, targeting high-risk patients with a large enough sample size. Such a trial could potentially disrupt the current state of clinical equipoise and provide robust evidence of the efficacy of CP.

The principle of clinical equipoise, however, faces challenges in immunocompromised individuals with B cell deficiencies and/or inability to produce antibodies, in whom a strong biological rationale for benefit exists alongside difficulties in generating well-powered RCT evidence. These individuals may experience a prolonged viral phase without an endogenous antibody response, creating ideal conditions for high titer CP therapy to be effective, regardless of symptom duration (Senefeld et al. 2023; Hueso et al. 2020; Thompson et al. 2021; Denkinger et al. 2023). Different levels of evidence in totality support this biological rationale (Senefeld et al. 2023; Thompson et al. 2021), raising ethical questions about the appropriateness of placebo-controlled RCT in immunocompromised individuals for whom small molecule antiviral agents are often contraindicated.

A solution is to employ alternative study designs to address the remaining knowledge gaps. Examining the interaction of CP with other antiviral agents within an RCT design (Ullah et al. 2024), or to retrospectively analyze the dose–response relationship of CP (Joyner et al. 2021), where the dose is not controlled prospectively, could investigate causality and add rigor to the accumulating evidence. Once effectiveness is established, a comprehensive exploration of the characteristics and biological mechanisms of antibody activity—including titer, isotypes, affinity, avidity, and Fc-effector activity—could provide deeper insights into the optimal use of CP. Such approaches could yield meaningful data amid a pandemic, during which time prioritizing resource allocation is crucial. Study designs in a pandemic setting should be informed by an open-minded perspective, grounded in a biological rationale, and contextualized within available collective evidence, to efficiently generate findings that are clinically meaningful.

Building upon this rational approach to study designs of CP use, the following sections put forth a framework to conduct clinical trials for a new infectious disease, informed by lessons learned from CCP studies (see Table 1).

Efficient and Flexible Study Design: Since outbreaks of infectious diseases may spread and abate rapidly, clinical trials require tractable approaches that minimize

Table 1 Framework for conducting future CP studies

Area of focus	Key Points
Rapid collection and vetting of high-quality CP	• Strengthen community and blood center collaborations for CP collection • Leverage rapid availability of CP after the initial onset of a disease wave when no other therapeutics are available • Standardize the product for RCTs focusing on high antibody concentration and functional activity • Timely obtainment of Investigational New Drug application in the US, or its equivalent outside the US
Rational approach to study design	• Synthesize and utilize accumulated knowledge in assessment of clinical equipoise for CP use • Consider the use of a Bayesian framework that will allow more nuanced and adaptive decision-making • Non-placebo trials if clinical equipoise is resolved
Efficient and flexible study design and conduct	• Adopt adaptive design for efficiency and flexibility • Ensure continuous communication with the DSMB for ongoing trial adjustments • Implement a centralized and streamlined trial process with uniform protocols and procedures • Integrate trials into standard care • Limit the use of CP on compassionate grounds without supporting evidence • Automate data extraction from electronic medical records to minimize manual data collection efforts
Targeted disease spectrum	• Focus on an early disease state considered to be driven by viral replication rather than inflammation • Incorporate biomarkers to identify high-risk individuals with high viral loads and insufficient endogenous antibody production • Achieve a balance between a pragmatic approach and excessive heterogeneity to yield clinically applicable evidence
Meaningful outcome measures	• Pre-define clinically meaningful outcome measures, including surrogate biomarkers that inform biological mechanism • Include long-term outcomes to account for post-acute infection syndromes
Recruitment innovation	• Expedite recruitment processes • Innovate accessibility strategies (mobile trial unit, mail-order) • Utilize educational material and social media for wider outreach and participant engagement
Diversity and inclusivity	• Include immunocompromised, pregnant, pediatric, rural and minority populations • Include individuals most affected by the initial wave of disease to improve the generalizability of findings • Build a diverse study team

(continued)

Table 1 (continued)

Area of focus	Key Points
	• Develop strategies to reach individuals with limited access to healthcare due to barriers (age, financial, language, cultural, and/or health literacy)
Learning from experience	• Embrace flexibility to adapt to unpredictable pandemic trends • Value the insights from all trial outcomes, including null results • Stay vigilant and prepared for shifts in pandemic dynamics, such as new virus variants, emerging therapeutics, and evolving population immunity. Make timely adjustments in study protocols and focus
Foster stakeholder collaboration	• Effective and respectful collaboration among a wide range of stakeholders including trialists, clinicians, patients, regulatory bodies, blood banks, hospital leadership, DSMB, and research scientists • Open communication and ongoing dialog regarding any obstacles, new discoveries, and ideas • Build trust among stakeholders, especially with community groups most affected by the disease

logistical barriers and optimize patient recruitment and diversity. Some well-designed CCP RCTs terminated early when pandemic waves abated, highlighting this point (Libster et al. 2021; Li et al. 2020). For CP trials, this will require an infrastructure to support multicenter clinical trials, oversight of CP qualification for use, trial governance, and universal trial and IRB protocols and consent documents. In the US, the NIH-sponsored ACTT (NIAID) and ACTIV (NHLBI) supported multicenter trials that were facilitated by dedicated teams responsible for trial design and oversight.

The use of a Bayesian approach to study design and analysis is particularly beneficial in a pandemic setting, as it facilitates the iterative updating of prior beliefs with newly observed data, enabling rapid conclusions to guide clinical care and/or to adjust trial parameters. This approach is well-suited for an adaptive trial design, where modifications in trial parameters are allowed according to pre-defined rules and evolving evidence. The role of the DSMB is particularly important in this context, to facilitate timely adaptations to the trial protocol. During the COVID-19 pandemic, real-time patient-level meta-analysis emerged as an innovative method to accelerate evidence acquisition (Troxel et al. 2022; Petkova et al. 2020). Platform trials, exemplified by the successful evaluation of corticosteroids in the RECOVERY trial (RECOVERY Collaborative Group et al. 2021) and interleukin-6 antagonist in the REMAP-CAP trial (Investigators 2021), along with factorial design trials such as the COVID-OUT trial (Bramante et al. 2022), can rapidly generate efficacy data. Leveraging this approach for future studies of CP with careful consideration of the target disease spectrum shared with other simultaneously tested agents could expedite research progress. Once clinical equipoise is resolved, late-phase trials can explore various dosing strategies of CP (number of units or dosing frequency) or its combination with other therapeutic agents.

Targeted Disease Spectrum: The use of biomarkers to establish objective inclusion criteria, target the viral replication phase of the disease, and monitor the CP response is important. Although imperfect, the cycle threshold of a polymerase chain reaction test obtained using a uniform platform may serve as a proxy of viral load. Caution should be exercised with the use of shorter symptom duration as a manifestation of early disease for inpatient RCTs, as it may select patients with rapid disease progression, which may be misinterpreted as "early" disease. While there is no efficacy data to support the use of CCP for prophylaxis (Shoham et al. 2023), it remains an important area for future investigation, given the temporal principle of antibody therapy (Casadevall et al. 2021a), especially when vaccination is not available or effective.

Clinically Meaningful Outcome Measures: The selection of clinical outcomes in trials should be tailored to the study population, clinical relevance, and pandemic phase to ensure an adequate sample size. For inpatient studies of respiratory diseases, outcomes such as oxygenation and mortality are crucial, but caution must be exercised to separate effects on mortality and ventilation as competing risks. Outpatient studies might focus on preventing hospitalization or the onset of hypoxia or other evidence of clinical deterioration. Surrogate outcomes, such as viral clearance and changes in antibody profiles and/or inflammatory markers can provide valuable biological insights and should be included in future studies, although their correlation with clinical efficacy of CCP was challenging due to patient and disease spectrum heterogeneity. As more individuals survive the acute phase of a disease, protracted symptoms may become significant and drive patient-oriented outcome measures (Yoon et al. 2024). These long-term outcome measures should include a wide spectrum of clinical symptoms and biomarkers collected systematically.

Collaboration and Recruitment Innovation: A collaborative approach, involving stakeholders such as trial investigators, clinicians, nurses, regulatory bodies, and patients, is crucial to refine study designs and ensure efficient workflow. Utilization of shared trial infrastructure and a centralized randomization process can streamline RCTs. In addition, integrating RCTs into standard clinical care and leveraging health information technology should be considered. The ability of the RECOVERY trial to rapidly test multiple interventions was partly due to the platform on which it was built, which included flexibility in trial procedures and efficient data collection (RECOVERY Collaborative Group et al. 2021). In outpatient settings, innovative recruitment strategies, such as in-home recruitment (Libster et al. 2021) and mail-order enrollment (as seen in the ACTIV trial (Lindsell 2023)), utilization of telemedicine, and social media engagement (Boulware et al. 2020) can be effective mechanisms to reach broader participant groups.

Diversity and Inclusivity: In conducting clinical studies during an infectious disease outbreak, it is vital to include those disproportionately affected, especially racial and ethnic minority groups and individuals with limited access to healthcare. Additionally, a thoughtful approach to including other vulnerable groups, such as immunocompromised, pregnant persons, those without capacity, and children is important, especially when the product has demonstrated safety, to ensure the generation of knowledge that benefits a diverse population (Trøseid et al. 2022).

Learning from History and Collective Experience: As shown in Chap. 5 on the history of CP use, early administration of the product, attention to the match of serum (or plasma) antibody to disease variant, and level of serum (or plasma) antibody, i.e., the principles of antibody therapy (Casadevall et al. 2021a), were emphasized repeatedly in the early twentieth century when CP emerged as the inaugural antimicrobial therapy and was used extensively. Unfortunately, these historical lessons were not taken into consideration in the design of many CCP trials. The CCP RCTs during the COVID-19 pandemic, while unequivocally demonstrating safety, faced specific challenges such as ensuring CCP product quality and standardization, identifying the most relevant and uniform disease spectrum for investigation, and rapidly adapting to the pandemic's evolving dynamics. These factors contributed to heterogeneity in treatment effects and insufficient power to test the hypothesized mechanism, thereby complicating the assessment of clinical equipoise and the translation of findings to clinical guidelines. The study of CCP in patients with a later phase of COVID-19 in the RECOVERY trial (RECOVERY Collaborative Group et al. 2021) underscores that even the most exemplary trial design can generate uninformative results and risk the dissemination of misleading generalizations about an agent's efficacy if the study did not adhere to biological plausibility (Casadevall et al. 2021b; Casadevall and Pirofski 2023).

As the COVID-19 pandemic evolved, RCT findings often became less applicable, necessitating a shift toward utilizing evidence from well-executed observational studies that better represented vulnerable patient groups, such as immunocompromised patients (Senefeld et al. 2023). RCTs are not without flaws; their focus on internal validity to demonstrate efficacy can inadvertently yield evidence less generalizable to real-world patients, especially those in vulnerable groups who fail to meet the narrow inclusion criteria of "well-designed" RCTs. Systematic reviews and meta-analyses play an essential role in filling evidence gaps. Despite these challenges, RCTs that tested pre-defined hypotheses have deepened our understanding of COVID-19 pathogenesis, the outcomes of viral damage that CP do not address, and the critical need for early intervention among high-risk populations. They also reinforced the principles of antibody therapy and provided explanatory power for some null results. These insights now serve to validate and extend earlier historical lessons to new lessons learned from COVID-19, thereby providing a roadmap for better study designs that incorporate biologically plausible, pragmatic, flexible, and creative approaches to study therapies during future pandemics.

Conflict of Interest Neither author has any conflict of interest to declare.

References

Agarwal A, Mukherjee A, Kumar G, Chatterjee P, Bhatnagar T, Malhotra P (2020) Convalescent plasma in the management of moderate covid-19 in adults in India: open label phase II multicentre randomised controlled trial (PLACID Trial). BMJ 371:m3939. https://doi.org/10.1136/bmj.m3939

Bandopadhyay P, D'Rozario R, Lahiri A et al (2021) Nature and dimensions of systemic hyperinflammation and its attenuation by convalescent plasma in severe COVID-19. J Infect Dis 224(4):565–574. https://doi.org/10.1093/infdis/jiab010

Bégin P, Callum J, Jamula E et al (2021) Convalescent plasma for hospitalized patients with COVID-19: an open-label, randomized controlled trial. Nat Med. https://doi.org/10.1038/s41591-021-01488-2

Boulware DR, Pullen MF, Bangdiwala AS et al (2020) A randomized trial of hydroxychloroquine as postexposure prophylaxis for Covid-19. N Engl J Med 383(6):517–525. https://doi.org/10.1056/NEJMoa2016638

Bramante CT, Huling JD, Tignanelli CJ et al (2022) Randomized trial of metformin, ivermectin, and fluvoxamine for Covid-19. N Engl J Med 387(7):599–610. https://doi.org/10.1056/NEJMoa2201662

Casadevall A, Pirofski LA (2020) The convalescent sera option for containing COVID-19. J Clin Invest 130(4):1545–1548. https://doi.org/10.1172/jci138003

Casadevall A, Pirofski LA (2023) Misinterpretation of clinical research findings and COVID-19 mortality. Ann Intern Med 176(7):993–994. https://doi.org/10.7326/m23-0737%m37186922

Casadevall A, Joyner MJ, Pirofski LA (2020) SARS-CoV-2 viral load and antibody responses: the case for convalescent plasma therapy. J Clin Invest 130(10):5112–5114. https://doi.org/10.1172/jci139760

Casadevall A, Pirofski LA, Joyner MJ (2021) The principles of antibody therapy for infectious diseases with relevance for COVID-19. mBio 12(2). https://doi.org/10.1128/mBio.03372-20

Casadevall A, Dragotakes Q, Johnson PW et al (2021) Convalescent plasma use in the USA was inversely correlated with COVID-19 mortality. eLife 10:e69866. https://doi.org/10.7554/eLife.69866

Cecil RL (1939) Present status of serum therapy in pneumonia. Bull N Y Acad Med 15(2):104–115

Denkinger CM, Janssen M, Schäkel U et al (2023) Anti-SARS-CoV-2 antibody-containing plasma improves outcome in patients with hematologic or solid cancer and severe COVID-19: a randomized clinical trial. Nat Cancer 4(1):96–107. https://doi.org/10.1038/s43018-022-00503-w

Dougan M, Nirula A, Azizad M et al (2021) Bamlanivimab plus Etesevimab in mild or moderate Covid-19. N Engl J Med 385(15):1382–1392. https://doi.org/10.1056/NEJMoa2102685

Focosi D, Franchini M (2021) COVID-19 convalescent plasma therapy: hit fast, hit hard! Vox Sang 116(9):935–942. https://doi.org/10.1111/vox.13091

Gasser R, Cloutier M, Prévost J et al (2021) Major role of IgM in the neutralizing activity of convalescent plasma against SARS-CoV-2. Cell Rep 34(9):108790. https://doi.org/10.1016/j.celrep.2021.108790

Herman JD, Wang C, Burke JS et al (2022) Nucleocapsid-specific antibody function is associated with therapeutic benefits from COVID-19 convalescent plasma therapy. Cell Rep Med 3(11):100811. https://doi.org/10.1016/j.xcrm.2022.100811

Hueso T, Pouderoux C, Péré H et al (2020) Convalescent plasma therapy for B-cell-depleted patients with protracted COVID-19. Blood 136(20):2290–2295. https://doi.org/10.1182/blood.2020008423

The REMAP-CAP Investigators (2021) Interleukin-6 receptor antagonists in critically ill patients with Covid-19. N Engl J Med 384(16):1491–1502. https://doi.org/10.1056/NEJMoa2100433

Writing Committee for the REMAP-CAP Investigators (2021) Effect of convalescent plasma on organ support–free days in critically ill patients with COVID-19: a randomized clinical trial. JAMA 326(17):1690–1702. https://doi.org/10.1001/jama.2021.18178

Joyner MJ, Carter RE, Senefeld JW et al (2021) Convalescent plasma antibody levels and the risk of death from Covid-19. N Engl J Med. https://doi.org/10.1056/NEJMoa2031893

Kunze KL, Johnson PW, van Helmond N et al (2021) Mortality in individuals treated with COVID-19 convalescent plasma varies with the geographic provenance of donors. Nat Commun 12(1):4864. https://doi.org/10.1038/s41467-021-25113-5

Li M, Beck EJ, Laeyendecker O et al (2022) Convalescent plasma with a high level of virus-specific antibody effectively neutralizes SARS-CoV-2 variants of concern. Blood Adv 6 (12):3678–3683. https://doi.org/10.1182/bloodadvances.2022007410

Li L, Zhang W, Hu Y et al (2020) Effect of convalescent plasma therapy on time to clinical improvement in patients with severe and life-threatening COVID-19: a randomized clinical trial. JAMA. https://doi.org/10.1001/jama.2020.10044

Libster R, Pérez Marc G, Wappner D et al (2021) Early high-titer plasma therapy to prevent severe Covid-19 in older adults. N Engl J Med. https://doi.org/10.1056/NEJMoa2033700

Lindsell CJ (2023) ACTIV-6: Operationalizing a decentralized, outpatient randomized platform trial to evaluate efficacy of repurposed medicines for COVID-19. J Clin Transl Sci 7(1):e221. https://doi.org/10.1017/cts.2023.644

McAdam AJ (2020) Special issue on diagnostic testing for severe acute respiratory syndrome Coronavirus 2 and lessons from this pandemic. J Clin Microbiol 58(8). https://doi.org/10.1128/jcm.01324-20

Misset B, Piagnerelli M, Hoste E et al (2023) Convalescent plasma for Covid-19-induced ARDS in mechanically ventilated patients. N Engl J Med. 389(17):1590–1600. https://doi.org/10.1056/NEJMoa2209502, https://pubmed.ncbi.nlm.nih.gov/37889107/

Natarajan H, Crowley AR, Butler SE et al (2021) Markers of polyfunctional SARS-CoV-2 antibodies in convalescent plasma. mBio 12(2). https://doi.org/10.1128/mbio.00765-21

Ordaya EE, Abu Saleh OM, Stubbs JR, Joyner MJ (2022) Vax-plasma in patients with refractory COVID-19. Mayo Clin Proc 97(1):186–189. https://doi.org/10.1016/j.mayocp.2021.11.001

Ortigoza MB, Yoon H, Goldfeld KS et al (2022) Efficacy and safety of COVID-19 convalescent plasma in hospitalized patients: a randomized clinical trial. JAMA Intern Med 182(2):115–126. https://doi.org/10.1001/jamainternmed.2021.6850

Peeling RW, Heymann DL, Teo YY, Garcia PJ (2022) Diagnostics for COVID-19: moving from pandemic response to control. Lancet 399(10326):757–768. https://doi.org/10.1016/s0140-6736(21)02346-1

Petkova E, Antman EM, Troxel AB (2020) Pooling data from individual clinical trials in the COVID-19 era. JAMA 324(6):543–545. https://doi.org/10.1001/jama.2020.13042

RECOVERY Collaborative Group (2021) Convalescent plasma in patients admitted to hospital with COVID-19 (RECOVERY): a randomised controlled, open-label, platform trial. Lancet 397(10289):2049–2059. https://doi.org/10.1016/s0140-6736(21)00897-7

RECOVERY Collaborative Group, Horby P, Lim WS, et al. (2021) Dexamethasone in hospitalized patients with Covid-19. N Engl J Med. 384(8):693–704. https://doi.org/10.1056/NEJMoa2021436, https://pubmed.ncbi.nlm.nih.gov/32678530/

Ripoll JG, Gorman EK, Juskewitch JE et al (2022) Vaccine-boosted convalescent plasma therapy for patients with immunosuppression and COVID-19. Blood Adv 6(23):5951–5955. https://doi.org/10.1182/bloodadvances.2022008932

Salazar E, Kuchipudi SV, Christensen PA et al (2020) Convalescent plasma anti-SARS-CoV-2 spike protein ectodomain and receptor-binding domain IgG correlate with virus neutralization. J Clin Invest 130(12):6728–6738. https://doi.org/10.1172/jci141206

Self WH, Tenforde MW, Stubblefield WB et al (2020) Decline in SARS-CoV-2 antibodies after mild infection among frontline health care personnel in a multistate hospital network—12 States, April-August 2020. MMWR Morb Mortal Wkly Rep 69(47):1762–1766. https://doi.org/10.15585/mmwr.mm6947a2

Self WH, Wheeler AP, Stewart TG et al (2022) Neutralizing COVID-19 convalescent plasma in adults hospitalized with COVID-19: a blinded, randomized, placebo-controlled trial. Chest 162 (5):982–994. https://doi.org/10.1016/j.chest.2022.06.029

Senefeld JW, Franchini M, Mengoli C et al (2023) COVID-19 convalescent plasma for the treatment of immunocompromised patients: a systematic review and meta-analysis. JAMA Netw Open 6(1):e2250647–e2250647. https://doi.org/10.1001/jamanetworkopen.2022.50647

Shoham S, Bloch EM, Casadevall A et al (2023) Transfusing convalescent plasma as post-exposure prophylaxis against severe acute respiratory syndrome coronavirus 2 (SARS-CoV-2) infection: a double-blinded, phase 2 randomized, controlled trial. Clin Infect Dis 76(3):e477–e486. https://doi.org/10.1093/cid/ciac372

Simonovich VA, Burgos Pratx LD, Scibona P et al (2020) A randomized trial of convalescent plasma in Covid-19 severe pneumonia. N Engl J Med 384(7):619–629. https://doi.org/10.1056/NEJMoa2031304

Sterlin D, Mathian A, Miyara M et al (2021) IgA dominates the early neutralizing antibody response to SARS-CoV-2. Sci Transl Med 13(577). https://doi.org/10.1126/scitranslmed.abd2223

Sullivan DJ, Gebo KA, Shoham S et al (2022) Early outpatient treatment for Covid-19 with convalescent plasma. N Engl J Med 386(18):1700–1711. https://doi.org/10.1056/NEJMoa2119657

Sullivan D, Casadevall A (2023) COVID-19 serology data provide guidance for future deployments of convalescent plasma. mBio 14(3):e0042823. https://doi.org/10.1128/mbio.00428-23

Thompson MA, Henderson JP, Shah PK et al (2021) Association of convalescent plasma therapy with survival in patients with hematologic cancers and COVID-19. JAMA Oncol. https://doi.org/10.1001/jamaoncol.2021.1799

Trøseid M, Hentzien M, Ader F et al (2022) Immunocompromised patients have been neglected in COVID-19 trials: a call for action. Clin Microbiol Infect Offl Publ Eur Soc Clin Microbiol Infect Dis 28(9):1182–1183. https://doi.org/10.1016/j.cmi.2022.05.005

Troxel AB, Petkova E, Goldfeld K et al (2022) Association of convalescent plasma treatment with clinical status in patients hospitalized with COVID-19: a meta-analysis. JAMA Netw Open 5 (1):e2147331–e2147331. https://doi.org/10.1001/jamanetworkopen.2021.47331

Ullah I, Escudie F, Scandale I et al (2024) Bioluminescence imaging reveals enhanced SARS-CoV-2 clearance in mice with combinatorial regimens. iScience 27(3):109049. https://doi.org/10.1016/j.isci.2024.109049

Vickers MA, Sariol A, Leon J et al (2021) Exponential increase in neutralizing and spike specific antibodies following vaccination of COVID-19 convalescent plasma donors. Transfusion 61 (7):2099–2106. https://doi.org/10.1111/trf.16401

Weinreich DM, Sivapalasingam S, Norton T et al (2020) REGN-COV2, a neutralizing antibody cocktail, in outpatients with Covid-19. N Engl J Med 384(3):238–251. https://doi.org/10.1056/NEJMoa2035002

Weissleder R, Lee H, Ko J, Pittet MJ (2020) COVID-19 diagnostics in context. Sci Transl Med 12 (546):eabc1931. https://doi.org/10.1126/scitranslmed.abc1931

World Health Organization (2020) Laboratory testing for coronavirus disease 2019 (COVID-19) in suspected human cases: interim guidance. https://iris.who.int/handle/10665/331329

Wu F, Zhao S, Yu B et al (2020) A new coronavirus associated with human respiratory disease in China. Nature 579(7798):265–269. https://doi.org/10.1038/s41586-020-2008-3
Yoon HA, Bartash R, Gendlina I et al (2021) Treatment of severe COVID-19 with convalescent plasma in Bronx, NYC. JCI Insight. https://doi.org/10.1172/jci.insight.142270
Yoon H, Li Y, Goldfeld KS et al (2024) COVID-19 convalescent plasma therapy: long-term implications. Open Forum Infect Dis 11(1):ofad686. https://doi.org/10.1093/ofid/ofad686
Zhu N, Zhang D, Wang W et al (2020) A novel coronavirus from patients with pneumonia in China, 2019. N Engl J Med 382(8):727–733. https://doi.org/10.1056/NEJMoa2001017

Hyunah Yoon is an infectious diseases physician-scientist at Albert Einstein College of Medicine and Montefiore Medical Center. She is developing a career in clinical and translational research, investigating antibody immunity to cryptococcosis and COVID-19.

Liise-anne Pirofski is an infectious diseases physician-scientist, the Selma and Dr. Jacques Mitrani Chair in Biomedical Research at Albert Einstein College of Medicine, and chief of the division of infectious diseases at Albert Einstein College of Medicine and Montefiore Medical Center. Her research program is focused on antibody immunity, host–pathogen interactions, and mechanisms of vaccine protection.

Convalescent Plasma and Other Antibody Therapies for Infectious Diseases—Lessons Learned from COVID-19 and Future Prospects

David J. Sullivan Jr.

Contents

Abstract Antiviral passive antibody therapy includes convalescent plasma, hyperimmune globulin, and monoclonal antibodies. Passive antibodies have proven effective in reducing morbidity and mortality for SARS-CoV-2 and other infectious diseases when given early in the disease course with sufficiently high specific total and neutralizing antibody levels. Convalescent plasma can be delivered to patients before vaccination implementation or novel drug production. Carefully designed and executed randomized controlled trials near the pandemic outset are important for regulatory bodies, healthcare workers, guideline committees, the public, and the government. Unfortunately, many otherwise well-designed antibody-based clinical trials in COVID-19 were futile, either because they intervened too late in the disease or provided plasma with insufficient antibodies. The need for early

D. J. Sullivan Jr. (✉)
Department of Molecular Microbiology and Immunology, Johns Hopkins Bloomberg School of Public Health, 615 N. Wolfe St Rm W4606, Baltimore, MD 21205, USA
e-mail: dsulliv7@jhmi.edu

Current Topics in Microbiology and Immunology (2025) 443: 219–242
https://doi.org/10.1007/82_2024_273

Published Online: 9 August 2024

treatment mandates outpatient clinical trials in parallel with inpatient trials. Early outpatient COVID-19 convalescent plasma transfusion with high antibody content within 9 days of symptom onset has proven effective in blunting disease progression and reducing hospitalization, thus reducing hospital overcrowding in a pandemic. Convalescent plasma offers the opportunity for hope by enabling community participation in outpatient curative therapy while monoclonal therapies, vaccines, and drugs are being developed. Maintaining the appropriate infrastructure for antibody infusion in both outpatient and inpatient facilities is critical for future pandemic readiness.

Glossary of Abbreviations

AHF	Argentine Hemorrhagic Fever
AU	Arbitrary unit
BARDA	Biomedical Advanced Research and Development Authority
BAU	Binding antibody unit
B-Cell	Bursa of Fabricius-like lymphocyte-producing antibodies
BNT	BioNtech
CCP	COVID-19 convalescent plasma
CD4	Cluster of differentiation 4
CMV	Cytomegalovirus
COVID-19	Coronavirus disease 2019
CP	Convalescent plasma
DNA	Deoxynucleic acid
EAP	Expanded Access Program
EUA	Emergency use authorization
Fc	Fragment, crystallizable
FDA	Food and Drug Administration
FFP	Fresh frozen plasma
HIV	Human immunodeficiency virus
IgG	Immunoglobulin G
IMV	Invasive mechanical ventilation
IVIG	Intravenous immunoglobulin
RCT	Randomized controlled trial
RNA	Ribonucleic acid
RSV	Respiratory syncytial virus
SARS-CoV-2	Severe acute respiratory syndrome coronavirus 2
SeroNet	Serological Sciences Network
STRIVE	Strategies and Treatments for Respiratory Infections and Viral Emergencies
TU	Therapeutic units
VZV	Varicella zoster virus

1 Antibody Therapy Types

Specific antiviral antibody therapy—convalescent plasma (CP), hyperimmune gamma globulin fractions, or monoclonal antibodies—is passive, in contrast to vaccines which actively generate antibodies. COVID-19 convalescent plasma (CCP) from survivors is no different from fresh frozen plasma (FFP), except that it contains large amounts of total and neutralizing viral-specific antibodies (Bloch et al. 2020). Convalescent plasma is ready within weeks of a pandemic and can be piggybacked on to the existing blood banking infrastructure, relying on the regulatory framework that covers FFP. However, FFP intended to provide antibodies to patients with an infection needs to be carefully assessed to ensure that it contains sufficient amounts of antibodies appropriate to the specific infection. CCP is an old treatment used in the early 1900s before antibiotics, with an updated understanding after its massive mobilization during COVID-19.

Hyperimmune globulin or gamma globulin is principally immunoglobulin G (IgG) which has been fractionated from thousands of plasma units. This product does not have to be blood group antigen A, B, or O typed and its preparation eliminates the possibility of hepatitis, HIV, or other viral transmission (Huygens et al. 2023). Gamma globulin fractionation was first accomplished in the 1930s by Arne Tiselius who developed electrophoresis, a development for which he was awarded the Nobel Prize in Chemistry in 1948 (Tiselius 1937). Cohn and colleagues later scaled up gamma globulin fractionation for clinical use (Cohn and Strong 1946). The fractionation and safety evaluation takes several months before the product can be provided to humans, requiring significant batch safety testing and addressing regulatory hurdles (Looney and Huggins 2006). Monoclonal antibodies require cloning, lead selection, and large 25,000 L batch production requiring 4–8 months before human clinical testing (Lehman et al. 2022).

All three passive antibody modalities have the goal of providing the patient with high antibody levels capable of viral-specific neutralization. Early in the COVID-19 pandemic, CCP was initially intended to be a bridge to other therapies or vaccine prevention. The potential impact of CCP is very high in a naive population early in a new pandemic. Later, when most patients are immunized or convalescent, CCP's impact may be more specific to certain immunocompromised populations. Vaccinations require an active response that takes weeks in the vaccinated to create viral-specific antibodies to microbial proteins originating from the protein, DNA, or RNA construct of the target virus. SARS-CoV-2 RNA vaccines were designed in January 2020, tested within months, and validated before a year had passed from pandemic onset. Healthcare workers were the first to receive the COVID-19 vaccine —in December 2020. In 2019, before COVID-19, approximately 5 billion doses of all forms of vaccine were manufactured. But by the end of 2021, more than 10 billion SARS-CoV-2 vaccine doses had been manufactured (Kreier 2022).

2 Antibody Works Against Multiple Infectious Diseases

In the 1890s, Behring and Kitasato produced immunity to tetanus and diphtheria toxins in experimental animals after toxin injection, thus producing active immunity, but most importantly, were able to produce passive immunity in other animals by injecting them with serum taken from the immunized animals (see Chap. X). At about the same time, Ehrlich generalized acquired serum immunity to non-infectious toxins such as plant ricin and abrin (Browning 1955). Within a decade, the understanding of early, high-titer transfer of specific immunity to individual toxins in humans was elucidated (Browning 1955; Addetia et al. 2020).

In the 1940s and 1950s, gamma globulin was used to prevent measles, rubella during pregnancy, hepatitis A, and poliomyelitis and also to provide disease attenuation when administered within a week of exposure or symptoms (Gross et al. 1959) (Table 1). Doses varied by microbial infection type (Gross et al. 1959). In measles, 0.1 mL/pound of bodyweight of gamma globulin completely prevented infection when given within 6 days of exposure. Attenuation of symptoms could be achieved with one-tenth of that dose. Doubling the dose was necessary for both prevention and disease attenuation if more than 6 days from exposure had elapsed. For Hepatitis A, a dose of 0.01 mL/pound consistently prevented infection, with evidence that even lower doses of 0.005 mL/pound limited the disease trajectory while preserving full active convalescent immunity to Hepatitis A.

Currently, hyperimmune globulin is used for post-exposure prophylaxis of varicella zoster virus (VZV), cytomegalovirus (CMV), hepatitis B, and rabies. For the rare cases of infant botulism and tetanus, hyperimmune globulins neutralize unbound toxins. The effective antibody doses required for preventing or treating novel viral, bacterial, or protozoal microbes will have to be determined for each agent.

More than 170 monoclonal antibody molecules have been approved by the FDA or European agencies for cancer ($\sim$80), hemostasis ($\sim$10), immunomodulation ($\sim$50), and infectious diseases ($\sim$15) (Therapeutic monoclonal antibodies approved or in review in the EU or US 2023). Among the infectious diseases, monoclonal efficacy has been shown for prophylaxis of anthrax, recurrent *Clostridium difficile*, rabies, and SARS-CoV-2 (see Chap. Y). Clinically effective monoclonals limit disease in anthrax, gram-negative endotoxemia, HIV that is resistant to drug treatments (Ibalizumab-CD4 target), respiratory syncytial virus (RSV), and pre-omicron SARS-CoV-2 (Therapeutic monoclonal antibodies approved or in review in the EU or US 2023) (Table 1). Unfortunately, SARS-CoV-2 evolved mutations that rendered monoclonals resistant to clinically relevant antibody doses during the omicron variant period (Jaki et al. 2023; Focosi et al. 2023; He et al. 2023).

Drugs and antibodies work together to limit microbes with many -static (i.e., microbe growth inhibiting) rather than -cidal (i.e., microbe killing) antimicrobials reliant on antibodies to clear infections. For example, during bone marrow transplants, targeted antibiotic therapy must sometimes continue for weeks until immune

Table 1 Clinically approved antibody therapies for infectious diseases (excluding vaccines) hyperimmune globulin (gamma)

Antibody type	Period of use	Infection protection (days post-exposure)	Disease attenuation	Notes
Measles (Gross et al. 1959)	Historic	6	Effective < 6 d	Double the mL/pound dose for disease attenuation 7–15 days
German measles (Gross et al. 1959)	Historic	3–15		Lots of variations in antibody content
Hepatitis A (Gross et al. 1959)	Historic	14	Effective 5–7 d	Half of infection prevention dose attenuates disease
Poliomyelitis (Gross et al. 1959)	Historic	Before exposure or	Effective 5-7d	Effectiveness in family contacts at the time of primary case
Variola and Vaccinia (Gross et al. 1959; Kempe et al. 1956, 1961)	Historic	7–19 d	Effective 2–5 d	In immunosuppressed reduced fatal disease after vaccinia
Mumps (Gross et al. 1959)	Historic		1 day after parotitis	Lowered orchitis
VZV (Keller and Stiehm 2000)	Current	Post-exposure		Prophylaxis in immunodeficient individuals
CMV (Keller and Stiehm 2000)	Current	Post-exposure		Prophylaxis in immunodeficient individuals
Hepatitis B (Keller and Stiehm 2000)	Current	Post-exposure		Neonate, unvaccinated exposed, or liver transplant utility
Rabies (Keller and Stiehm 2000)	Current	Post-exposure		Human or horse only to unvaccinated
Botulism (Keller and Stiehm 2000)	Current		Effective	Use in infants as soon as possible after diagnosis
Tetanus (Keller and Stiehm 2000)	Current		Effective	Neutralizes only unbound toxin
SARS-CoV-2 (Group IS 2022)	none		Not effective	RCT in hospitalized not effective and outpatient trials in the process

(continued)

Table 1 (continued)

Antibody type	Period of use	Infection protection (days post-exposure)	Disease attenuation	Notes
Convalescent plasma				
Junin Virus AHF	Historic/ current		Effective 0–8 d	Mortality reduced from 16 to 1%
SARS-CoV-2 (Sullivan et al. 2022b; Shoham et al. 2023)	Current	Not effective	Effective 0–9 d	50% within 9d and 80% within 5d to prevent outpatient hospitalization
Monoclonals				
Anthrax (Mazumdar 2009)	Current	Effective	Effective	Raxibacumab/ Obiltoxaximab- serum effective in the pre-antibiotic era. 2 mAb are FDA approved on limited data
C. difficile (Wilcox et al. 2017)	Current	Effective		Bezlotoxumab-used to prevent recurrence
Endotoxin (GNR) (Marks 2012)	Current		Effective-denied FDA approval	Centoxin-
Ebola (Markham 2021)	Current		Effective	Atoltivimab, maftivimab, and odesivimab
HIV (Chahine and Durham 2021)	Current		Effective	Ibalizumab targets CD4 in treatment-resistant patients
Rabies (Melo et al. 2022)	Current	Effective		Ormutivimab/ Docaravimab and Miromavimab
RSV (Keam 2023; Messina et al. 2022)	Current	Effective	Effective	Palivizumab/Nirsevimab
SARS-CoV-2 (Group A-TL-CS et al. 2021; Weinreich et al. 2021a, b)	Historic	Effective	Effective < 5 days	Variant evolution made mAb ineffective

reconstitution restores leukocytes (Toor et al. 2001). This lesson has repeated itself in B-cell deficient COVID-19 patients who were receiving rituximab (B-cell ablation therapy), had cancer, or were transplant recipients. In these situations, both before and after drug availability, patients often took months to clear the virus. Multiple rounds of CCP or combinations with interferons or later antivirals were often required to eliminate the virus to halt disease progression (Thornton et al. 2022; Thompson et al. 2021).

In the first two years of the pandemic, antibody therapy, in the form of monoclonals or CCP, was the only available therapeutic option for COVID-19 outpatients. Antibodies continued to play a role later even after vaccination had been established, especially when intravenous remdesivir was not available. Virus-specific antibodies wane over time after new infection, vaccination, or receipt of monoclonals or CCP. Moreover, viral variant evolution decreased the efficacy of natural convalescent SARS-CoV-2 antibodies, enabling breakthrough infections characteristic of coronavirus (Li et al. 2022; Sullivan et al. 2022a, 2023). As shown in Fig. 1, when immunity and antibodies are present, there is less reliance on small molecule drugs, but in the absence of active production of antibodies—often because of immunosuppression—small molecule drug therapy is not always able to clear infections.

Fig. 1 Antibody and drug interactions—relationship between 100% antibody dependence when effective drugs are not available and 100% drug dependence when antibody levels are low in naïve seronegative individuals or immunocompromised individuals

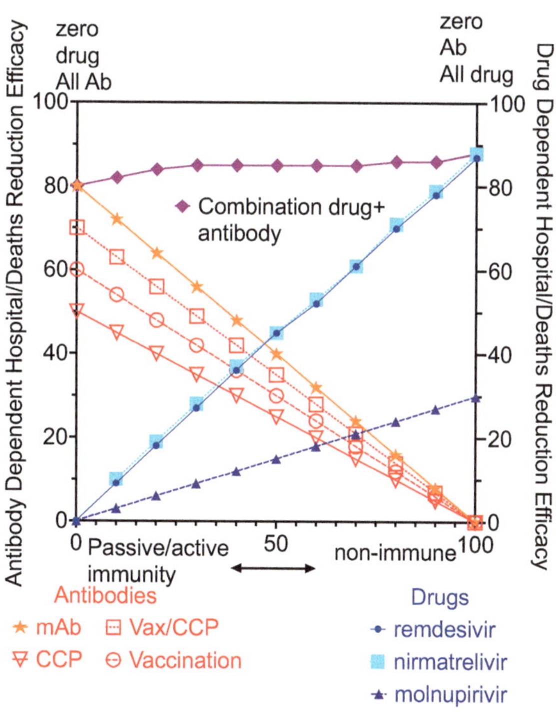

Both the fraction of the population who are seropositive and average antibody levels have risen during the COVID-19 pandemic as a result of both infection and vaccination, as illustrated in Table 2, which is based on data from CDC's SeroNet program and its national blood donor database. Documented COVID-19 infections, including reinfections, total more than 100 million or 33% of the population. Approximately 70% of the US population has been vaccinated twice or more. Seropositivity to COVID-19 in the general population as of June 2022 was at 58% in the SeroNet population of more than 1.4 million tested specimens per month (Wiegand et al. 2022). In blood donors, seroprevalence was reported to be 97% by February 2023, with 78% of this total estimated to come from infections (Miller et al. 2020; Jones et al. 2022).

3 Antibody Therapeutic Principles—Early Treatment with High Specific Neutralizing Dosage

For the treatment of infectious diseases, both drug and antibody therapy require a neutralizing dose for the causal organism, which is time-dependent. Higher doses of both are required later in disease progression. Two convalescent plasma trials conducted before COVID-19 with contrasting findings illustrate the need to provide sufficient antibodies in a timely fashion.

The first trial studied the Arena virus which causes Argentine Hemorrhagic Fever (AHF), a disease characterized by a mild febrile illness for a week followed by a severe hemorrhagic component that leads to a case-fatality rate of at least 15% (Enria et al. 1986). A trial of convalescent plasma in 188 participants studied immunologically naïve ambulatory participants within the first week of illness, reducing the death rate from 16% in recipients of control plasma to 1% in those receiving 500 mLs of AHF-convalescent plasma (Maiztegui et al. 1979). However, transfusions after a week of illness did not reduce deaths (Maiztegui et al. 1979). Subsequent trials of AHF-convalescent plasma were able to determine an effective neutralizing dose of 3,500 TU/kg-neutralizing antibody dose per kg of body weight (Enria et al. 1984, 2008). Doses much lower than that did not reduce mortality.

In contrast, the second trial, a phase 3 RCT of influenza convalescent plasma in severe influenza, selected a population to study 50% of whom had recently been vaccinated and 76% of whom were receiving antiviral drugs (Beigel et al. 2019). Many were already in the intensive care unit (44%) and 30% were being mechanically ventilated. Under those conditions, influenza convalescent plasma did not change the outcome in this trial, despite treatment within 5 days of symptom onset (median 3 days).

Convalescent plasma appears to have been effective when high levels of antibodies were provided early in immunologic naïve patients in a disease where progression to severe disease takes more than a week (Maiztegui et al. 1979). But in the influenza trial, transfused patients were largely immunologically experienced—

Table 2 USA data on COVID-19 reported cases, percent vaccinated, percent positive for spike antibodies in SeroNet (seroprevalence), and percent positive blood donors. The USA has an estimated population size of 330 million as of 2023

Study period	End date	Cumulative COVID-19 cases[a] millions	Received at least one dose[b] Millions (%)	Completed primary series[b] Millions (%)	Received updated bivalent dose[b] Millions (%)	Sero-Prevalence (SeroNet,) (%)[c]	Blood donors (%)[d]	Sero-Prevalence (Infection-Induced NucCap)
Apr–Dec 2020	12-31-2020	19.58 (6)	5.57 (0)	0.04 (0)	0 (0)	13	7	7
Jan–Jun 2021	06-30-2021	33.31 (10)	183.54 (55)	163.82 (49)	0	22	82	15
Jul–Nov 2021	11-30-2021	48.03 (15)	233.81 (70)	202.53 (61)	0	32	90	22
Dec 21–Jun 22	06-30-2022	86.31 (26)	259.52 (78)	223.52 (67)	0	58	94	55
Jul 22–Feb 23	02-28-2023	102.02 (31)	269.55 (81)	230.24 (69)	53.92 (16)	No data	97	78
Mar 23–Aug 23	08-31-2023	103.44 (31)	270.22 (81)	230.63 (69)	56.48 (17)	No data	No data	No data

[a]Cumulative cases were obtained at—https://ourworldindata.org/coronavirus/country/united-states. [b]USA cumulative vaccine history at https://covid.cdc.gov/covid-data-tracker/#vaccination-trends. [c]COVID-19 seroprevalence from CDC Seronet https://covid.cdc.gov/covid-data-tracker/#national-lab through February 17, 2022. [d]Blood donor seroprevalence estimates from https://covid.cdc.gov/covid-data-tracker/#nationwide-blood-donor-seroprevalence-2022 through December 2022

they were seropositive, receiving antiviral drugs, and had a disease with rapid progression (Beigel et al. 2019). Another possibility is that the influenza titer threshold for the plasma used in treatment (hemagglutination inhibition titer 1:80) was based upon outdated information (Hobson et al. 1972; Potter and Oxford 1979). The neutralizing level of antibodies to prevent infection might be higher or lower than that needed to attenuate the progression of the disease and must be worked out for each type of viral illness.

4 Crisis Therapeutic Antibody Timeline

To paraphrase Joshua Sharfstein "The first order of business in a crisis is to declare a crisis and act upon the ramifications of the crisis" (Sharfstein and Sharfstein 2018). SARS-CoV-2 spread into an immunologically naïve population with no proven available, effective chemotherapeutic options. Passive antibody therapies were critical, early in the first two years of the pandemic. On January 10, 2020, the sequence of SARS-CoV-2 was shared with the world by Chinese researchers (GenBank-Accession number MN988668) (Carvalho et al. 2021). In a matter of days, both Moderna and BNT started to design spike vaccines with introduced amino acid changes (prolines) to promote closed fist spike immunogenic conformation which interacts with the human SARS-CoV-2 receptor, the angiotensin-converting enzyme 2 (Carvalho et al. 2021). Within 2 months, phase 1 trials began; within 4 months, phase 2 trials were underway; and by 6 months, the final phase of vaccine testing—phase 3 trials—were underway. Operation Warp Speed—a US federal program—provided guaranteed funding on May 15, 2020 for two candidates in each of three different platforms—mRNA, replication-defective adenovirus vectored, and adjunctive recombinant protein. Preliminary effectiveness data was reviewed by the FDA after just 10 months of development in October 2020, and the first vaccine was administered to Sandra Lindsay on Dec 14, 2020 (Carvalho et al. 2021) (Fig. 2).

Convalescent plasma was first used in China on January 20, 2020 in severely ill hospitalized patients (Shen et al. 2020) and in the USA on March 29 (Salazar et al. 2021). The FDA created an Expanded Access Program (EAP) at the Mayo Clinic which allowed CCP treatment of hospitalized patients but under the condition that a registry of this use was created (see Chap. Z). The findings of the EAP led to an emergency use authorization (EUA) in hospitalized COVID-19 patients on August 23, 2020. Because the findings that supported CCP use were based on lower mortality when higher antibody levels in the transfused plasma were found (Joyner et al. 2021), the FDA authorized CCP treatment when a specific antibody threshold in the plasma was met (O'Shaughnessy 2021). In December 2021, FDA authorized the outpatient use of CCP in immunocompromised patients (O'Shaughnessy 2021). Hyperimmune globulin (ITAC trial) was found ineffective in trials of hospitalized patients with severe disease. An outpatient randomized controlled trial (RCT) of hyperimmune globulin is currently in progress (Group IS 2022) (Fig. 2).

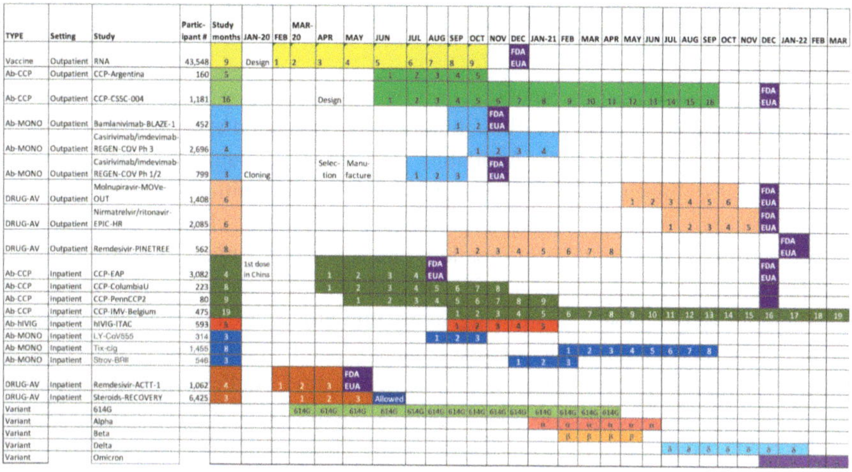

Fig. 2 COVID-19 therapy RCT study timelines—study duration and start and end dates in the context of dominant variant(s) of concern

The development of effective monoclonal antibodies was remarkably rapid. In the development of casirivimab and imdevimab, hybridization cloning from previously infected human patients to immunized mouse cells on the Velocimmune platform (immunoglobin mouse heavy and kappa light chain variable regions are replaced with human variable region DNA) was initiated in February 2020 and by April 15, 2020 had produced selected leads. This was followed by outpatient human trials initiated on two of these leads just two months later—on June 10—after both 10,000 L and 25,000 L scale production runs. The EUA for outpatient use of casirivimab and imdevimab was issued 5 months later, on November 21, 2020 (Release Fp 2020), meaning that the entire process took just over 9 months. Monoclonals saved many lives with outpatient infusion to high-risk COVID-19 patients, but acquired monoclonal antibody resistance effectively removed this entire class of therapeutics during the omicron SARS-CoV-2 variant period (He et al. 2023; Lin et al. 2022; Takashita et al. 2022).

While antibody therapies were sanctioned under EUAs within a year of the beginning of the SARS-CoV-2 pandemic, oral antiviral therapy was not allowed by the FDA for nearly 2 years, with licensing of nirmatrelvir/ritonavir (Harris 2023) and molnupiravir (Molnupiravir for treatment of COVID-19 2022) in December 2021 along with outpatient intravenous remdesivir in January 2022 (COVID-19 updates 2022) (Fig. 2). Inpatient intravenous remdesivir had been approved by the FDA in May 2020 (Remdesivir (Veklury) for COVID-19 2020).

5 Trial Design in Outpatients Needs to Be Priority One

Outpatient trials of new therapeutics in a pandemic need to be designed from the outset rather than being an afterthought and only conducted after inpatient trials, as was the case for COVID-19. Available safe, effective outpatient therapy is the foundation for limiting hospital overcrowding and permitting a rapid return to normal societal function during a pandemic. Outpatient clinical trials must be performed in parallel with early inpatient efficacy trials. Although everyone wants results immediately during pandemics, the only result valued by regulatory bodies is a randomized clinical trial showing safety and efficacy. Clinicians and guideline committees place high value on RCT data regardless of context and biologic plausibility, although retrospective or matched control studies can yield important early information. During COVID-19, many academic centers first ran antibody and repurposed drug trials in hospitalized individuals. Initially, monoclonals and hyperimmune globulins did not perform well in the hospital setting trials (Group IS 2022; ACTIV-3 2023; Activ-3-Therapeutics 2022; Group A-TL-CS et al. 2021). Monoclonals were validated in the outpatient setting and were subsequently widely successful in developed countries with access to monoclonals. The BARDA-funded emergency use of CCP in hospitalized patients documented the safety of this product in more than 20,000 patients (Joyner et al. 2020), as well as showing that high-titer plasma provided before the use of mechanical ventilation prevented deaths (Joyner et al. 2021). In hospitalized patients, many studies (e.g., RECOVERY 2020) transfused late in disease progression or provided CCP with low levels of viral-specific antibodies (e.g., Agarwal et al. 2020).

The choice by drug companies to test nirmatrelivir (Hammond et al. 2022) and molnupiravir (Jayk Bernal et al. 2022) in the outpatient setting was wise. Antimicrobials have the greatest potential for reducing morbidity if given before disease progression leads to hospitalization. While outpatient trials may be more difficult to conduct than studies in bed-ridden inpatients, they have greater potential to illustrate differences between intervention and control patients because they operate earlier in the disease process.

Future pandemics require preparation on many operational levels (Williams et al. 2023). The Strategies and Treatments for Respiratory Infections and Viral Emergencies (STRIVE) platform maintains a constant inpatient infrastructure for trials (NIAID 2022; Montejano et al. 2019). But having a similar ready-to-go infrastructure for *outpatient* clinical trials during pandemics may be more important than having the inpatient infrastructure.

6 Convalescent Plasma RCTs and Matched Controls

In early 2024, CCP remains clinically relevant alongside repeated vaccinations. Both approaches generate polyclonal antibodies which can keep up with changing SARS-CoV-2 variants. Pre-omicron CCP with high antibody levels, unlike the monoclonals, neutralizes omicron variants (Focosi et al. 2023; Li et al. 2022; Sullivan et al. 2022a, 2023; Takashita et al. 2022). The processing requirements of intravenous immunoglobulin (IVIG) require at least 6–12 months from the point of plasma collection, increasing the possibility of a mismatch to current viral variants. This time delay and limited clinical trial efficacy hinder hyperimmunoglobulin development in a fast-moving pandemic. Clinical trial effectiveness and correlates of protection for vaccines (Pormohammad et al. 2021; Gilbert et al. 2022), monoclonals (Khoury et al. 2021, 2023; Tao et al. 2022), and CCP for hospital use (Klassen et al. 2021; Senefeld et al. 2023) have been described elsewhere and in other chapters of this volume. Herein follows descriptions of outpatient CCP trials and selected hospital-based trials.

Randomized control trials demonstrate effectiveness at three stages of COVID-19.

(1) mild disease to reduce hospitalizations,
(2) already hospitalized moderate disease to reduce deaths, and
(3) within 48 h of mechanical ventilation to reduce deaths.

Two well-conducted outpatient RCTs have shown that CCP can reduce hospitalizations (Sullivan et al. 2022b; Libster et al. 2021) as do three outpatient CCP studies that were stopped near the midpoint of projected enrollment (Gharbharan et al. 2022; Korley et al. 2021; Alemany et al. 2022). Combining both the complete and incomplete trials in an individualized participant meta-analysis noted the significance of all five studies in hospital risk reduction, an effect that was strengthened when restricted to participants who received plasma units in the upper half of antibody level given early (Levine et al. 2023). Two RCTs in hospitalized participants achieved significant reductions in deaths in recipients of early transfusion with high antibody levels (Bar et al. 2021; O'Donnell et al. 2021) (Fig. 3). As discussed in Chaps. A and B, a comprehensive meta-analysis has confirmed significant reductions in mortality in hospitalized patients given CCP (Senefeld et al. 2023).

One trial in severely ill patients—The Acute Respiratory Distress Syndrome CONFIDENT trial—transfused patients with high SARS-CoV-2 antibody level plasma (top 15% of all pre-delta donor units collected) late in the disease course, but within 5 days of invasive mechanical ventilation (IMV). Maximum mortality reduction was seen when two units were transfused within 48 h of IMV (Misset et al. 2023) (Fig. 3). Stratification by neutralizing IC_{50} antibody levels at titer measurements of 160, 320, 640, or greater (all within the top 15%) did not further reduce deaths, even though there were more deaths (>50%) in participants in the lowest pretransfusion antibody level quartile, suggesting partial antiviral action.

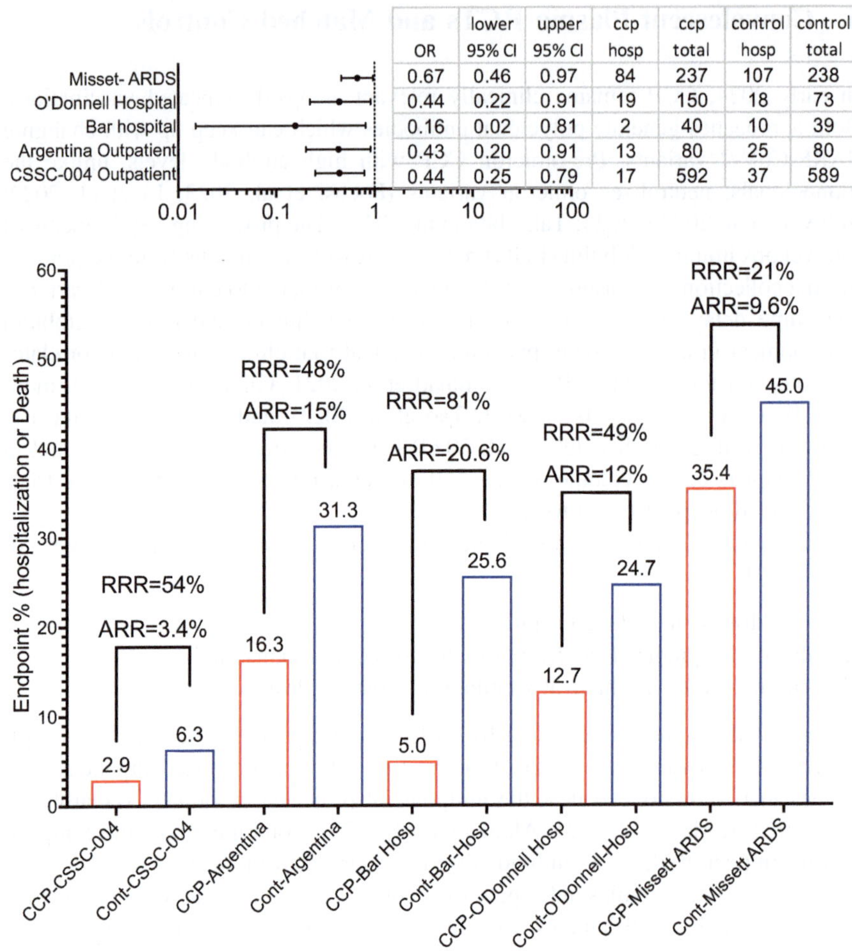

	OR	lower 95% CI	upper 95% CI	ccp hosp	ccp total	control hosp	control total
Misset-ARDS	0.67	0.46	0.97	84	237	107	238
O'Donnell Hospital	0.44	0.22	0.91	19	150	18	73
Bar hospital	0.16	0.02	0.81	2	40	10	39
Argentina Outpatient	0.43	0.20	0.91	13	80	25	80
CSSC-004 Outpatient	0.44	0.25	0.79	17	592	37	589

Fig. 3 Effective CCP RCTs. RCTs in outpatients, hospitalized, and newly mechanically ventilated are summarized graphically with odds ratios on the upper panel and percent hospital or death reduction on the lower panel (CCP red and control blue)

More than 97% of participants received anti-inflammatory steroids, but fewer than 7% received IL-6 antagonists or remdesivir. Previous convalescent plasma trials or studies with high SARS-CoV-2 antibodies support early antiviral (Joyner et al. 2021; Sullivan et al. 2022c) and anti-inflammatory (Habtehyimer et al. 2023) effects. This large well-powered CONFIDENT trial demarcates the critical period earlier than 2 days of IMV for biologically plausible potent convalescent plasma antiviral and anti-inflammatory effects (Pirofski and Casadevall 2020).

Examining mortality rates in hospitalized COVID-19 populations, we find two notable matched control studies, one in immunocompetent patients with an 80%

relative risk reduction in deaths (Salazar et al. 2021) and another in patients with hematologic cancers with 46% relative risk reduction (Thompson et al. 2021), and two large study analyses of the EAP which looked at recipients of high-titer compared to low-titer plasma with smaller, but statistically significant reductions of deaths (Joyner et al. 2021; Belov et al. 2022) (Fig. 4). Another matched control trial in the immunocompromised is ongoing at Mayo Clinic with COVID-19 high-titer plasma obtained after recent donor vaccination. Preliminary data suggests findings

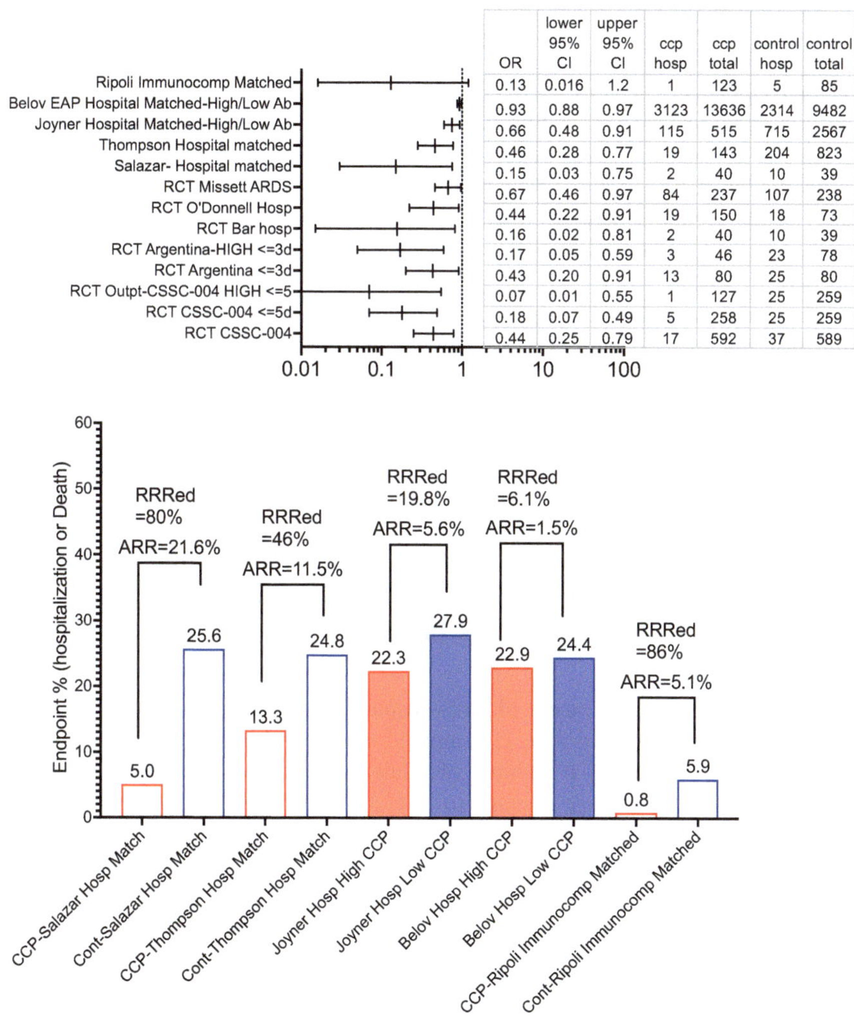

	OR	lower 95% CI	upper 95% CI	ccp hosp	ccp total	control hosp	control total
Ripoli Immunocomp Matched	0.13	0.016	1.2	1	123	5	85
Belov EAP Hospital Matched-High/Low Ab	0.93	0.88	0.97	3123	13636	2314	9482
Joyner Hospital Matched-High/Low Ab	0.66	0.48	0.91	115	515	715	2567
Thompson Hospital matched	0.46	0.28	0.77	19	143	204	823
Salazar- Hospital matched	0.15	0.03	0.75	2	40	10	39
RCT Missett ARDS	0.67	0.46	0.97	84	237	107	238
RCT O'Donnell Hosp	0.44	0.22	0.91	19	150	18	73
RCT Bar hosp	0.16	0.02	0.81	2	40	10	39
RCT Argentina-HIGH <=3d	0.17	0.05	0.59	3	46	23	78
RCT Argentina <=3d	0.43	0.20	0.91	13	80	25	80
RCT Outpt-CSSC-004 HIGH <=5	0.07	0.01	0.55	1	127	25	259
RCT CSSC-004 <=5d	0.18	0.07	0.49	5	258	25	259
RCT CSSC-004	0.44	0.25	0.79	17	592	37	589

Fig. 4 Effective matched control studies. Matched control studies with early and late transfusion of the two outpatient CCP RCTs and the three inpatient RCTs are summarized graphically with odds ratios on the upper panel and percent hospital or death reduction on the lower panel (CCP red and control blue)

approaching traditional significance levels, with further data analyses in progress (Ripoll et al. 2023).

The large outpatient trial—with 1181 participants—robustly demonstrated more than a 50% reduction in relative risk of hospitalization when CCP was transfused within 9 days of symptom onset (Sullivan et al. 2022c). When transfusion was within 5 days of symptom onset, the relative risk of hospitalization was reduced by 80%. High-risk subgroups—older, obese, diabetic, and hypertensive participants—all showed a similar 50% hospital risk reduction.

A week after the public release of this outpatient trial result, on December 27, 2021, the FDA made an emergency use allowance for CCP with high titers of anti-SARS-CoV-2 antibodies "*for the treatment of COVID-19 in patients with immunosuppressive disease or receiving immunosuppressive treatment, in either the outpatient or inpatient setting.*" "*Based on the totality of the scientific evidence available, it is reasonable to believe that CCP with high titers of anti-SARS-CoV-2 antibodies may be effective when it is used in the treatment of COVID-19 either early in the course of disease (e.g., prior to hospitalization in immunocompetent patients) or in patients with immunosuppressive disease or receiving immunosuppressive treatment.*"

This large outpatient trial helped to change clinical practice, even though the trial was not focused especially on immunosuppressed patients. Additionally, in October of 2023, more than a dozen international guideline committees endorsed CCP use when other therapies were not available or for use in the immunocompromised (Franchini et al. 2023).

7 The Upper Deciles of Donor Convalescent Plasma Must Contain Enough Antibodies so that After Dilution It Still Provides Protective Levels in Recipients

The antibodies in convalescent plasma—usually provided in a volume of 200 to 500 mL—are diluted 10 to 20-fold compared to the level in the donor when mixed into the approximately 6 L volume of adult whole blood (Gordon et al. 2021; Shoham et al. 2023). Thus, average to low antibody levels in donor units might easily be diluted to ineffective levels as measured by viral-specific total or neutralization amounts. The large effective outpatient CCP study excluded the lower third of units based on low spike antibody levels in the convalescent plasma (Sullivan et al. 2022c). Comparing the remaining middle and upper third (in terms of antibody content) of all units, recipients of the upper third given within 5 days of symptom onset had no hospitalizations, compared with 5 hospitalizations in participants who received the infusion later or containing antibodies in the middle third (Levine et al. 2023; Sullivan et al. 2022c). Similar findings were found in the other four outpatient convalescent plasma trials when donors' plasma was divided into upper and lower halves of antibody content. When given early, the higher antibody levels greatly reduced hospitalizations (Levine et al. 2023) or, as in the Argentina outpatient trial, progression of the respiratory illness (Libster et al. 2021).

The fundamental goal for plasma is to ensure that sufficient SARS-CoV-2 antibodies are available so that viral neutralization is above a pre-determined threshold even after the dilution described above. CCP offers extra viral clearance as neither the constant region function (also called the fragment, crystallizable (Fc) region), which augments phagocyte- or complement-dependent function, nor antibodies to the viral nucleocapsid, which can clear infected cells, are measured in the virus neutralization assays. In many publications reporting SARS-CoV-2 variant neutralizations, virologists describe virus neutralization assays, but without reporting the important associated total antibody levels to the spike protein, and also finding anywhere from low to high neutralization capacity in convalescent populations with uncharacterized total antibody levels (Sullivan et al. 2022a, 2023). We have found that donor units obtained in the pre-omicron era with total IgG antibodies above Eurimmun AU 10 or 140 RU/mL correlated with > 95% omicron neutralization, especially when antibodies to Spike or S1 protein were high (Li et al. 2022). Recently (within 6 months of infection or vaccination) collected CCP was able to retain virus neutralizations of future variants if antibody levels were high enough. High-titer CCP neutralizes SARS-CoV-2 variants despite no previous donor exposure to the new variants (Sullivan et al. 2022a, 2023).

Over the three years of the pandemic with repeated SARS-CoV-2 infections and vaccinations, antibody levels in the population have risen almost 100-fold (Fig. 5). The present qualifications of donor plasma would thus include more than 90% of all individuals. This threshold should, however, be raised to only include the top 20% of antibody levels, because at those high antibody levels, activity is retained against future variants (Sullivan et al. 2022a, 2023).

Early in a pandemic, when antibody levels may not be determinable, using plasma units from different donors can maximize the potential to deliver neutralizing antibodies (Sullivan and Casadevall 2023). After the establishment of assays for antibody level determination, a single CP unit in the top quintile of antibody content and neutralizing capacity will be ideal for therapeutic plasma as a single high antibody donor unit with ten times the levels contributes to more than doubling low antibody level donor units. An early registry to save aliquots of donor units for later antibody level determinations when diagnostic tests become available is important to create at the outset of the pandemic.

8 Conclusion

Antibodies work! Therapeutic convalescent plasma units containing high levels of specific viral neutralization antibodies need to be distinguished from nontherapeutic plasma, which, while still protective in the donor, will not protect recipients after the 10-to-20-fold dilution that occurs when transfused into a recipient. A cup of water can quench a small fire and high-titer plasma can stem early viral disease

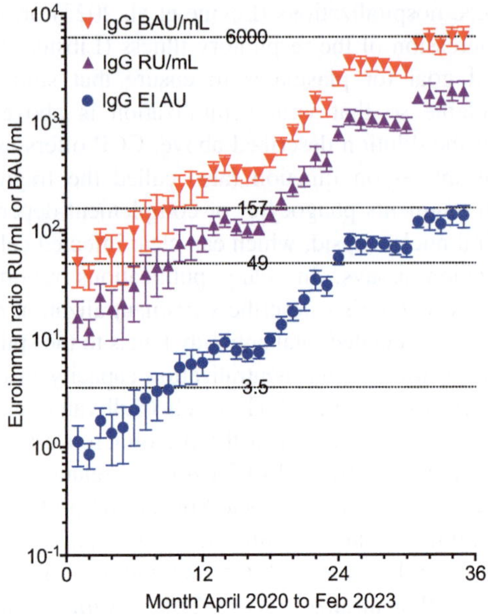

Fig. 5 Increasing trend of spike IgG antibodies in a tertiary hospital patient population. Over a 3-year period (April 2020–February 2023), a retrospective analysis measured spike IgG antibodies in a tertiary hospital clinical immunology laboratory using the Euroimmun SARS-CoV-2 IgG ELISA (Caturegli et al. 2023). The 3-year interval was classified into five periods based on the SARS-CoV-2 strain variant epidemiology matching the above periods. A total of 15,820 sera, derived from 11,022 individuals (6,362 females and 4,660 males) ranging from severe immunocompromised state to routine health visits, were tested. Spike IgG levels rose significantly over time, from a median ELISA 0.13 Arbitrary Units (AU) in the first period to 48.7 in the last period ($p < 0.0001$). The 80th percentile of the spike IgG distribution was 0.55, 8.1, 9.6, 64.9, and 151 AU (7.7, 113, 134, 910, and 2114 RU/mL; 25, 362, 429, 2912, and 6765 BAU/mL) in the five periods. Using 3.5 AU which approximates 49–55 RU/mL, the FDA threshold for qualifying CCP donors with this Euroimmun assay, the percentage of subjects eligible for CCP donation would have been 11%, 44%, 61%, 81%, and 91% in the five time periods. The conversion of Euroimmun AU 1 = 14 RU/mL = 45 BAU/mL such that on the low range 3.5 AU ~ 49 RU/mL ~ 157 BAU/mL and on the high range 150 AU ~ 2100 RU/mL ~ 6720 BAU/mL

progression. Buckets of water may not be able to quench a raging fire and much higher plasma doses may be needed if the target is a severe respiratory distress. In both circumstances, promptness, and not waiting until things get out of hand, is crucial. Convalescent plasma offers the opportunity for hope by enabling community participation in outpatient curative therapy while monoclonal therapies, vaccines, and drugs are being developed. "Damn the spikes, full antibodies ahead" (McPherson 2012).

Declaration of Competing Interest DJS reports AliquantumRx Founder and Board member with stock options (macrolide for malaria), Hemex Health malaria diagnostics consulting, and royalties for malaria diagnostic test control standards to Alere—all outside of submitted work.

Funding This work was supported by the U.S. Department of Defense's Joint Program Executive Office for Chemical, Biological, Radiological and Nuclear Defense (JPEO-CBRND), in collaboration with the Defense Health Agency (DHA) (contract number: W911QY2090012), with additional support from Bloomberg Philanthropies, State of Maryland, the National Institutes of Health (NIH), and National Institute of Allergy and Infectious Diseases 3R01AI152078-01S1. The study sponsors did not contribute to the manuscript design; to the collection, analysis, and interpretation of data; to manuscript preparation, nor to the decision to submit the work for publication.

References

ACTIV-3 (2022) Efficacy and safety of two neutralising monoclonal antibody therapies, sotrovimab and BRII-196 plus BRII-198, for adults hospitalised with COVID-19 (TICO): a randomised controlled trial. Lancet Infect Dis 22(5):622–635

Activ-3-Therapeutics (2022) Tixagevimab-cilgavimab for treatment of patients hospitalised with COVID-19: a randomised, double-blind, phase 3 trial. Lancet Respir Med 10(10):972–984

Addetia A, Crawford KHD, Dingens A, Zhu H, Roychoudhury P, Huang ML et al (2020) Neutralizing antibodies correlate with protection from SARS-CoV-2 in humans during a fishery vessel outbreak with a high attack rate. J Clin Microbiol 58(11)

Agarwal A, Mukherjee A, Kumar G, Chatterjee P, Bhatnagar T, Malhotra P et al (2020) Convalescent plasma in the management of moderate covid-19 in adults in India: open label phase II multicentre randomised controlled trial (PLACID Trial). BMJ 371:m3939

Alemany A, Millat-Martinez P, Corbacho-Monne M, Malchair P, Ouchi D, Ruiz-Comellas A et al (2022) High-titre methylene blue-treated convalescent plasma as an early treatment for outpatients with COVID-19: a randomised, placebo-controlled trial. Lancet Respir Med 10 (3):278–288

Bar KJ, Shaw PA, Choi GH, Aqui N, Fesnak A, Yang JB et al (2021) A randomized controlled study of convalescent plasma for individuals hospitalized with COVID-19 pneumonia. J Clin Invest 131(24)

Beigel JH, Aga E, Elie-Turenne MC, Cho J, Tebas P, Clark CL et al (2019) Anti-influenza immune plasma for the treatment of patients with severe influenza A: a randomised, double-blind, phase 3 trial. Lancet Respir Med 7(11):941–950

Belov A, Huang Y, Villa CH, Whitaker BI, Forshee R, Anderson SA et al (2022) Early administration of COVID-19 convalescent plasma with high titer antibody content by live viral neutralization assay is associated with modest clinical efficacy. Am J Hematol 97(6):770–779

Bloch EM, Shoham S, Casadevall A, Sachais BS, Shaz B, Winters JL et al (2020) Deployment of convalescent plasma for the prevention and treatment of COVID-19. J Clin Invest 130 (6):2757–2765

Browning CH (1955) Emil Behring and Paul Ehrlich: their contributions to science. Nature 175 (4457):570–575

Carvalho T, Krammer F, Iwasaki A (2021) The first 12 months of COVID-19: a timeline of immunological insights. Nat Rev Immunol 21(4):245–256

Caturegli P, Laeyendecker O, Tobian AAR, Sullivan D (2023) Hundred-fold increase in SARS-CoV-2 spike antibody levels over three years in a hospital clinical laboratory. Microbiol Spectr. https://doi.org/10.1128/spectrum.02183-23

Chahine EB, Durham SH (2021) Ibalizumab: the first monoclonal antibody for the treatment of HIV-1 infection. Ann Pharmacother 55(2):230–239

Cohn EJ, Strong LE et al (1946) Preparation and properties of serum and plasma proteins; a system for the separation into fractions of the protein and lipoprotein components of biological tissues and fluids. J Am Chem Soc 68:459–475

COVID-19 Updates (2022) Remdesivir (Veklury) in high-risk outpatients with COVID-19. Med Lett Drugs Ther 64(1644):31–32

de Melo GD, Hellert J, Gupta R, Corti D, Bourhy H (2022) Monoclonal antibodies against rabies: current uses in prophylaxis and in therapy. Curr Opin Virol 53:101204

Enria DA, Briggiler AM, Fernandez NJ, Levis SC, Maiztegui JI (1984) Importance of dose of neutralising antibodies in treatment of Argentine haemorrhagic fever with immune plasma. Lancet 2(8397):255–256

Enria D, Franco SG, Ambrosio A, Vallejos D, Levis S, Maiztegui J (1986) Current status of the treatment of Argentine hemorrhagic fever. Med Microbiol Immunol 175(2–3):173–176

Enria DA, Briggiler AM, Sanchez Z (2008) Treatment of Argentine hemorrhagic fever. Antiviral Res 78(1):132–139

Focosi D, McConnell S, Sullivan DJ, Casadevall A (2023) Analysis of SARS-CoV-2 mutations associated with resistance to therapeutic monoclonal antibodies that emerge after treatment. Drug Resist Updat 71:100991

Franchini M, Casadevall A, Senefeld JW, Joyner MJ, Sullivan DJ, Focosi D (2023) Recommendations on the use of COVID-19 convalescent plasma to treat immunocompromised patients. Semin Thromb Hemost

Gharbharan A, Jordans C, Zwaginga L, Papageorgiou G, van Geloven N, van Wijngaarden P et al (2022) Outpatient convalescent plasma therapy for high-risk patients with early COVID-19. A randomized placebo-controlled trial. Clin Microbiol Infect 29:208–214

Gilbert PB, Donis RO, Koup RA, Fong Y, Plotkin SA, Follmann D (2022) A Covid-19 milestone attained—a correlate of protection for vaccines. N Engl J Med 387(24):2203–2206

Gordon O, Brosnan MK, Yoon S, Jung D, Littlefield K, Ganesan A et al (2021) Pharmacokinetics of high-titer anti-SARS-CoV-2 human convalescent plasma in high-risk children. JCI Insight

Gross PA, Gitlin D, Janeway CA (1959) The gamma globulins and their clinical significance. IV. Therapeutic uses of gamma globulins. N Engl J Med 260(4):170–178

Group A-TL-CS, Lundgren JD, Grund B, Barkauskas CE, Holland TL, Gottlieb RL et al (2021) A neutralizing monoclonal antibody for hospitalized patients with Covid-19. N Engl J Med 384 (10):905–914

Group IS (2022) Hyperimmune immunoglobulin for hospitalised patients with COVID-19 (ITAC): a double-blind, placebo-controlled, phase 3, randomised trial. Lancet 399(10324):530–540

Habtehyimer F, Zhu X, Redd A, Gebo K, Abraham A, Patel E, Laeyendecker O, Gniadek T, Fernandez R, Baker O, Ram M, Cachay E, Currier J, Fukuta Y, Gerber J, Heath S, Meisenberg B, Huaman M, Levine A, Shenoy A, Anjan S, Blair J, Cruser D, Forthal D, Hammitt L, Kassaye S, Mosnaim G, Patel B, Paxton J, Raval J, Sutcliffe C, Abinante M, Oei K, Cluzet V, Cordisco M, Greenblatt B, Rausch W, Shade D, Gawad A, Klein S, Pekosz A, Shoham S, Casadevall A, Bloch E, Hanley D, Tobian A, Sullivan D (2023) COVID-19 convalescent plasma therapy decreases inflammatory cytokines: a randomized controlled tria. Microbiol Spectr. accepted

Hammond J, Leister-Tebbe H, Gardner A, Abreu P, Bao W, Wisemandle W et al (2022) Oral nirmatrelvir for high-risk, nonhospitalized adults with Covid-19. N Engl J Med 386(15):1397–1408

Harris E (2023) FDA grants full approval to Paxlovid, COVID-19 antiviral treatment. JAMA 329 (24):2118

He Q, Wu L, Xu Z, Wang X, Xie Y, Chai Y et al (2023) An updated atlas of antibody evasion by SARS-CoV-2 Omicron sub-variants including BQ.1.1 and XBB. Cell Rep Med 4(4):100991

Hobson D, Curry RL, Beare AS, Ward-Gardner A (1972) The role of serum haemagglutination-inhibiting antibody in protection against challenge infection with influenza A2 and B viruses. J Hyg (lond) 70(4):767–777

Huygens S, Hofsink Q, Nijhof IS, Goorhuis A, Kater AP, Te Boekhorst PAW et al (2023) Hyperimmune globulin for severely immunocompromised patients hospitalized with coronavirus disease 2019: a randomized, controlled trial. J Infect Dis 227(2):206–210

Jaki L, Weigang S, Kern L, Kramme S, Wrobel AG, Grawitz AB et al (2023) Total escape of SARS-CoV-2 from dual monoclonal antibody therapy in an immunocompromised patient. Nat Commun 14(1):1999

Jayk Bernal A, Gomes da Silva MM, Musungaie DB, Kovalchuk E, Gonzalez A, Delos Reyes V et al (2022) Molnupiravir for oral treatment of Covid-19 in nonhospitalized patients. N Engl J Med 386(6):509–520

Jones JM, Opsomer JD, Stone M, Benoit T, Ferg RA, Stramer SL et al (2022) Updated US infection- and vaccine-induced SARS-CoV-2 seroprevalence estimates based on blood donations, July 2020–December 2021. JAMA 328(3):298–301

Joyner MJ, Bruno KA, Klassen SA, Kunze KL, Johnson PW, Lesser ER et al (2020) Safety update: COVID-19 convalescent plasma in 20,000 hospitalized patients. Mayo Clin Proc 95 (9):1888–1897

Joyner MJ, Carter RE, Senefeld JW, Klassen SA, Mills JR, Johnson PW et al (2021) Convalescent plasma antibody levels and the risk of death from Covid-19. N Engl J Med 384(11):1015–1027

Keam SJ (2023) Nirsevimab: first approval. Drugs 83(2):181–187

Keller MA, Stiehm ER (2000) Passive immunity in prevention and treatment of infectious diseases. Clin Microbiol Rev 13(4):602–614

Kempe CH, Berge TO, England B (1956) Hyperimmune vaccinal gamma globulin; source, evaluation, and use in prophylaxis and therapy. Pediatrics 18(2):177–188

Kempe CH, Bowles C, Meiklejohn G, Berge TO, St Vincent L, Babu BV et al (1961) The use of vaccinia hyperimmune gamma-globulin in the prophylaxis of smallpox. Bull World Health Organ 25:41–48

Khoury DS, Cromer D, Reynaldi A, Schlub TE, Wheatley AK, Juno JA et al (2021) Neutralizing antibody levels are highly predictive of immune protection from symptomatic SARS-CoV-2 infection. Nat Med 27(7):1205–1211

Khoury DS, Schlub TE, Cromer D, Steain M, Fong Y, Gilbert PB et al (2023) Correlates of protection, thresholds of protection, and immunobridging among persons with SARS-CoV-2 infection. Emerg Infect Dis 29(2):381–388

Klassen SA, Senefeld JW, Johnson PW, Carter RE, Wiggins CC, Shoham S et al (2021) The effect of convalescent plasma therapy on mortality among patients with COVID-19: systematic review and meta-analysis. Mayo Clin Proc 96(5):1262–1275

Korley FK, Durkalski-Mauldin V, Yeatts SD, Schulman K, Davenport RD, Dumont LJ et al (2021) Early convalescent plasma for high-risk outpatients with Covid-19. N Engl J Med 385 (21):1951–1960

Kreier F (2022) Ten billion COVID vaccinations: world hits new milestone. Nature

Lehman A, Muniz VA, Chaney R, Pimentel J, Mattila J, Lawrence S (2022) Speed and need: twin development challenges in rapid response for a SARS-CoV-2 antibody cocktail. Curr Opin Biotechnol 76:102715

Levine AC, Fukuta Y, Huaman MA, Ou J, Meisenberg BR, Patel B et al (2023) Coronavirus disease 2019 convalescent plasma outpatient therapy to prevent outpatient hospitalization: a meta-analysis of individual participant data from 5 randomized trials. Clin Infect Dis 76 (12):2077–2086

Li M, Beck EJ, Laeyendecker O, Eby Y, Tobian AAR, Caturegli P et al (2022) Convalescent plasma with a high level of virus-specific antibody effectively neutralizes SARS-CoV-2 variants of concern. Blood Adv 6(12):3678–3683

Libster R, Perez Marc G, Wappner D, Coviello S, Bianchi A, Braem V et al (2021) Early high-titer plasma therapy to prevent severe Covid-19 in older adults. N Engl J Med 384(7):610–618

Lin WT, Hung SH, Lai CC, Wang CY, Chen CH (2022) The impact of neutralizing monoclonal antibodies on the outcomes of COVID-19 outpatients: a systematic review and meta-analysis of randomized controlled trials. J Med Virol 94(5):2222–2229

Looney RJ, Huggins J (2006) Use of intravenous immunoglobulin G (IVIG). Best Pract Res Clin Haematol 19(1):3–25

Maiztegui JI, Fernandez NJ, de Damilano AJ (1979) Efficacy of immune plasma in treatment of Argentine haemorrhagic fever and association between treatment and a late neurological syndrome. Lancet 2(8154):1216–1217

Markham A (2021) REGN-EB3: first approval. Drugs 81(1):175–178

Marks L (2012) The birth pangs of monoclonal antibody therapeutics: the failure and legacy of Centoxin. Mabs 4(3):403–412

Mazumdar S (2009) Raxibacumab. MAbs 1(6):531–538

McPherson JM (2012) Damn the torpedoes. War on the waters. The union and confederate navies, 1861–1865. University of North Carolina Press, pp 207–223

Messina A, Germano C, Avellis V, Tavella E, Dodaro V, Massaro A et al (2022) New strategies for the prevention of respiratory syncytial virus (RSV). Early Hum Dev 174:105666

Miller MJ, Himschoot A, Fitch N, Jawalkar S, Freeman D, Hilton C et al (2020) Association of trends in severe acute respiratory syndrome coronavirus 2 (SARS-CoV-2) seroprevalence and state-issued nonpharmaceutical interventions: United States, 1 August 2020 to 30 March 2021. Clin Infect Dis 75(Supplement_2):S264–S270

Misset B, Piagnerelli M, Hoste E, Dardenne N, Grimaldi D, Michaux I et al (2023) Convalescent plasma for Covid-19–induced ARDS in mechanically ventilated patients. N Engl J Med 389 (17):1590–1600

Molnupiravir for treatment of COVID-19 (2022). Med Lett Drugs Ther 64(1642):10–11

Montejano R, Soler-Carracedo A, Borobia AM, Martin-Quiros A, Grandioso D, Cañada C et al (2019) The current mismatch between coronavirus disease 2019 clinical trial design and the evolving profile of hospitalized patients: a retrospective cohort analysis. Clin Infect Dis

NIAID (2022) Strategies and treatments for respiratory infections and viral emergencies (STRIVE) NIAID clinical trials2023 [updated 22 Dec 2022]. Available from: https://www.niaid.nih.gov/clinical-trials/strategies-treatments-respiratory-infections-viral-emergencies

O'Donnell MR, Grinsztejn B, Cummings MJ, Justman JE, Lamb MR, Eckhardt CM et al (2021) A randomized double-blind controlled trial of convalescent plasma in adults with severe COVID-19. J Clin Invest 131(13)

O'Shaughnessy JA (2021) Convalescent plasma EUA letter of authorization 12282021. FDA website. https://www.fda.gov/media/141477/download

Pirofski LA, Casadevall A (2020) Pathogenesis of COVID-19 from the perspective of the damage-response framework. mBio 11(4)

Pormohammad A, Zarei M, Ghorbani S, Mohammadi M, Razizadeh MH, Turner DL et al (2021) Efficacy and safety of COVID-19 vaccines: a systematic review and meta-analysis of randomized clinical trials. Vaccines (Basel) 9(5)

Potter CW, Oxford JS (1979) Determinants of immunity to influenza infection in man. Br Med Bull 35(1):69–75

Recovery (2020) Dexamethasone in hospitalized patients with Covid-19. N Engl J Med 384 (8):693–704

Release Fp (2020) Coronavirus (COVID-19) update: FDA authorizes monoclonal antibodies for treatment of COVID-19. https://www.fdagov/news-events/press-announcements/coronavirus-covid-19-update-fda-authorizes-monoclonal-antibodies-treatment-covid-19

Remdesivir (Veklury) for COVID-19 (2020). Med Lett Drugs Ther 62(1612):186–188

Ripoll JG, Tulledge-Scheitel SM, Ford S, Pike ML, Gorman EK, Hanson SN et al (2023) Outpatient treatment with concomitant vaccine-boosted convalescent plasma for patients with immunosuppression and COVID-19. medRxiv 2023.08.29.23293790

Salazar E, Christensen PA, Graviss EA, Nguyen DT, Castillo B, Chen J et al (2021) Significantly decreased mortality in a large cohort of coronavirus disease 2019 (COVID-19) patients transfused early with convalescent plasma containing high-titer anti-severe acute respiratory syndrome coronavirus 2 (SARS-CoV-2) spike protein IgG. Am J Pathol 191(1):90–107

Senefeld JW, Gorman EK, Johnson PW, Moir ME, Klassen SA, Carter RE et al (2023) Rates among hospitalized patients with COVID-19 treated with convalescent plasma: a systematic review and meta-analysis. Mayo Clin Proc Innov Qual Outcomes 7(5):499–513

Sharfstein JM, Sharfstein JM (2018) 6 Recognizing a crisis. The public health crisis survival guide: leadership and management in trying times. Oxford University Press

Shen C, Wang Z, Zhao F, Yang Y, Li J, Yuan J et al (2020) Treatment of 5 critically ill patients with COVID-19 with convalescent plasma. JAMA 323(16):1582–1589

Shoham S, Bloch EM, Casadevall A, Hanley D, Lau B, Gebo K et al (2023) Transfusing convalescent plasma as post-exposure prophylaxis against severe acute respiratory syndrome coronavirus 2 (SARS-CoV-2) infection: a double-blinded, phase 2 randomized, controlled trial. Clin Infect Dis 76(3):e477–e486

Sullivan DJ, Franchini M, Joyner MJ, Casadevall A, Focosi D (2022a) Analysis of anti-SARS-CoV-2 Omicron-neutralizing antibody titers in different vaccinated and unvaccinated convalescent plasma sources. Nat Commun 13(1):6478

Sullivan DJ, Gebo KA, Shoham S, Bloch EM, Lau B, Shenoy AG et al (2022b) Early outpatient treatment for Covid-19 with convalescent plasma. N Engl J Med 386(18):1700–1711

Sullivan D, Casadevall A (2023) COVID-19 serology data provide guidance for future deployments of convalescent plasma. mBio 14(3):e0042823

Sullivan DJ, Gebo KA, Shoham S, Bloch EM, Lau B, Shenoy AG et al (2022c) Early outpatient treatment for Covid-19 with convalescent plasma. N Engl J Med

Sullivan DJ, Franchini M, Senefeld JW, Joyner MJ, Casadevall A, Focosi D (2023) Plasma after both SARS-CoV-2 boosted vaccination and COVID-19 potently neutralizes BQ.1.1 and XBB.1. J Gen Virol 104(5)

Takashita E, Kinoshita N, Yamayoshi S, Sakai-Tagawa Y, Fujisaki S, Ito M et al (2022) Efficacy of antibodies and antiviral drugs against Covid-19 Omicron variant. N Engl J Med 386 (10):995–998

Tao K, Tzou PL, Kosakovsky Pond SL, Ioannidis JPA, Shafer RW (2022) Susceptibility of SARS-CoV-2 Omicron variants to therapeutic monoclonal antibodies: systematic review and meta-analysis. Microbiol Spectr 10(4):e0092622

Therapeutic monoclonal antibodies approved or in review in the EU or US [Internet] (2023). [cited 10/17/2023]

Thompson MA, Henderson JP, Shah PK, Rubinstein SM, Joyner MJ, Choueiri TK et al (2021) Association of convalescent plasma therapy with survival in patients with hematologic cancers and COVID-19. JAMA Oncol 7(8):1167–1175

Thornton CS, Huntley K, Berenger BM, Bristow M, Evans DH, Fonseca K et al (2022) Prolonged SARS-CoV-2 infection following rituximab treatment: clinical course and response to therapeutic interventions correlated with quantitative viral cultures and cycle threshold values. Antimicrob Resist Infect Control 11(1):28

Tiselius A (1937) Electrophoresis of serum globulin: electrophoretic analysis of normal and immune sera. Biochem J 31(9):1464–1477

Toor AA, van Burik JA, Weisdorf DJ (2001) Infections during mobilizing chemotherapy and following autologous stem cell transplantation. Bone Marrow Transplant 28(12):1129–1134

Weinreich DM, Sivapalasingam S, Norton T, Ali S, Gao H, Bhore R et al (2021a) REGN-COV2, a neutralizing antibody cocktail, in outpatients with Covid-19. N Engl J Med 384(3):238–251

Weinreich DM, Sivapalasingam S, Norton T, Ali S, Gao H, Bhore R et al (2021b) REGEN-COV antibody combination and outcomes in outpatients with Covid-19. N Engl J Med 385(23):e81

Wiegand RE, Deng Y, Deng X, Lee A, Meyer WA, III, Letovsky S et al (2022) Estimated SARS-CoV-2 antibody seroprevalence trends and relationship to reported case prevalence from a repeated, cross-sectional study in the 50 states and the District of Columbia, United States 2014; October 25, 2020 and 2013; February 26, 2022. The Lancet Regional Health – Americas 18

Wilcox MH, Gerding DN, Poxton IR, Kelly C, Nathan R, Birch T et al (2017) Bezlotoxumab for prevention of recurrent clostridium difficile infection. N Engl J Med 376(4):305–317
Williams BA, Jones CH, Welch V, True JM (2023) Outlook of pandemic preparedness in a post-COVID-19 world. NPJ Vaccines 8(1):178

David J. Sullivan , MD is a Johns Hopkins Malaria Research Institute faculty member in the Bloomberg School of Public Health where he has focused on malaria diagnostics and drugs revolving around heme, hemozoin, and metal metabolism. As an Infectious Diseases physician, he pivoted to COVID-19 clinical and bench research focused on convalescent antibody correlates of protection to reduce COVID-19 morbidity and mortality. He led the outpatient randomized clinical trial comparing early COVID-19 convalescent plasma (CCP) transfusion within 9 days from symptom onset which demonstrated effectiveness in reducing progression to hospitalization. This outpatient trial was cited by the FDA in its emergency use allowance of CCP for immunocompromised COVID-19 patients. This outpatient plasma research provides a blueprint of hope for early high antibody dose convalescent plasma as rapidly deployable therapy for future pandemics.

Monoclonal Antibodies and Hyperimmune Immunoglobulins in the Next Pandemic

Massimo Franchini and Daniele Focosi

Contents

Abstract Pandemics are highly unpredictable events that are generally caused by novel viruses. There is a high likelihood that such novel pathogens belong to entirely novel viral families for which no targeted small-molecule antivirals exist. In addition, small-molecule antivirals often have pharmacokinetic properties that make them contraindicated for the frail patients who are often the most susceptible to a novel virus. Passive immunotherapies—available from the first convalescent patients—can then play a key role in controlling pandemics. Convalescent plasma is immediately available, but if manufacturers have fast platforms to generate marketable drugs, other forms of passive antibody treatment can be produced. In this chapter, we will review the technological platforms for generating monoclonal antibodies and hyperimmune immunoglobulins, the current experience on their use for treatment of COVID-19, and the pipeline for pandemic candidates.

M. Franchini
Department of Transfusion Medicine and Hematology, Carlo Poma Hospital, Mantua, Italy

D. Focosi (✉)
North-Western Tuscany Blood Bank, Pisa University Hospital, Pisa, Italy
e-mail: daniele.focosi@gmail.com

Current Topics in Microbiology and Immunology (2025) 443: 243–257
https://doi.org/10.1007/82_2024_274

Published Online: 15 June 2024

Abbreviations

CCP	COVID19 convalescent plasma
EMA	European Medicines Agency
mAb	Monoclonal antibody
nAb	Neutralizing antibody
RBD	Receptor-binding domain
RCT	Randomized controlled trial

1 Introduction

A novel pandemic invariably represents a dramatic societal event with enormous negative social, economic, and healthcare consequences. Pandemic preparedness, including nonpharmaceutical interventions (NPIs) and supportive treatments, can do much to mitigate these consequences, but novel pathogens are very likely to be unresponsive to all the existing therapeutic classes of small-molecule antivirals. In this scenario, medicine must turn to passive immunotherapies, available from the first convalescent patients. While convalescent plasma will invariably represent the first line of reaction, convalescent plasma derivatives such as monoclonal antibodies (mAbs) and hyperimmune immunoglobulins (HIG) represent more standardized solutions which can provide immediate protection during the time it takes for vaccines to be developed and for individuals to mount effective vaccine responses. Passive antibody therapies are especially important for immunocompromised populations who may not be able to mount an adequate immune response.

Optimal target product profiles outlining use cases, intended populations, and desirable product characteristics are needed (Gupta et al. 2023). To facilitate the administration of mAbs in nonspecialized healthcare settings, development of intramuscular or subcutaneous formulations may need to occur in tandem with advancing alternative delivery technologies (e.g., viral vectors, RNA, DNA, and oral or respiratory administration (Focosi and Maggi 2023)).

2 mAb Manufacturing Platforms

The ideal pandemic preparedness mAbs should target highly conserved epitopes shared across multiple viruses and should act through a combination of mechanisms. The product development timeline from lead mAb identification to phase 1 investigational new drug application (IND) was 18 months in 2015 but was reduced to 10–12 months in 2020 in light of the COVID pandemic, and thanks to public and private COVID-19 investments. Ideally, this process should be reduced to 5–6 months to appropriately meet pandemic needs (Kelley 2020). Rapid clinical production capacity has benefited from the development of highly productive cell lines and larger bioreactors with single-use technology, enabling the production of thousands of doses from a single batch of over 5 kg of mAb. One way of increasing the speed of identifying leads is to screen prospectively isolated panels against new pathogens and many viral strains. A rapid development model should focus on IgG_1 and the already widely used Chinese hamster ovary (CHO) cell lines (as opposed to the promising but still much more rarely used *Escherichia coli*, yeasts, or plants), for which safety has been well assessed, risk is low, and scalability of manufacturing is high (Kelley 2020).

Several companies have developed cell lines using targeted integration of mAb expressing vectors (as opposed to historical random integration), which provide more consistent expression through integration of low copy numbers in highly active transcriptional hotspots. A transfectant pool is generated that provides gram quantities of proteins in 2 months, which can be used as a representative material (Ye 2012). Cell sorting is later implemented to enrich the transfection pool for high producers, improving productivity. Transfection is transient, and the DNA is lost during cell divisions. What is needed here is the rare event of DNA integration into the genome, in what is called a stable transfection pool. Moving directly from the stable transfectant pool to cloning (a single clone origin being required by regulatory authorities for biotherapeutics approval) is becoming a standard practice today. Multiple candidate clones can be screened with very small bioreactors using small-volume tubes or ambr15 bioreactors, instead of screening using 5-L bioreactors.

The initial clinical entry product profile would likely be a liquid drug product (to avoid delays related to lyophilization) at a moderate concentration (50–75 mg), stored cold or frozen. Parallel production trains could provide both small-volume batches quickly and larger volume batches soon afterward. The portability of the commercial process to enable production at 15,000–25,000 L is an important requirement of a process that must be capable of supplying millions of doses in a short time, and consortia of manufacturers sharing production capacity and commercialization capacities should be encouraged and pre-defined for better preparedness.

3 mAb Insights from the COVID-19 Experience

The COVID-19 pandemic has represented an unprecedented boost to the mAb field, and we have learned a great deal about supply chains, delivery routes, logistical hurdles, baseline resistance, and treatment-emergent resistance. SARS-CoV-2 has represented a tricky enemy, with very fast viral evolution, and prolonged circulation in the population (3 years and counting so far).

The first point to address is expectations. Early in the pandemics, mAbs were advertised as magic bullets for all patients. The failure of the first RCTs in hospitalized patients (Gottlieb et al. 2021) reminded us of the rational use of antiviral agents, which requires treatment in the early viral replication phase of the disease. Manufacturers were smart at retargeting early disease stages, i.e., in outpatients, where the mAbs proved successful. Moving to outpatients was a logical step for manufacturers (larger eligible populations mean higher profits) and for healthcare systems (fewer patients progress to hospitalization). It was only later that companies focused on pre-exposure prophylaxis in immunocompromised patients, with fewer patients, and requiring repeated infusions.

The second point is resilience. The authorization of the first mAb candidates has discouraged many small companies from finalizing their promising candidates, thinking that a marketing delay would have compromised revenues. Only in 2023, after the withdrawal of ineffective mAbs by the FDA (FDS 2022, 2023, 2024), several small but enlightened companies (e.g., Invivyd VYD222 and Aerium Therapeutics AER-800) reboosted their pipelines to reach the stage of performing clinical trials of their products. The EMA has been surprisingly passive, leaving mAbs without any clinical efficacy to be permitted in the European market, and this might have discouraged innovation by small companies. Similarly, the availability of cheaper and more scalable small-molecule antivirals—together with viral evolution leading to baseline resistance—discouraged many manufacturers (e.g., Eli Lilly) from reinvesting in novel candidates. AstraZeneca (AZD3152 and AZD3959) and Regeneron (REGN17092) have invested in COVID-19 mAb updates, but their slow pace in running clinical trials suggests that this market area is not of sufficient priority, especially with the cessation of public funding.

The third point is that cocktails are key to prevent failures. Cocktails such as REGN-CoV-2 remained in clinical use even after imdevimab was totally resistant, and Evusheld™ remained in clinical use even after tixagevimab was totally resistant. Cocktails of two mAbs are, however, suboptimal, because when only one agent remains active the chances of treatment-emergent immune escape approach what is seen with monotherapies. Cocktails of three mAbs should be encouraged. Although this invariably comes with increased costs, increased market persistence benefiting the vendor should limit costs. Notably, this approach has been already in place with Ebola virus mAb cocktails (see below) and should be encouraged with fast-evolving pathogens (such as RNA viruses). Another alternative is to create bispecific antibodies (e.g., 14-H-06 (Ku et al. 2022)) combining two Fab regions binding to different epitopes. Bispecific antibodies may be a cost-effective alternative to mAb cocktails and are a promising strategy to improve potency and breadth.

The fourth point is that an extended half-life is needed for preparations intended for pre-exposure prophylaxis, which is the main unmet need for immunocompromised patients who cannot mount an endogenous immune response. This longer half-life can be achieved by engineering the Fc fragment of antibodies, e.g., with Met428Leu/Asn434Ser (LS) mutation without affecting ADCC function (Zalevsky et al. 2010; Schäfer et al. 2020) (as was achieved with nirsevimab and Evusheld) or with the M252Y/S254T/T256E (YTE) mutation (as was achieved with sotrovimab and adintrevimab).

4 mAb Pipeline for Selected Pandemic Agents

Viruses with pandemic potential are almost exclusively RNA viruses. The list includes flaviviruses (e.g., West Nile Virus (WNV), dengue virus, Zika virus (Guarner and Hale 2019)), chikungunya virus (Vairo et al. 2019), influenza viruses A, e.g., A(H_1N_1), A(H_5N_1) (Vogel and Manicassamy 2020), Ebola virus (EBOV) (Hoenen et al. 2019), and respiratory betacoronaviruses (SARS-CoV (Booth et al. 2003) and MERS-CoV (Bermingham et al. 2012)). Given the unpredictability of what will be the next pandemic agent, it is hard for manufacturers to invest on specific viral families. Thus far, they have mostly focused on viruses classified as bioweapons and on viruses without any effective small-molecule antivirals.

As noted above, combining mAbs creates a polyclonal product. Thus, the creation of mAb cocktails, which have been generated and tested in humans for various pathogens in addition to SARS-CoV-2, will be a key element of the proposed pipeline for addressing new pandemics:

- Respiratory Syncytial Virus (RSV): Two mAbs against the prefusion conformation or the F glycoprotein have been individually licensed in the last decade.
  - palivizumab (Synagis™) is a humanized $IgG_1\kappa$ approved for infants below the age of 2
  - nirsevimab/MEDI8897 (Beyfortus™) is an IgG_1 with an engineered Fc (M252Y/S254T/T256E (YTE)) improving pH-dependent FcRn interactions, and then improving half-life.
- Influenza virus: Methods for rapid cloning of high-affinity human mAbs against influenza virus have been developed (Wrammert et al. 2008), making it likely that many will be developed during a new influenza pandemic. In particular, transchromosomic bovines have been used as a platform for mAbs against H_1 (53C10 (Gao et al. 2020)). At least, 11 mAbs against viral proteins expressed on the virion surface (H and M2) or other virus-induced human cell surface markers have been tested in clinical trials (Table 1). While controlled human challenge trials have provided encouraging results (Han et al. 2021), studies in naturally occurring infections have provided mixed results (Beigel 2018). A relevant pipeline of heterotypic broadly neutralizing mAb is awaiting to be tested in clinical trials, targeting either group 1 (F10, 6F12, C179) or group 2 (CR8043, 9H10) or both (FI6, CT149, 31.a.83, 56.a.09, 31.b.09, 16.a.26, 16.g.07, 41-5E04, C05, CR9114) (Corti et al. 2017). emphasis on antibody cocktails for treatments (Bimler et al. 2022) prioritized early in the pandemic response.
- Ebola virus:
  - ZMapp™ (Mulangu et al. 2019) is a cocktail of three chimeric mAbs:

    c13C6 is a chimeric mAb that comes from a previously existing antibody cocktail called “MB-003” (LeafBio) (c13C6, h13F6, and c6D8) produced in Chinese hamster ovary or in whole plant (*Nicotiana benthamiana*)

Table 1 Monoclonal antibodies in clinical trials for the treatment of influenza

Target	Cocktail	Ingredients	Specificity	Manufacturer	Prior clinical studies
Anti-H	CT-P27 (fully human IgG with half-life of only 6 days)	CT-P22/ CT120/ firivumab	H_1, H_2, H_5, and H_9	Celltrion (Korea)	KCT0001179 NCT02071914 KCT0001617 NCT03511066
		CT-P23/ CT149/ navivumab	H_1, H_2, H_5, and H_9 (stem fusion domain in HA2)		Preclinical only (Yi et al. 2020)
	CR6261/diridavumab		Group 1 influenza viruses (El-Ekiaby et al. 2015), and binds a highly conserved epitope in the HA stalk/stem)	Crucell/Janssen Vaccines (Netherlands)	NCT01406418 phase II NCT02371668 challenge (Han et al. 2021)
	CR8020		Group 2 viruses (HA stem near the viral membrane)		NCT01756950 NCT01938352 NCT02015533
	MHAA4549A/39.29		H_1, H_2, H_3, H_5, and H_7 (stalk)	Genentech (USA)	NCT01877785 and NCT02284607 phase I (Lim et al. 2016) NCT01980966 (Deng et al. 2018) NCT02293863 (Deng et al. 2020) NCT02623322
	MHAB5553A (IgG_1)		Influenza B		NCT02528903 phase I (Lim et al. 2017)
	MEDI8852/46B8 (FY1-derived IgG_1 kappa) (Kallewaard et al. 2016; Paules et al. 2017)		Inhibits the host cell protease cleavage of H_1 and H_3 HA0 to prevent membrane fusion	MedImmune/ AstraZeneca (USA)	NCT02350751 (Mallory et al. 2017) NCT02603952 (Ali et al. 2018)
	VIS410		Broad IgG_1 (stem) against group 1 and group 2, including H_7N_9	Visterra (USA)	NCT02045472 (Wollacott et al. 2016) NCT02468115 (Sloan et al. 2020)

(continued)

Table 1 (continued)

Target	Cocktail	Ingredients	Specificity	Manufacturer	Prior clinical studies
			(Tharakaraman et al. 2015)		NCT02989194 (Hershberger et al.2019) NCT03040141
	VIR-2482		Fully human IgG_1 with extended half-life for prophylaxis (Sager et al. 2020)	Vir Biotechnology, Inc. (USA)	NCT04033406 NCT05567783
Ectodomain of the matrix protein 2	TCN-032			Theraclone Sciences (USA)	NCT01390025 NCT01719874 (Ramos et al. 2015)
TSG101 (human protein flipped on surface of influenza-infected cells)	FGI-101-1A6 (fully human IgG_1)			Functional Genetics (USA)	NCT01299142

cells, that protected four out of six rhesus macaques when delivered 24–48 h post-infection (Olinger et al. 2012) and does not impair the development of protective endogenous immune responses (Qiu et al. 2013). c2G4 and c4G7 are chimeric mAbs coming from a different murine antibody cocktail called ZMab (Defyrus) (m1H3, m2G4, and m4G7 (Qiu et al. 2014)).

- REGN-EB3 (Inmazeb™, Regeneron): a coformulated mixture of three human IgG_1 monoclonal antibodies (atoltivimab/maftivimab/odesivimab), approved in October 2020 by the FDA for Zaire Ebola virus.
- Ansuvimab/Mab114 (Ebanga™, Ridgeback Biotherapeutics) is a recombinant human IgG_1 isolated from a survivor of the 1995 pandemic, which binds to the portion of the envelope glycoprotein that binds NPC1 receptor (neutralizing infection).

In the PALM, open-label RCT conducted during the 2018–2020 Ebola virus outbreak in the Democratic Republic of Congo, both ansuvimab and REGN-EB3 were superior to ZMapp™ in reducing mortality from EVD: overall mortality of patients in the ZMapp and remdesivir groups were 49% and 53% compared to 34% and 29% for ansuvimab and REGN-EB3 (Mulangu et al. 2019), respectively.

Clearly, the pandemic pipeline is much restricted to date, and pandemic preparedness should include further pathogens.

5 Hyperimmune Immunoglobulin Manufacturing Platforms

As the pandemic evolved, several countries began to direct CCP donations toward plasma manufacturers to investigate CCP fractionation to produce polyclonal immunoglobulin (Ig) formulations (Focosi et al. 2021). Polyclonal Ig can be administered intravenously (IVIg), intramuscularly, subcutaneously, or, more rarely, intranasally. It can be produced from both immunized humans and animals (Cao et al. 2019). Hyperimmune immunoglobulins have been successfully used in the last century to control both bacterial (e.g., *Streptococcus pneumoniae*) and viral infections (e.g., hepatitis A virus).

Anti-SARS-CoV-2 hyperimmune IVIg (hIVIg) is a plasma-derived product prepared from CCP unit pools (usually from a pool of more than 1000 donors) and requires dedicated production chains according to Good Manufacturing Practice [GMP] rules. They have the advantage over CCP in that they contain a high and standardized nAb titer against SARS-CoV-2 in a smaller volume (hIVIg are approximately tenfold more concentrated than CCP) (Yang 2023). Plasma fractionation results in IgG purification, with the consequent removal not only of undesirable prothrombotic factors but also of potentially useful IgG subclasses, in particular IgG_3 fraction, that could negatively impact the hIVIg neutralizing titer (Focosi et al. 2021). The mechanism of action of hIVIg could be in part independent of the antibody-dependent antiviral effect but related to anti-cytokine effects, a decrease in pro-inflammatory responses, inhibition of complement activation, and suppression of B- and T-cell functions (Yang 2023). In addition to the purification process, this plasma-derived product undergoes viral inactivating steps including solvent/detergent treatment (that inactivates SARS-CoV-2) and nanofiltration (that inactivates nonenveloped viruses) (Focosi et al. 2021).

To be considered economically sustainable, however, anti-SARS-CoV-2 hIVIg typically requires fractionation contracts exceeding 10,000 L of plasma per year (Radosevich and Burnouf 2010). The production on a large scale of such hIVIg encountered, therefore, the obstacle of needing a large number of CCP donations required to produce a single lot. These were very difficult to assemble in the pre-vaccine and pre-mAb era of the pandemic because most CCP units collected were utilized for emergency treatment of COVID-19. In addition, hIVIg manufacturing requires several months between initiation of CCP collection and distribution of lots (creation of dedicated CCP production chains, conduction of well-designed phase I-III trials, and acquisition of all necessary certifications for marketing by regulatory agencies). Another, not less important, issue is the economic sustainability of the high costs of IVIg production, whose price is noticeably higher than that of CCP. These economic considerations induced many manufacturers to view investments in hIVIg as economically non-advantageous, requiring several years (well beyond the duration of the pandemic) to be amortized

(Radosevich and Burnouf 2010). These obstacles induced many manufacturers to privilege mAb development, production, and marketing, considering that their production is free from plasma collection and was supported by public funding. To overcome these limitations, methods for Ig purification from pools of a few CCP donations have been implemented at single transfusion centers during the late phase of pandemic in a few developing countries (El-Ekiaby et al. 2015; Li et al. 2002).

6 Hyperimmune Immunoglobulins Insights from the COVID-19 Pandemic

The results of RCTs on the clinical use of hIVIg against COVID-19 have been contradictory (Focosi et al. 2021; Franchini and Focosi 2024). In a phase I/II RCT by Ali and colleagues in 50 patients with severe and critical COVID-19 and randomized to receive hIVIg or standard of care, the administration of anti-SARS-CoV-2 specific immunoglobulins was associated with a mortality reduction (25% in the intervention group versus 60% in the control group) (Ali et al. 2020). The administration of hIVIg in 60 hospitalized COVID-19 patients, randomized to receive hIVIg or standard care, was found to be safe and well-tolerated being characterized with respect to control arm by a shorter time to viral clearance, and an early reduction of inflammatory biomarkers (Parikh et al. 2021). Another RCT enrolled 18 severely immunocompromised patients hospitalized for COVID-19 to receive either hIVIg (10 patients) or standard IVIg (8 patients) and showed that hIVIg significantly reduced the incidence of severe COVID-19 (20% versus 88%, $P = 0.015$) (Huygens et al. 2023). By contrast, the ITAC RCT enrolled 593 participants of whom 301 received hIVIg and 292 placebo (Polizzotto et al. 2022): compared with placebo, the hIVIg group did not have significantly greater odds of a more favorable outcome at day 7 (adjusted OR: 1.06; 95% CI 0.77–1.45; $P = 0.72$). A recent Cochrane systematic review identified 5 RCTs with 947 participants (688 treated with different formulations of hIVIg) hospitalized with moderate or severe COVID-19 (Kimber et al. 2023). Given the limitations of the great heterogeneity (difference in dosing and human or animal origin) of marketed hIVIg formulations, the author found no impact of this treatment on patients' mortality or clinical improvement. Notably, in a recent head-to-head comparative RCT (hVIg versus CCP), treatment outcomes were better in patients treated with hIVIg on day 28 but not on day 14. Overall, although limited, the available literature data seem to indicate that hIVIg have no beneficial effect in immunocompetent hospitalized COVID-19 patients, but a potential role in improving outcome in those severely immunocompromised. In the current post-vaccine era, however, the hybrid (i.e., vaccine and infection) exposure of a large part of the population has led to high-titer heterologous immunity against

SARS-CoV-2 providing an efficient cross-protection against most variants, including Omicron sub-lineages (Booth et al. 2003). In this new context, the currently available standard IVIg, originating mostly from CCP donations, is equivalent to hIVIg in terms of anti-SARS-CoV-2 nAb content, thus rendering the creation of a dedicated manufacturing chain poorly cost-effective and even obsolete (Focosi and Franchini 2022).

7 Hyperimmune Immunoglobulin Candidates for Selected Pandemic Agents

Previously published studies pointed out the encouraging clinical benefits and safety of hIVIg therapy in patients affected by several pandemic pathogens, including SARS, MERS, influenza virus, WNV, anthrax, and Ebola virus (Yang 2023).

Regarding the influenza virus, a recent (n = 329) randomized, double-blind, placebo-controlled, phase III clinical trial (INSIGHT 006; ClinicalTrials.gov NCT02287467) revealed that the robust rise in hemagglutination inhibition titers following Flu-IVIg treatment did not result in any clinical benefit for influenza A virus-infected participants (Davey et al. 2019). In contrast, a beneficial effect of Flu-IVIg in influenza B virus-infected patients was evident. In another multicenter double-blind randomized controlled trial for patients with severe 2009 influenza A (H_1N_1) infection, the treatment with hIVIg within 5 days of symptom onset was associated with a lower viral load and reduced mortality (Hung et al. 2013).

Regarding the Ebola virus, following the first successful infusion in 1977 of a (heat-treated) convalescent disease "serum", together with human interferon, into a research investigator who had accidentally been infected by this virus (Emond et al. 1977), several trials have investigated the role of convalescent plasma for this severe hemorrhagic viral disease (Garraud 2017). The demonstration of the efficacy of mAbs derived from a single antibody renewed the interest toward antibody-based therapy of Ebola and recent the production of hIVIg derived from pooled convalescent plasma with a five- to sixfold increased potency opens new scenarios for the treatment of persons infected by Ebola virus (Ciencewicki et al. 2022).

Regarding anthrax, a purified hIVIg (Anthrasil™) produced with plasma obtained from healthy donors vaccinated with Anthrax Vaccine was approved in 2015 by the FDA for the treatment (in combination with an antibiotic) of patients with inhalational anthrax (US Department of HHS FDA 2015). The drug has been studied under the Animal Efficacy Rule since human efficacy trials are not feasible or ethical. Survival ranged from 36 to 70% in monkeys and 26% in rabbits versus 0%–2% in control animals. Safety was evaluated in 74 healthy human volunteers with side effects including headache, back pain, nausea, infusion site pain, and swelling (Couse et al. 2021).

8 Conclusions

MAbs represent an effective passive antibody-based therapy against viral infections developed from nAbs in convalescent plasma. However, as mAb development is very complex and requires several months for production and marketing, it is not used as the front-line therapy but in the middle or late phases of pandemics. mAb cocktails are definitely what we need to counteract basal and treatment-emergent resistance. Nonetheless, mAb cocktail approaches have limitations, including heightened manufacturing costs, time-consuming processes, and high-dose infusions for patients: all of those limitations could be overcome by bispecific Abs, which should be the focus of further research (Radić et al. 2023).

Finally, hIVIg, whose clinical effectiveness has been investigated in several previous infective diseases including COVID-19, has the potential to be a useful therapeutic tool in the case of future pandemic outbreaks, such as those caused by influenza virus and WNV. Stockpiling HIG seems desirable but unless the pathogen has very low mutation rates (which is not common for RNA viruses), this invariably comes with the risk of reduced efficacy at times of deployment. The same concerns apply to programs that produce HIG after the pandemic gets started, given viral evolution in the meanwhile of the manufacturing process.

Acknowledgements None.

Author Contributions D.F. wrote the sections on monoclonal antibodies. M.F. wrote the sections on hyperimmune immunoglobulins.

Funding None.

Conflict of Interest Declaration None.

References

Ali S, Luxmi S, Anjum F, Muhaymin SM, Uddin SM, Ali A et al (2020) Hyperimmune anti-COVID-19 IVIG (C-IVIG) therapy for passive immunization of severe and critically ill COVID-19 patients: a structured summary of a study protocol for a randomised controlled trial. Trials 21(1):905

Ali SO, Takas T, Nyborg A, Shoemaker K, Kallewaard NL, Chiong R et al (2018) Evaluation of MEDI8852, an anti-influenza A monoclonal antibody, in treating acute uncomplicated influenza. Antimicrob Agents Chemother 62(11)

Beigel JH (2018) Polyclonal and monoclonal antibodies for the treatment of influenza. Curr Opin Infect Dis 31(6):527–534

Bermingham A, Chand MA, Brown CS, Aarons E, Tong C, Langrish C et al (2012) Severe respiratory illness caused by a novel coronavirus, in a patient transferred to the United Kingdom from the Middle East, September 2012. Euro Surveill Bulletin European sur les maladies transmissibles = Eur Commun Dis Bull 17(40):20290

Bimler L, Kim T, Ronzulli SL, Song AY, Johnson SK, Jones CA et al (2022) M2e-specific monoclonal antibody cocktails against influenza A virus are superior to individual monoclonal antibody treatments, universally effective, and viral escape mutant resistant. bioRxiv 2022.04.02.486847. A low-dose triple antibody cocktail against the ectodomain of the matrix protein 2 of influenza A is a universally effective and viral escape mutant resistant therapeutic agent. bioRxiv

Booth CM, Matukas LM, Tomlinson GA, Rachlis AR, Rose DB, Dwosh HA et al (2003) Clinical features and short-term outcomes of 144 patients with SARS in the greater Toronto area. JAMA 289(21):2801–2809

Cao W, Liu X, Bai T, Fan H, Hong K, Song H et al (2020) High-dose intravenous immunoglobulin as a therapeutic option for deteriorating patients with coronavirus disease 2019. Open Forum Infect Dis 7(3)

Ciencewicki JM, Herbert AS, Storm N, Josleyn NM, Huie KE, McKay LGA et al (2022) Characterization of an anti-Ebola virus hyperimmune globulin derived from convalescent plasma. J Infect Dis 225(4):733–740

Corti D, Cameroni E, Guarino B, Kallewaard NL, Zhu Q, Lanzavecchia A (2017) Tackling influenza with broadly neutralizing antibodies. Curr Opin Virol 24:60–69

Couse Z, Cui X, Li Y, Moayeri M, Leppla S, Eichacker PQ (2021) A review of the efficacy of FDA-approved *B. anthracis* anti-toxin agents when combined with antibiotic or hemodynamic support in infection- or toxin-challenged preclinical models. Toxins 13(1)

Davey RT Jr, Fernández-Cruz E, Markowitz N, Pett S, Babiker AG, Wentworth D et al (2019) Anti-influenza hyperimmune intravenous immunoglobulin for adults with influenza A or B infection (FLU-IVIG): a double-blind, randomised, placebo-controlled trial. Lancet Respir Med 7(11):951–963

Deng R, Lee AP, Maia M, Lim JJ, Burgess T, Horn P et al (2018) Pharmacokinetics of MHAA4549A, an anti-influenza A monoclonal antibody, in healthy subjects challenged with influenza A virus in a phase IIa randomized trial. Clin Pharmacokinet 57(3):367–377

Deng R, She G, Maia M, Lim JJ, Peck MC, McBride JM et al (2020) Pharmacokinetics of the monoclonal antibody MHAA4549A administered in combination with oseltamivir in patients hospitalized with severe influenza A infection. J Clin Pharmacol 60(11):1509–1518

El-Ekiaby M, Vargas M, Sayed M, Gorgy G, Goubran H, Radosevic M et al (2015) Minipool caprylic acid fractionation of plasma using disposable equipment: a practical method to enhance immunoglobulin supply in developing countries. PLoS Neglected Trop Dis 9(2): e0003501-e

Emond RT, Evans B, Bowen ET, Lloyd G (1977) A case of Ebola virus infection. Br Med J 2 (6086):541–544

FDA Announces Bebtelovimab is Not Currently Authorized in Any US Region (2022). Accessed online at https://www.fda.gov/drugs/drug-safety-and-availability/fda-announces-bebtelovimab-not-currently-authorized-any-us-region on 1 Dec 2022

FDA announces Evusheld is not currently authorized for emergency use in the U.S. (2023). Accessed online at https://www.fda.gov/drugs/drug-safety-and-availability/fda-announces-evusheld-not-currently-authorized-emergency-use-us on 3 Feb 2023

FDA authorizes revisions to fact sheets to address SARS-CoV-2 variants for monoclonal antibody products under emergency use authorization (2024). Accessed on 12 Feb 2024 at https://www.fda.gov/drugs/drug-safety-and-availability/fda-authorizes-revisions-fact-sheets-address-sars-cov-2-variants-monoclonal-antibody-products-under

Focosi D, Maggi F (2023) Respiratory delivery of passive immunotherapies for SARS-CoV-2 prophylaxis and therapy. Hum Vaccin Immunother 19(2):2260040

Focosi D, Franchini M, Tuccori M (2021) The road towards polyclonal anti-SARS-CoV-2 immunoglobulins (hyperimmune serum) for passive immunization in COVID19. Life (basel, Switzerland) 11:144

Focosi D, Franchini M (2022) Passive immunotherapies for COVID-19: the subtle line between standard and hyperimmune immunoglobulins is getting invisible. Rev Med Virol e2341

Franchini M, Focosi D (2024) Hyperimmune Plasma and Immunoglobulins against COVID-19: a narrative review. Life 14(2):214

Gao R, Sreenivasan CC, Sheng Z, Hause BM, Zhou B, Wentworth DE et al (2020) Human monoclonal antibody derived from transchromosomic cattle neutralizes multiple H1 clades of influenza A virus by recognizing a novel conformational epitope in the hemagglutinin head domain. J Virol 94(22)

Garraud O (2017) Use of convalescent plasma in Ebola virus infection. Transfus Apheresis Sci off J World Apheresis Assoc off J Eur Soc Haemapheresis 56(1):31–34

Gottlieb RL, Nirula A, Chen P, Boscia J, Heller B, Morris J et al (2021) Effect of Bamlanivimab as monotherapy or in combination with Etesevimab on viral load in patients with mild to moderate COVID-19: a randomized clinical trial. JAMA 325(7):632–644

Guarner J, Hale GL (2019) Four human diseases with significant public health impact caused by mosquito-borne flaviviruses: West Nile, Zika, dengue and yellow fever. Semin Diagn Pathol 36 (3):170–176

Gupta R, Purcell LA, Corti D, Virgin HW (2023) Pandemic preparedness strategies must go beyond vaccines. Sci Transl Med 15(690):eadd3055

Han A, Czajkowski L, Rosas LA, Cervantes-Medina A, Xiao Y, Gouzoulis M et al (2021) Safety and efficacy of CR6261 in an influenza A H1N1 healthy human challenge model. Clin Infect Dis 73(11):e4260–e4268

Hershberger E, Sloan S, Narayan K, Hay CA, Smith P, Engler F et al (2019) Safety and efficacy of monoclonal antibody VIS410 in adults with uncomplicated influenza A infection: results from a randomized, double-blind, phase-2, placebo-controlled study. EBioMedicine 40:574–582

Hoenen T, Groseth A, Feldmann H (2019) Therapeutic strategies to target the Ebola virus life cycle. Nat Rev Microbiol 17(10):593–606

Hung IFN, To KKW, Lee CK, Lee KL, Yan WW, Chan K et al (2013) Hyperimmune IV immunoglobulin treatment: a multicenter double-blind randomized controlled trial for patients with severe 2009 influenza A(H1N1) infection. Chest 144(2):464–473

Huygens S, Hofsink Q, Nijhof IS, Goorhuis A, Kater AP, Te Boekhorst PAW et al (2023) Hyperimmune globulin for severely immunocompromised patients hospitalized with coronavirus disease 2019: a randomized, controlled trail. J Infect Dis 227(2):206–210

Kallewaard NL, Corti D, Collins PJ, Neu U, McAuliffe JM, Benjamin E et al (2016) Structure and function analysis of an antibody recognizing all influenza A subtypes. Cell 166(3):596–608

Kelley B (2020) Developing therapeutic monoclonal antibodies at pandemic pace. Nat Biotechnol 38(5):540–545

Kimber C, Valk SJ, Chai KL, Piechotta V, Iannizzi C, Monsef I et al (2023) Hyperimmune immunoglobulin for people with COVID-19. Cochrane Database Syst Rev 1(1):Cd015167

Ku Z, Xie X, Lin J, Gao P, Sahili AE, Su H et al (2022) Engineering SARS-CoV-2 cocktail antibodies into a bispecific format improves neutralizing potency and breadth. Nat Commun 13 (1):5552

Li G, Stewart R, Conlan B, Gilbert A, Roeth P, Nair H (2002) Purification of human immunoglobulin G: a new approach to plasma fractionation. Vox Sang 83(4):332–338

Lim JJ, Deng R, Derby MA, Larouche R, Horn P, Anderson M et al (2016) Two phase 1, randomized, double-blind, placebo-controlled, single-ascending-dose studies to investigate the safety, tolerability, and pharmacokinetics of an anti-influenza a virus monoclonal antibody, MHAA4549A, in healthy volunteers. Antimicrob Agents Chemother 60(9):5437–5444

Lim JJ, Derby MA, Zhang Y, Deng R, Larouche R, Anderson M et al (2017) A phase 1, randomized, double-blind, placebo-controlled, single-ascending-dose study to investigate the safety, tolerability, and pharmacokinetics of an anti-influenza B virus monoclonal antibody, MHAB5553A, in healthy volunteers. Antimicrob Agents Chemother 61(8)

Mallory RM, Ali SO, Takas T, Kankam M, Dubovsky F, Tseng L (2017) A phase 1 study to evaluate the safety and pharmacokinetics of MEDI8852, an anti-influenza A monoclonal antibody, in healthy adult volunteers. Biologicals: J Int Assoc Biol Standard 50:81–86

Mulangu S, Dodd LE, Davey RT Jr, Tshiani Mbaya O, Proschan M, Mukadi D et al (2019) A randomized, controlled trial of Ebola virus disease therapeutics. N Engl J Med 381(24):2293–2303

Olinger GG, Pettitt J, Kim D, Working C, Bohorov O, Bratcher B et al (2012) Delayed treatment of Ebola virus infection with plant-derived monoclonal antibodies provides protection in rhesus macaques. Proc Natl Acad Sci 109(44):18030–18035

Parikh D, Chaturvedi A, Shah N, Patel P, Patel R, Ray S (2021) Safety and efficacy of COVID-19 hyperimmune globulin (HIG) solution in the treatment of active COVID-19 infection-findings from a prospective, randomized, controlled, multi-centric trial. medrXiv. 2021.07.26.21261119

Paules CI, Lakdawala S, McAuliffe JM, Paskel M, Vogel L, Kallewaard NL et al (2017) The hemagglutinin A stem antibody MEDI8852 prevents and controls disease and limits transmission of pandemic influenza viruses. J Infect Dis 216(3):356–365

Polizzotto MN, Nordwall J, Babiker AG, Phillips A, Vock DM, Eriobu N et al (2022) Hyperimmune immunoglobulin for hospitalised patients with COVID-19 (ITAC): a double-blind, placebo-controlled, phase 3, randomised trial. The Lancet 399(10324):530–540

Qiu X, Audet J, Wong G, Fernando L, Bello A, Pillet S et al (2013) Sustained protection against Ebola virus infection following treatment of infected nonhuman primates with ZMAb. Sci Rep 3:3365

Qiu X, Wong G, Audet J, Bello A, Fernando L, Alimonti JB et al (2014) Reversion of advanced Ebola virus disease in nonhuman primates with ZMapp. Nature 514(7520):47–53

Radić L, Sliepen K, Yin V, Brinkkemper M, Capella-Pujol J, Schriek AI et al (2023) Bispecific antibodies combine breadth, potency, and avidity of parental antibodies to neutralize sarbecoviruses. iScience 26(4):106540

Radosevich M, Burnouf T (2010) Intravenous immunoglobulin G: trends in production methods, quality control and quality assurance. Vox Sang 98(1):12–28

Ramos EL, Mitcham JL, Koller TD, Bonavia A, Usner DW, Balaratnam G et al (2015) Efficacy and safety of treatment with an anti-m2e monoclonal antibody in experimental human influenza. J Infect Dis 211(7):1038–1044

Sager J, Hong DK, Bonavia A, Connolly L, Cebrik D, Fanget MC et al (2020) Preliminary safety and pharmacokinetic profile of VIR-2482: a monoclonal antibody for the prevention of influenza A illness. Open Forum Infect Dis 7(Supplement_1):S376–S37S

Schäfer A, Muecksch F, Lorenzi JCC, Leist SR, Cipolla M, Bournazos S et al (2020) Antibody potency, effector function, and combinations in protection and therapy for SARS-CoV-2 infection in vivoIn vivo efficacy of anti–SARS-CoV-2 antibodies. J Exp Med 218(3)

Sloan SE, Szretter KJ, Sundaresh B, Narayan KM, Smith PF, Skurnik D et al (2020) Clinical and virological responses to a broad-spectrum human monoclonal antibody in an influenza virus challenge study. Antiviral Res 184:104763

Tharakaraman K, Subramanian V, Viswanathan K, Sloan S, Yen HL, Barnard DL et al (2015) A broadly neutralizing human monoclonal antibody is effective against H7N9. Proc Natl Acad Sci USA 112(35):10890–10895

US Dept. of HHS FDA (2023) ANTHRASIL (AIGIV) Injection prescribing information. March 2015. Available online: https://www.fda.gov/vaccines-blood-biologics/approved-blood-products/anthrasil. Accessed on 16 Dec 2023

Vairo F, Haider N, Kock R, Ntoumi F, Ippolito G, Zumla A (2019) Chikungunya: epidemiology, pathogenesis, clinical features, management, and prevention. Infect Dis Clin North Am 33 (4):1003–1025

Vogel OA, Manicassamy B (2020) Broadly protective strategies against influenza viruses: universal vaccines and therapeutics. Front Microbiol 11:135

Wollacott AM, Boni MF, Szretter KJ, Sloan SE, Yousofshahi M, Viswanathan K et al (2016) Safety and upper respiratory pharmacokinetics of the hemagglutinin stalk-binding antibody VIS410 support treatment and prophylaxis based on population modeling of seasonal influenza A outbreaks. EBioMedicine 5:147–155

Wrammert J, Smith K, Miller J, Langley WA, Kokko K, Larsen C et al (2008) Rapid cloning of high-affinity human monoclonal antibodies against influenza virus. Nature 453(7195):667–671
Yang X (2023) Passive antibody therapy in emerging infectious diseases. Front Med
Ye J (2012) Stable transfection pools for large quantity of protein production. Methods Mol Biol (clifton, NJ) 899:221–225
Yi KS, Choi JA, Kim P, Ryu DK, Yang E, Son D et al (2020) Broader neutralization of CT-P27 against influenza A subtypes by combining two human monoclonal antibodies. PLoS ONE 15 (7):e0236172
Zalevsky J, Chamberlain AK, Horton HM, Karki S, Leung IWL, Sproule TJ et al (2010) Enhanced antibody half-life improves in vivo activity. Nat Biotechnol 28(2):157–159

Using Passive Antibody Therapies in the Next Pandemic

Nigel Paneth, Michael J. Joyner and Arturo Casadevall

Contents

Disclaimer: all opinions expressed in this chapter are those of the authors and do not necessarily represent the views of any of their employing institutions.

N. Paneth (✉)
Departments of Epidemiology and Biostatistics and Pediatrics and Human Development, College of Human Medicine, Michigan State University, East Lansing, MI, USA
e-mail: paneth@msu.edu

M. J. Joyner
Department of Anesthesiology and Perioperative Medicine, Mayo Clinic, Rochester, MN, USA
e-mail: joyner.michael@mayo.edu

A. Casadevall
Department of Molecular Microbiology and Immunology, Johns Hopkins School of Medicine, Baltimore, MD, USA
e-mail: acasade1@jhu.edu

Current Topics in Microbiology and Immunology (2025) 443: 259–273
https://doi.org/10.1007/82_2024_283

Published Online: 19 December 2024

Abstract The twenty-first century has witnessed seven human viral pandemics. Approximately once every three to four years over the past quarter-century, the world has experienced a new viral epidemic that expanded well beyond its original national borders to become a pandemic. The probability that another pandemic caused by a previously unknown agent will occur in the near future is thus very high and public health agencies must prioritize mechanisms for detecting their first signals. At the onset of these recent pandemics, no specific therapeutic agent was available for any of the newly emergent pathogens. However, convalescent plasma therapy can be available as soon as there are survivors and is likely to be effective if used early and in sufficient strength. But for the three forms of passive antibody—convalescent plasma, monoclonal antibodies, and hyperimmune globulins—to be available and effective in a pandemic situation, careful strategic planning will be necessary. In the pre-pandemic period, we must reinforce the capacities of blood banks and plasma fractionating companies in the production and storage of their products; ensure that outpatient settings can provide intravenous products; educate providers in the proper use of plasma; and create a research infrastructure to examine the effectiveness of passive antibody products. Once a pandemic is underway, regulatory bodies should simplify the approval of research and emergency use protocols and develop treatment registries. Incentives for the rapid production of monoclonal antibodies and hyperimmune globulins will likely be required. A national resource to link providers with passive antibody products and national databases to monitor pandemic progress and pandemic treatment will permit the most effective allocation of pandemic-fighting resources. We cannot afford to wait until the next pandemic is upon us to respond. The time to strengthen clinical, research, and manufacturing infrastructure to permit us to be ready to confront the next new virulent pathogen is now.

Glossary of Abbreviations

BARDA	Biomedical Advanced Research and Development Authority
CP	Convalescent plasma
DOD	Department of Defenses
EAP	Expanded Access Program
FDA	Food and Drug Administration
hIG	Hyperimmune immunoglobulins
IRB	Institutional Review Board
NIH	National Institutes of Health
WHO	World Health Organization

1 Introduction

In this chapter, we address the many critically important concepts that must guide the use of antibody therapies in the future, avoiding mistakes that were made during the most recent pandemic. And, recognizing the extent to which we were caught flat-footed when COVID-19 hit with ferocity in 2020, we emphasize the need to develop an infrastructure for the efficient acquisition, distribution, and administration of CP and other antibody therapies as well as a parallel infrastructure for collecting data to assess the effectiveness of treatment. Advance planning is necessary if we are to minimize the mortality and morbidity of the next pandemic.

2 Recognizing the Next Pandemic

Although much preparatory work must be conducted during inter-epidemic periods, prompt recognition of the earliest signs of a potential pandemic is crucial if passive antibody therapies are to be used at scale. Since many recent pandemics owe their origins to the spread of a novel infectious agent from wildlife or large domestic animal agricultural operations, coordination between wildlife, agricultural, and (human) public health officials is essential. Systems for communication between these often very separate public agencies must be developed and be in place well in advance of any potential microbial threat.

In high-income countries disease surveillance capabilities are usually adequate, but bureaucratic territorialism can hamper their effectiveness. However, resources and expertise must be allocated to low-and middle-income countries whose surveillance capacities for animal and human infectious diseases are at times inadequate (Sasie et al. 2024). Because infectious disease threats transcend geopolitical boundaries and rivalries, it is in the best interest of all to minimize barriers and maximize cooperation. Economic or other considerations that can limit transparent reporting of likely outbreaks in many parts of the world must be resisted firmly at all administrative levels of government and international organizations, avoiding incentivizing under or late reporting.

A range of data sources has been proposed for surveillance for new pandemics, each of which has advantages and disadvantages. Traditional surveillance modalities include deaths recorded in vital data (Mays and Cochran 2021), reportable disease requirements (Centers for Disease Control and Prevention (CDC) 1996), and the use of sentinel practices and laboratories for detecting influenza and other respiratory diseases (Hay and McCauley 2018). Since 1952, the WHO has conducted influenza virological surveillance through its Global Influenza Surveillance and Response System (GISRS) (World Health Organization 2024).

New technologies have been exploited in recent years to increase public health capacity in infectious disease surveillance. Thus, the WHO surveillance program noted above added a web-based system in 1997 entitled FLUNET (https://www.who.int/tools/flunet). A recent review described several newer forms of infectious disease surveillance including the use of electronic medical records and publicly available sources (school and work absenteeism, medication sales, social media such as Twitter, online searches, telephone helplines, telephone or electronic surveys, and media reports) finding that participatory surveys and electronic health records were most likely to provide timely data that correlated well with information from traditional sources (Hammond et al. 2022). Conflicting results were found for the timeliness and validity of the other web-based surveillance approaches.

Emergency Department Syndromic Surveillance (EDSyS), in which signs, symptoms, and preliminary diagnoses are uncovered through the search of records, was viewed as "successful" in a recent review of more than 500 studies (Hughes et al. 2020). Surveillance of wastewater and other sources for viral agents developed greatly with the COVID-19 pandemic and current technologies including new absorbents for capturing agents, and molecular methods to categorize them have considerable promise, but technological challenges remain (Hayes and Gagnon 2024). Although whole genome pathogen sequencing has been proposed as a surveillance tool, it may be economically prohibitive to use as a front-line tool (Tran et al. 2023). Perhaps the most important principle in using the several surveillance mechanisms is their integration at a national and even international level permitting coordinated action across countries and continents.

In recent decades, accentuated by the COVID-19 pandemic, trust in public health agencies has eroded, and resources to sustain public health surveillance may have been weakened. Recognition that a new pandemic comparable to COVID-19, which killed at least 7 million people worldwide, is nearly inevitable in the future, should signal to all nations the need to enhance public health infrastructure, especially in the arena of infectious disease surveillance.

3 Monoclonal and Polyclonal Forms of Passive Antibody Treatment

Passive antibody therapies involve the administration of pre-formed antibodies to a host for the purpose of preventing or treating disease. They differ from vaccination in that the administration of pre-formed antibodies provides immediate immunity (Casadevall 2002), although this form of immunity is of relatively short duration. Antibody preparations can be polyclonal or monoclonal.

Polyclonal preparations are generally obtained from an immune or vaccinated host (usually human, but animals have been used) and were the only type of antibody therapy available until the development of hybridoma technology in 1975, which allowed the production of monoclonal antibodies.

CP and hIG are the two forms of passive antibody therapy that are polyclonal. CP is collected from previously infected persons by blood banks based on voluntary contributions of plasma. Only blood banks are legally permitted to provide plasma for direct infusion into patients. hIG, by contrast, is obtained from presumably immune individuals and then fractionated and purified through stringent chromatographic processing to produce a product with very high immunoglobulin G content (which is why the product is often referred to as gamma globulin). hIG is prepared from plasma collected by commercial operations that in the United States usually pay for plasma. The dosage volume is much smaller than CP and the product is usually administered intramuscularly. Historically, hIG has been used both for prevention and treatment, for example for rabies, hepatitis A and B viruses, measles, tetanus, and varicella-zoster virus among other applications. Used only sporadically in COVID, its record of effectiveness was mixed, but, as noted in Maor and Zimhony (2024), at least one trial found hIG superior to CP in lowering 28-day mortality in COVID. CP has the advantage, in a pandemic, of being immediately available, whereas hIG requires a stringent manufacturing process that usually takes many months. The delay in producing hIG is a major disadvantage if the infectious agent mutates rapidly, as has been the case with SARS-CoV-2, since antigenic changes can potentially nullify the efficacy of hIG. By contrast, CP obtained shortly after recovery should be effective in treating the most recent antigenic variant.

Polyclonal and monoclonal antibody preparations differ in antibody composition and specific antibody content. In polyclonal preparations, the amount of specific antibody available is small relative to the total immunoglobulin content. In contrast, for mAb preparations all the immunoglobulin is directed toward a relevant antigen. However, the monoclonal nature of mAbs constitutes its Achilles heel in viral treatment. Since antiviral activity is targeted to a single antigenic site on the viral pathogen—for the mAbs for COVID, the target was the SARS-CoV2 spike protein —viral evolution can allow the agent to bypass that site and remain pathogenic. mAb preparations, when targeting antigenically variable pathogens, will select for survival antibody-escape variants that are resistant to the original therapies. This phenomenon was predicted more than 20 years ago (Casadevall 2002) and unfortunately was demonstrated by the successive failure of each monoclonal antibody used in the treatment of COVID-19 by the end of 2022 (Focosi et al. 2023; Ren et al. 2023). In the future, designing variant-resistant mAbs together with cocktails that target several epitopes may be useful principles to extend the clinical utility of mAbs in the future (Casadevall et al. 2024).

By contrast, polyclonal preparations such as CP tend to have low specific activity relative to their total immunoglobulin content, but this is compensated by the diversity of antibodies available, which can target different epitopes and provide multiple effector functions in the form of IgM, IgG, and IgA classes. Polyclonal preparations are also heterogeneous and can manifest considerable lot-to-lot variation requiring careful testing of the antibody potency of each unit of convalescent plasma or lot of hyperimmune immunoglobulin (hIG).

3.1 *The Critical Need for Early Treatment with Sufficient Antibody*

All antibody therapies work best when given early in the course of the disease. In a sense, this is no different than many anti-infective therapies that work best when given early but for antibody therapies, there appears a sharp reduction in efficacy when antibody administration is delayed after symptoms appear. For example, serum therapy was most effective against infections with *Streptococcus pneumoniae* when given within three days of hospitalization (Cecil 1937) and CP is most effective if administered within 5 days of COVID-19 symptoms (Sullivan et al. 2023). A meta-analysis of results from RCTs of CP in COVID-19 involving 18,877 patients found that efficacy in hospitalized patients was limited to those who received therapy in the first 7 days after the onset of symptoms (Franchini et al. 2024). The COVID-19 pandemic thus provided rigorous RCT confirmation of the absolute requirement for early treatment in CP therapy, validating the historically documented experience dating back more than a century.

The requirement for early administration of antibody therapies is not completely understood and may have different explanations for different diseases. One argument proposed for the temporal dependence of efficacy is that antibody protects the host by neutralizing the inoculum and there may be insufficient antibodies to modify the course of infection once the pathogen has established itself and begun to proliferate (Robbins et al. 1995). Another explanation has to do with the pathogenesis of the disease being treated. Pulmonary disease caused by pathogens such as Streptococcus pneumoniae and SARS-Cov-2 is driven by an exuberant inflammatory response that once established cannot be easily modified by the provision of opsonic or neutralizing antibody, respectively. Many hIGs are used even before the disease is manifest as a mode of disease prevention, especially in outbreak situations or when an exposure has taken place.

Timely administration must be accompanied by appropriate dosing. Antibody therapies are only effective when the antibody preparation contains sufficient specific immunoglobulin to mediate a therapeutic effect. For mAb preparations, dosing is simpler since the amount of immunoglobulin is known and the pharmacokinetics of antibody half-life and volume of distribution are well understood. Such considerations guided the mAb randomized controlled trials of mAbs to SARS-CoV-2, which established that they were effective in preventing hospitalization of high-risk individuals with COVID-19. The dosing of polyclonal preparations such as CP, however, is more challenging since the amount of specific antibody is not always known. While the antibody titer provides a measure of antibody activity in a preparation, therapeutic effectiveness depends on both antibody amount and affinity to the pathogen. Consequently, the preparation is harder to standardize. Two preparations may have the same titer and yet differ in antibody amount, so that a preparation with high specific antibody content with low average affinity could have the same titer as another preparation with low specific antibody content that has high average affinity. This concern notwithstanding the availability of sufficient antibody in infused plasma can be ensured by selecting high titer units. One practical way to do this is to screen donated plasma and to use plasma in the upper quintile of antibody titers for treatment (Sullivan and Casadevall 2023).

4 Infrastructure Required for Passive Antibody Treatment

4.1 The Required Research Infrastructure

Yoon and Pirofski (2024), have set out a detailed research prospectus for the use of CP in the next pandemic. Some key features of their proposal are support for a national trial infrastructure that can be mobilized in the face of a pandemic; recognition that the clinical equipoise required for ethically including a placebo/ standard treatment arm may not hold in all circumstances; the need for flexibility in trial design and conduct; and the need for openness to trial designs other than the RCT in evaluating effectiveness.

As noted by Yoon and Pirofski, the NIH did in fact support two COVID trial networks (ACTIV and ACTT) but these began their work later in the pandemic than would have been ideal. Now is the time for NIH and other key supporters of biomedical research such as BARDA, the FDA, and even the DOD, to prospectively establish a large multicenter trial network that can be activated as soon as the first indicators of the next pandemic are recognized. But for appropriate trials of CP to actually begin, it must first be established in the laboratory that newly collected CP contains a sufficient level of neutralizing antibody to the agent of concern. This information can then ensure that all human studies of CP, both experimental and observational, assess a truly therapeutic dose.

Trials should not be started in a new epidemic with a new pathogen until solid information is available on dosage and timing. Starting trials before this key information is known risks obtaining false negative findings from trials treating participants too late with plasma of weak antibody content, and false negative findings can damage the reputational efficacy of this approach, as happened with several trials during the COVID-19 pandemic. In fact, conclusive evidence of the efficacy of CCP was not available until the second year of the pandemic when the result of outpatient trials became available but by then CCP had been abandoned in many countries thus removing an effective tool for combating COVID-19. In the USA alone, the abandonment of CCP was associated with an excess of at least 30,000 deaths (Casadevall et al. 2021; Dragotakes et al. 2024b).

The alternative to trials is to initiate a registry such as the COVID-19 EAP (see Joyner 2024) as soon as information on CP neutralizing capacity is available. Such a registry, focused on patients treated with antibody therapies, can collect information on efficacy and timing that can then inform the design of RCTs (Senefeld et al. 2023; Perichon et al. 2023). The EAP mechanism allowed FDA regulators to ensure that virtually all US treatment with CP was available for analysis prior to the emergency authorization of COVID-19 CP of August 2020, just five months after the EAP was initiated and to confirm evidence of its effectiveness. If this kind of registry is initiated in the next pandemic, investigators will be able to harness the statistical power of all treated patients to rapidly assess safety and effectiveness.

Agent-neutralizing CP can then be used to treat immunosuppressed patients, reserving RCTs for circumstances where equipoise exists and that are designed with sufficient power to examine pre-specified patient sub-groups and that are flexible enough to accommodate design changes based on new therapeutic information or agent mutation.

4.2 The Clinical Treatment Infrastructure

4.2.1 The Capacity to Obtain, Move, Process, and Use Convalescent Plasma and Immunoglobulins

In the US, the blood banking infrastructure is composed of a network of not-for-profit organizations that collect blood products from volunteers and distribute them to hospitals. In most high- and middle-income countries there is a centralized blood banking regulatory authority. Developing blood banking capabilities in less developed countries is a priority as part of the provision and expansion of basic surgical services in these countries, as surgery often requires access to blood products.

In the US, it will be critically important to work with organizations such as the Association for the Advancement of Blood and Biotherapies (aabb.org) and America's Blood Centers (americasblood.org) to ensure that the blood banking world is well prepared to obtain and distribute CP as early as possible in a pandemic. We must recognize that the demand for plasma may be in competition with regular blood products and that ramping up plasma capacity must consider the need to continue to distribute red blood cells, platelets, fresh frozen plasma, and other blood bank products for use in clinical care. Furthermore, it is important to take into account in planning the fact that CP—a product based on human donation—will be in competition with for-profit therapies. We must guard against any discouragement of its use in favor of the lucrative testing of more profitable therapies, especially in situations where its use would disqualify the recipient from participating in other studies. This is an ethical issue that needs to be addressed in anticipation of future pandemics.

If early production of hIG is contemplated—as it should be—the plasma fractionating industry must also be included. Most large plasma manufacturers are headquartered overseas (Takeda, CSL, Grifols, Octapharma) but all have substantial presence in the USA. An effort must be made to include these and other fractionating companies in the planning process for making an hIG that can prevent or treat the pandemic infection.

All use of plasma and its products will depend upon antibody assays that will enable blood banks and fractionating companies to determine appropriate donors and effective products. Prototype assay and clinical tests should be developed in advance for families of pathogens so that it will be possible to rapidly generate the reagents for specific infectious agents once identified. However, in situations involving a novel agent for which no prior experience is available, physicians could hedge their bets in providing sufficient antibodies by transfusing more than one unit of plasma from different donors (Sullivan and Casadevall 2023). Hardware and disposable supplies needed to perform large scale apheresis will need to be made available via either repurposing from other uses or de novo. In less resource-rich environments, simple collection and processing systems such as those suggested in Bloch (2024) and Nierich et al. (2024) may be useful.

The use of CP in a timely fashion depends upon the capacity to provide it on an outpatient basis to high-risk patients. Outpatient clinics will need to adapt their facilities, staffing, and equipment to administer CP and monoclonals. Preventing inpatient services from being overwhelmed, as they were in early 2020, will be a high priority for the next pandemic, and this cannot be accomplished without adapting outpatient settings to efficiently administer intravenous products.

Healthcare staffing during a pandemic may be bottlenecked as a result of illness among technicians and nurses on the front lines of donor centers and processing facilities. Repurposing classes of healthcare workers may have to be considered, along with thinking about how to streamline and perhaps automate some of the processes involved in plasma procurement and distribution. As noted in Lebovits (2024), community partnerships in the mobilization of plasma donation are essential. Expertise in logistics and supply chain management and distribution coming from the business world could help streamline the complex process of acquiring and distributing plasma.

No less important is the education of healthcare professionals. It was apparent during the COVID pandemic that many physicians did not understand the key principles of passive antibody use which led to inappropriate designs for many trials and subsequent disuse of a valuable therapy. This knowledge had been forgotten in the decades since 1940, when serum therapy was abandoned in light of the emergence of antibiotics. Educational programs will be needed to explain the rationale for passive antibody therapy use as well as the modes of administration that assure the best patient outcomes.

4.2.2 The Need for Plasma Banks

Access to plasma proved to be an important problem in the early days of the COVID pandemic. Thus, the development of a system of regional plasma banks so that plasma is available in sufficient quantities throughout the country and can be transported rapidly to fill local needs will be an essential element of the pandemic response. As long ago as 1947, the Nobelist in literature, Albert Camus, recognized, even as a layman, that variation in strains of pathogens might necessitate local development of plasma products.[1]

This early insight was validated for COVID by Kunze et al. (2021) and Kunze and Johnson (2024) who found mortality to be 27% lower if plasma came from within 150 miles of the recipient. Nonetheless, there may have to be some inter-regional transportation of plasma to handle emergency needs and the use of high titer units may possibly compensate for distance considerations. Decisions as to the siting of regional plasma banks should be part of pandemic planning.

Major plasma fractionating companies are widely distributed in the US, and hIG is more easily transportable than plasma. With appropriate planning and organization, we assume that plasma fractionating companies can store and distribute their products without additional facilities. However, they will have to be ready to place their manufacturing capacity on a pandemic footing, prioritizing the production and distribution of hIG directed toward the pathogen of concern.

4.2.3 The Regulatory Infrastructure and Its Challenges

The flexibility shown by the US FDA in regulating CP during the COVID-19 pandemic has been vindicated by the outcomes data on COVID treatment as described in chapter 8 that recapitulated the substantial evidence of effectiveness on the value of antibody therapy of the past 130 years described in Paneth et al. (2024). The FDA's decision to make CP available to American citizens early in the pandemic, using such regulatory measures as compassionate use, the EAP, and eventual emergency use authorization, is estimated to have saved tens of thousands of lives in the United States, and might have saved more lives had CP been more widely and effectively used (Dragotakes et al. 2024a). As noted in analyses performed by the FDA (Marks 2024) and by the Mayo team directing the EAP (Joyner et al. 2021), in data collected in just 20 weeks, higher levels of antibody in administered CP lowered mortality rates in unventilated patients. It is noteworthy that for subjects enrolled in the EAP neither the patient nor the physician had any information on the antibody titer of CCP units, which ranged from absent to high, and this analysis thus reflected a measure of the randomization and blinding that minimizes bias and confounding in RCTs. The findings of the EAP provided a marker of the effectiveness of CP well before the findings of trials were available and allowed the FDA to permit emergency access to CP in August 2020, no doubt saving many lives. Unlike most clinical trials, an FDA-mandated patient registry allows CP to be administered at scale outside of research-intensive academic hospitals and this was achieved in the United States (Senefeld et al. 2021).

---

[1]Rieux hoped that a vaccine made with cultures of the bacilli obtained locally would take effect more actively than serum coming from the outside, since the local bacillus differed slightly from the normal plague bacillus as defined in textbooks of tropical diseases. Albert Camus: The Plague. Cited from the first Vintage International Edition, 1991. P. 133 (translated from the French by Stuart Gilbert).

A major strength of the EAP, which should be emulated in any new pandemic, is that it had a single central IRB, that of the Mayo Clinic. In any new pandemic, this function might be taken over as a federal responsibility or delegated to a major research institution as was done during the COVID-19 pandemic. To ensure the capacity to begin a program of treatment early, standard pre-developed consent forms in multiple languages should be prepared, along with provisions for consent procedures in patients too sick or otherwise unable to provide consent. Online enrollment should include information on the treating physicians, treated patients (with documentation of consent), and a substantial data matrix recorded online. This dataset should include patient demographics, clinical status on a daily basis (Ma et al. 2022), and key unambiguous outcomes, such as mortality, mechanical ventilation, oxygen dependence, and complications of therapy. Centralized data collation and analysis with timely adjudication of serious adverse events is also a necessity.

5 Conclusions and Recommendations

In this section, we set out in brief our recommendations for making the most of passive antibody therapies in the next pandemic. Clearly, each of the recommendations will require fleshing out, including determining who is responsible for each component of planning. We first describe pre-pandemic preparation, then consider the application of passive antibody principles to a developing pandemic, and finally, we consider how to sustain the effort, evaluating its efficacy on a continuing basis.

5.1 Pre-Pandemic Planning

Just as no nation can afford to wait until the onset of combat to develop a military, no nation can afford to wait until the pandemic occurs to develop the capacity to address it. In the first 21 years of the twenty-first century, humanity experienced seven major viral outbreaks (SARS, H1N1, MERS, Ebola, Zika, SARS-CoV-2, and mPox). In light of this history, we can safely conclude that more outbreaks are inevitable and planning for the next pandemic must begin now. For passive antibody treatments, we recommend the following pre-pandemic steps:

1. Organizing blood banks and plasma fractionating companies to work jointly to increase their capacity to collect convalescent plasma and produce hIG.
2. Developing a system of regional plasma banks to allow prompt plasma storage.
3. Developing outpatient capacity for transfusion of passive antibody products.

4. Developing a national protocol for assuring early access on an emergency basis to CP with monitoring of outcomes, in effect creating an early patient registry of treated cases, and the capacity to link patients in need to CP supply.
5. Developing a clinical trial network suitable for rapid testing of newly developed antibody products.
6. Developing and disseminating continuing medical education programs focused on the appropriate use of convalescent plasma and other passive antibody therapies.
7. Having a CDC task force specifically designated to address any pandemic threat, not just seasonal influenza, for which planning is well developed. (https://www.cdc.gov/pandemic-flu/php/national-strategy/index.html).

5.2 *Earliest Steps in Any New Pandemic*

1. Creating a single national IRB to oversee all functions involving passive antibody treatment that require informed consent, including registries, cohort studies, and trials. Hospitals and Universities must be prepared to cede oversight to this national IRB.
2. Setting up a registry of treated patients based on the protocol developed under recommendation #4 above.
3. Working with the pharmaceutical industry to rapidly develop monoclonal antibodies, with an emphasis on creating cocktails with more than one antibody that would be less vulnerable to viral evolution.
4. Working with plasma fractionating companies to rapidly develop hIG.
5. Constituting panels of physicians and ethicists to adjudicate how new for-profit lucrative therapies will be tested without depriving patients of potentially life-saving CP.

5.3 *Monitoring the Interventions*

1. Creating a national database that combines mortality and hospitalization data pertaining to the pandemic with treatment data on the use of passive antibody products.
2. Having a national center (with phone-in and online access) that can immediately direct providers to locations where CP is available.
3. Monitor the introduction of new therapies and ensure that if CP is withheld from patients in favor of testing of newer entities the decision is adjudicated ethically.

Although these recommendations may well be refined based upon additional experience during future outbreaks, their implementation could potentially save many lives moving forward.

Acknowledgements The helpful suggestions of Dr. Peter Marks are gratefully acknowledged.

Conflicts of Interest No conflicts to declare.

References

Bloch EM (2024) Blood banking capacity in low-and middle-income countries: Covid-19 convalescent plasma in context. Curr Top Microbiol Immunol. https://doi.org/10.1007/82_2024_266. PMID: 38772969

Casadevall A (2002) Passive antibody administration (immediate immunity) as a specific defense against biological weapons. Emerg Infect Dis 8(8):833–841

Casadevall A, Dragotakes Q, Johnson PW et al (2021) Convalescent plasma use in the USA was inversely correlated with COVID-19 mortality. Elife 4(10):e69866

Casadevall A, McConnell S, Focosi D (2024) Considerations for the development of monoclonal antibodies to address new viral variants in COVID-19. Expert Opin Biol Therapy. https://doi.org/10.1080/14712598.2024.2388186

Cecil RL (1937) Effects of very early serum treatment in pneumococcus type I pneumonia. JAMA 108:589–692

Centers for Disease Control and Prevention (CDC) (1996) Notifiable disease surveillance and notifiable disease statistics—United States, June 1946 and June 1996. Morbidity Mortality Weekly Rep 45(25):530+

Dragotakes Q, Johnson PW, Buras MR et al (2024a) Estimates of actual and potential lives saved in the United States from the use of COVID-19 convalescent plasma. Med Arch. https://doi.org/10.1101/2024.05.16.24307505

Dragotakes Q, Johnson PW, Buras MR et al (2024b) Estimates of actual and potential lives saved in the United States from the use of COVID-19 convalescent plasma. Proc Natl Acad Sci USA 121(41):e2414957121. PMID: 39352932

Focosi D, McConnell S, Sullivan DJ, Casadevall A (2023) Analysis of SARS-CoV-2 mutations associated with resistance to therapeutic monoclonal antibodies that emerge after treatment. Drug Resist Updates: Rev Comment Antimicrobial Anticancer Chemother 71:100991

Franchini M, Cruciani M, Mengoli C et al (2024) Convalescent plasma and predictors of mortality among hospitalized COVID-19 patients: a systematic review and meta-analysis. Clin Microbiol Infect

Hammond A, Kim JJ, Sadler H, Vandemaele K (2022) Influenza surveillance systems using traditional and alternative sources of data: a scoping review. Influenza Other Resp Viruses 16:965–974

Hay AJ, McCauley JM (2018) The WHO global influenza surveillance and response system (GISRS)—a future perspective. Influenza Other Respir Viruses 12(5):551–557

Hayes EK, Gagnon GA (2024) From capture to detection: a critical review of passive sampling techniques for pathogen surveillance in water and wastewater. Water Res 261:1–14

https://www.who.int/tools/flunet. Accessed 22 July 2024

Hughes HE, Edeghere O, O'Brien SJ et al (2020) Emergency department syndromic surveillance systems: a systematic review. BMC Public Health 20(1):1891

Joyner MJ, Carter RE, Senefeld JW et al (2021) Convalescent plasma antibody levels and the risk of death from Covid-19. N Engl J Med 384(11):1015–1027
Joyner MJ (2024) Convalescent plasma and the US expanded access program: a personal narrative. Curr Top Microbiol Immunol. https://doi.org/10.1007/82_2024_269. PMID: 38877204
Kunze KL, Johnson PW, van Helmond N, Senefeld JW, Petersen MM, Klassen SA et al (2021) Mortality in individuals treated with COVID-19 convalescent plasma varies with the geographic provenance of donors. Nat Commun 12(1):4864
Kunze KL, Johnson PW (2024) The importance of geographic proximity of convalescent plasma donors. Curr Top Microbiol Immunol. https://doi.org/10.1007/82_2024_270. PMID: 39117845
Lebovits C (2024) The role of the patient advocate during a pandemic: the case of convalescent plasma. Curr Top Microbiol Immunol. https://doi.org/10.1007/82_2024_271. PMID: 38877203
Ma T, Wiggins CC, Kornatowski BM et al (2022) The role of disease severity and demographics in the clinical course of COVID-19 patients treated with convalescent plasma. Front Med (Lausanne) 26(8):707895
Maor Y, Zimhony O (2024) Hyperimmune globulins in COVID-19. Curr Top Microbiol Immunol. PMID: 38877201
Marks PM (2024) The importance of antibody titer determination to the effective use of convalescent plasma. Curr Top Microbiol Immunol. (in press)
Mays VM, Cochran SD (2021) A look at when dying really counts: an *AJPH* supplement on mortality data in public health surveillance. Am J Public Health 111(Suppl 2):S47–S48
Nierich A, Bihariesingh R, Bansie R (2024) HemoClear: a practical and cost-effective alternative to conventional convalescent plasma retrieval methods. Curr Top Microbiol Immunol. https://doi.org/10.1007/82_2024_276. PMID: 39126485
Paneth N, Walsh M, Kornatowski B, Casadevall A (2024) A brief history of polyclonal antibody therapies against bacterial and viral diseases before COVID-19. Curr Top Microbiol Immunol. https://doi.org/10.1007/82_2024_279. PMID: 39207508
Perichon AM, Acosta A, Di Tulio L (2023) Factors associated with mortality among hospitalized patients with COVID-19 disease treated with convalescent plasma. mBio 14(6):e0177723
Ren Z, Shen C, Peng J (2023) Status and developing strategies for neutralizing monoclonal antibody therapy in the omicron era of COVID-19. Viruses 15(6):1297
Robbins JB, Schneerson R, Szu SC (1995) Perspective: hypothesis: serum IgG antibody is sufficient to confer protection against infectious diseases by inactivating the inoculum. J Infect Dis 171:1387–1398
Sasie SE, Ayano G, Mamo F et al (2024) Assessing the performance of the integrated disease surveillance and response systems: a systematic review of global evidence. Public Health 231:71–79
Senefeld JW, Johnson PW, Kunze KL (2021) Access to and safety of COVID-19 convalescent plasma in the United States expanded access program: a national registry. PLoS Med 18(12): e1003872
Senefeld JW, Marks P, Casadevall A, Joyner MJ (2023) The value of observational registry studies for the next infectious disease emergency. mBio 14(6):e0256523
Sullivan DJ, Focosi D, Hanley DF, Cruciani M, Franchini M, Ou J, Casadevall A, Paneth N (2023) Outpatient randomized controlled trials to reduce COVID-19 hospitalization: systematic review and meta-analysis. J Med Virol 95(12):e29310
Sullivan D, Casadevall A (2023) COVID-19 serology data provide guidance for future deployments of convalescent plasma. mBio e0042823
Tran M, Smurthwaite KS, Nghiem S (2023) Economic evaluations of whole-genome sequencing for pathogen identification in public health surveillance and health-care-associated infections: a systematic. Lancet Microbe 4:e953–e962
World Health Organization (2024) Global influenza surveillance and response system (GISRS). https://www.who.int/initiatives/global-influenza-surveillance-and-response-system. Accessed 22 July 2024

Yoon H, Pirofski L-a (2024) Generating the evidence base for convalescent plasma use for a new infectious disease. Curr Top Microbiol Immunol. PMID: 39117847

Nigel Paneth MD MPH is University Distinguished Professor Emeritus and founding chair of the Department of Epidemiology & Biostatistics at MSU. His principal research is in perinatal and child developmental epidemiology and he is a member of the leadership team of the National COVID-19 Convalescent Project (CCPP19). He is the author or editor of five books and 380 peer-reviewed publications.

Michael J. Joyner MD is the Caywood Professor of Anesthesiology and Perioperative Medicine at the Mayo Clinic. He is a physiologist who does basic research in human volunteers and patients. During the pandemic he repurposed his lab and led the US Expanded Access Program for convalescent plasma.

Arturo Casadevall MD PhD is a Bloomberg Distinguished Professor and Chair of the Department of Molecular Microbiology and Immunology at the Johns Hopkins School of Public Health. He has been involved in the development of antibody therapies for infectious diseases for more than three decades and chaired the CCPP19 group during the COVID-19 crisis.

Correction to: A Brief History of Polyclonal Antibody Therapies Against Bacterial and Viral Diseases Before COVID-19

Correction to:
Chapter 279 in: *Current Topics in Microbiology and Immunology* https://doi.org/10.1007/82_2024_279

The text 'a survival rate of 24.5%' in section 'THE ANTITOXIN ERA (1890–1930)' was published erroneously in the initially published version of this chapter. This has been corrected as 'a mortality rate of 24.5%'.

The updated version of this chapter can be found at
https://doi.org/10.1007/82_2024_279

Current Topics in Microbiology and Immunology 443: 275–275
https://doi.org/10.1007/82_2025_295

Published Online: 10 May 2025

MIX
Papier aus verantwortungsvollen Quellen
Paper from responsible sources
FSC® C105338

If you have any concerns about our products,
you can contact us on
ProductSafety@springernature.com

In case Publisher is established outside the EU,
the EU authorized representative is:
Springer Nature Customer Service Center GmbH
Europaplatz 3, 69115 Heidelberg, Germany

Printed by Libri Plureos GmbH
in Hamburg, Germany